THE SWORD OF DAMOCLES
OUR NUCLEAR AGE

By
Michael and James Hall

ISBN:

978-1-965384-57-2 (E-Book)
978-1-965384-58-9(Paperback)
978-1-965384-59-6(Hardcover)

Library of Congress Registration Number – TXu 2-423-213

(Effective Date of Registration-March 27, 2024 – *The Sword of Damocles, Our Nuclear Age*, Michael and James Hall)

i

Dedicated to Charles (Chuck) Francis Costa, a distinguished Nevada Test Site veteran, who, like so many men and women of his age, helped win the peace of the Cold War. May his kind come again to protect the peace once more. (See Appendix III.)

Cover photo by Eiichi Matsumoto, a Japanese photographer who photographed the aftermath of the nuclear bombing of Hiroshima and Nagasaki. Matsumoto served as a staff photographer in the Publications Division, Asahi Shimbun Tokyo Head Office, covering the firebombing of several Japanese cities. This image depicts a police station on September 15, 1945, in Shimoyanagi-cho, Hiroshima. The clock stopped at the time of the atomic bomb blast.

Table of Contents

INTRODUCTION

The sword of Damocles is an ancient parable that Roman philosopher Cicero passed down in his 45 BCE book *Tusculan Disputations*. He wrote of a tyrannical king of the Sicilian city of Syracuse named Dionysius II. In this ancient tale set four centuries before Christ, Dionysius became wealthy beyond his dreams at the expense of exploitive rule. This forced such a precarious situation upon him that he lived in fear of assassination day and night, so much so that he could not even trust his own servants and had to rely on his daughters to shave him.

One day, a court flatterer by the name of Damocles attracted Dionysius's attention with uncharacteristic words of admiration. This jester reasoned that such a powerful king must have a blissful life. Dionysius challenged his assumption. "Since this life delights you, do you wish to taste it yourself and make a trial of my good fortune?"[1] Damocles happily agreed.

Dionysius ordered his every comfort to be attended to by a host of servants providing succulent cuts of meat and scented perfumes. Damocles quickly felt like a king, that is, until he noticed that Dionysius had suspended a razor- sharp sword over his head. The unnerving weapon positioned above him maintained its stability only by a single strand of horsehair. By that point, Damocles understood how

---

[1] Cicero, *Tusculan Disputations* (Rome: Ancient Publishers, 45 BC); J. E. *King, Cicero. Tusculan Disputations, Loeb Classical Library* (Cambridge, MA: Harvard University Press, 1927), ISBN 0-674-99156-7.

dangerous Dionysius's life felt every moment, whether awake or asleep. Damocles no longer wished to be "so fortunate."[2]

On September 25, 1961, as the Cold War spawned nightmarish visions of nuclear war, President John F. Kennedy gave a chilling speech before the General Assembly of the United Nations. He stated, "Every man, woman, and child lives under a nuclear sword of Damocles, hanging by the slenderest of threads, capable of being cut at any moment by accident or miscalculation or by madness."[3] In those years, the possibility of a nuclear war breaking out was quite real.

I had the opportunity to study that perilous period in a unique way. As a museum director with four decades of experience behind me, I met and worked with some of the leaders in the field of nuclear weapons testing. During those years of my association with them, their heyday had long passed, but these "atomic vets" still conveyed many amazing tales. I collected their information during the years I managed, along with my coauthor and son James, a nationally designated Smithsonian-affiliated museum. The Department of Energy partially sponsored the museum, and we curated a history detailing over eighty years of nuclear weapons production and testing. During that time, I had access to stories and facility tours that few are ever privileged to. I learned from so many veterans of the nuclear weapons field, who, although long retired, were active in recounting the significance of their experiences.

I soon came to understand that the people who created the weapons of the nuclear age were not warmongers but brave, patriotic men and women who prided themselves on winning the peace of the Cold War. However, I also became aware that the great nuclear arsenals created over many decades, while reduced in numbers from the height of the Cold War, are still, to this day a sword of Damocles hanging over us all.

Our museum served an international audience, and I gave hundreds of tours to visitors from around the globe. I always cautioned our audience not to ignore the history of the nuclear age or take it lightly. I was especially cognizant of conveying that message to our younger audiences and school visitors.

I explained that once Pandora's Box had been opened, inevitably, other countries developed nuclear weapons as deterrents to counter each other's growing arsenals of bombs. They believed in the power of deterrence even though they had become hostages to their own mutually escalating nuclear arms race.

Admittedly, consensus tells us that, to date, nuclear deterrence has prevented a third world war, although it has only done so because rational leaders have continued to make sensible decisions. Will a day come when that could change? With growing tensions in today's world now outpacing what we saw even during the worst days of the Cold War, could a nuclear nightmare finally be realized?

This book contains a collection of independent and little-known vignettes detailing the horror of nuclear weapons and the culture inherent in their creation. The stories attempt to relate our present

---

[2] Ibid.

[3] John F. Kennedy Presidential Library and Museum, full transcript of President Kennedy's General Assembly speech of September 15, 1961, https://www.jfklibrary.org/.

times to our past lessons.

For decades, the world's population has lived in the dark shadow cast by the atomic age, which has correspondingly shaped many facets of daily life. We are reminded again of the ever-present danger of that nuclear sword of Damocles that President Kennedy warned us about not so long ago.

CHAPTER ONE
THE WORLD SET FREE

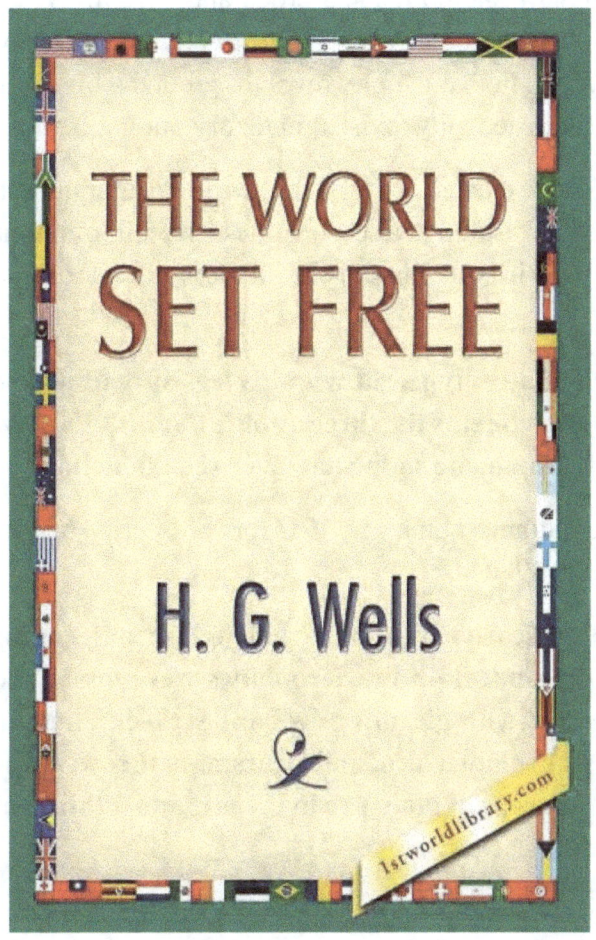

Public record image of the early cover of *The World Set Free*,
https://i.ebayimg.com/images/g/5WMAAOSwwrtZw8sC/s-l640.jpge.

We think of the atomic age beginning in the 1940s. Surprisingly, it actually goes back as far as the First World War, by which time the word "atomic bomb" had already made its way into our vocabulary—many thanks to science fiction writer H.G. Wells for that. Yet, we forget that Wells was not just a noted author but an influential personality of his age and closely connected with many key figures in the arenas of politics, science, and upper-class society. Wells was not a sci-fi footnote but a real player in the history of atomic theory.[4]

Many modern scientists and a significant number of the formative leaders of the 20th century freely admit they were inspired from childhood and into adulthood by H.G. Wells. A few great names on that list are world leaders like Winston Churchill and Franklin Roosevelt, and great physicists and engineers such

---

[4] Conversation with retired Los Alamos National Laboratory Director Siegfried S. Hecker, June 2015, National Atomic Testing Museum. Metallurgist and nuclear scientist Dr. Hecker served as Director of the Los Alamos National Laboratory from 1986 to 1997.

as Leo Szilárd and Robert H. Goddard. Nuclear physicist Dr. Robert R. Brownlee proudly traced his early fascination with science to the books of Wells.[5]

H.G. Wells is famous for iconic fictional stories such as *The War of the Worlds*. However, a much lesser-known book called *The World Set Free* did reach a broad audience in 1914. This story made print before the Great War broke out, and it told a tale involving a destructive conflict not unlike what the First World War became. It is unique, however, in that the plot revolves around a highly destructive and uncontrollable weapon of mass destruction that, in actuality, existed in theory but not reality, until the Second World War.

In *The World Set Free*, Wells created a story involving "a uranium-based hand grenade" that was dropped from planes and caused devastating effects on cities. Keep in mind that Wells conceived this story just before planes began to be used in war. His story continues with a French aviator on a bombing mission to Berlin:

His companion, a less imaginative type, sat with his legs spread wide over the long, coffin-shaped box which contained in its compartments the three atomic bombs, the new bombs that would continue to explode indefinitely and which no one so far had ever seen in action.[6]

Wells's description of the "atomic bomb" in 1914 proved eerily similar to accounts of the days of atmospheric atomic testing in the 1950s:

The bomb flashed blinding scarlet in mid-air and fell, a descending column of blaze eddying spirally in the midst of a whirlwind. Both the aeroplanes were tossed like shuttlecocks, hurled high and sideways, and the steersman, with gleaming eyes and set teeth, fought in great banking curves for a balance. Then that bomb had exploded also, and steersman, thrower, and aeroplane were just flying rags and splinters of metal and drops of moisture in the air, and a third column of fire rushed eddying down upon the doomed buildings below.[7]

In his book, Wells had an amazing intuition about what future nuclear war could look like and its repercussions. He explained that the bomb dropped on Berlin "never entirely exhausted… the battlefields and bomb-fields of that frantic time [were] sprinkled with radiant matter and so centers of inconvenient rays." The story continues:

Certainly it seems now that nothing could have been more obvious to the people of the early twentieth century than the rapidity with which war was becoming impossible. And as certainly they did not see it. They did not see it until the atomic bombs burst in their fumbling hands. Yet the broad facts must have glared upon any intelligent mind. All through the nineteenth and twentieth centuries the amount of energy that men were able to command was continually increasing. Applied to warfare, that meant that the power to inflict a blow, the power to destroy, was continually increasing. There was no increase whatever in the ability to escape. Every sort of passive defense…was being

---

[5] Annual Magazine and Annual National Atomic Testing Museum Report, 2018, pp. 16-21.

[6] H.G. Wells, The World Set Free (London: W. Collins Sons, 1924), Chapter Two, Section Three; Wells' book first appeared in a serial format that consisted of three books: A Trap to Catch the Sun, The Last War in the World and The World Set Free.

[7] Ibid, Chapter Two, Section Three; Online version at https://novelonline.ir/read/1781/The-World-Set-Free.

outmastered by this tremendous increase on the destructive side.[8]

In Wells's 1914 story, the fictionalized concept of an atomic bomb changed the way people looked at war, just as real nuclear weapons would in modern times:

Our mental setting had far more of the effect of a huge natural catastrophe. The atomic bomb had dwarfed the international issues to complete insignificance. When our minds wandered from the preoccupations of our immediate needs, we speculated upon the possibility of stopping the use of these frightful explosives before the world was utterly destroyed.[9]

In Wells's vision, the atomic age also brought the peaceful benefits of atomic power to society, eventually creating a type of socialist utopia or one "world republic." Wells, of course, developed many fictional works over the years expounding on the concept of mankind's mastery of energy. In *The Outline of History* in 1918, Wells utilized the scientific work of radiochemist Fredrick Soddy's study titled *Interpretation of Radium*, which dealt with the radioactive decay of elements like radium. Soddy had worked with renowned New Zealand physicist Ernest Rutherford during Rutherford's studies on radioactivity as Director of the Cavendish Laboratory at the University of Cambridge.[10] Rutherford was later considered the father of atomic theory and nuclear physics and influenced Wells's creative writing.

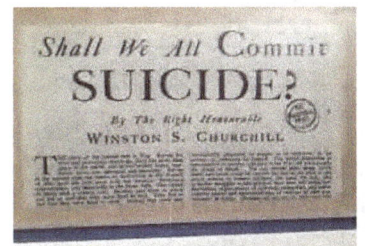

Authors' collection.

Winston Churchill became inspired by Wells when he served as First Lord of the British Admiralty from 1911 to 1915. His interest in Wells's writings continued when he served as an Army officer in the trenches in 1916, cabinet Minister of Munitions from 1917 to 1918, and Secretary of State for War in 1919. It is ironic because Wells, in his previous works, had envisioned such unbelievable weapons as tanks, submarines, and airplanes. In real life, Churchill inspired the creation of the first tank in modern (World War I) warfare. He and one of his future comrades, then an Assistant Secretary of the US Navy, Franklin Roosevelt, helped employ the first naval uses of airplanes and submarines. Years later, it would be Churchill and Roosevelt, lifelong fans of Wells, who joined the scientific forces of their two countries in the Manhattan Project.[11]

Churchill was inspired by Wells long before the days of the Manhattan Project. In 1924, the future World

---

[8] Ibid., Chapter One, Section One.

[9] Ibid., Chapter Three, Section One.

[10] Ernest Rutherford began his studies at the University of Cambridge and in 1896 pioneered research into radio waves just prior to Guglielmo Marconi's breakthroughs. He became Director of the Cavendish Laboratory at Cambridge in 1919 and under his tutelage the neutron was discovered by James Chadwick in 1932. In that same year, his students John Cockcroft and Ernest Walton split the nucleus of a lithium atom; David Wilson, *Rutherford, Simple Genius* (London: Hodder & Stoughton, 1983).

[11] Little known to this day, Churchill and Roosevelt signed a mutual consent agreement in 1943 for the use of the atomic bomb. Although Roosevelt did not live to see the bomb used, his successor, Harry Truman, formally honored that mutual consent agreement in a communiqué from Washington to Churchill on July 2, 1945, 14 days before the first bomb was even tested in New Mexico and 15 days before the Potsdam Conference began. Churchill and Truman, almost as in a scene from one of Wells' stories, thus jointly authorized the use of atomic weapons upon Japan. In his memoirs, Churchill confirms that fact citing an address to Parliament in November 1945.

War II-era Prime Minister wrote an article in the *Pall Mall Gazette* called "Shall We All Commit Suicide?"[12] He stated, "Might a bomb no bigger than an orange be found to possess a secret power to destroy a whole block of buildings—nay to concentrate the force of a thousand tons of cordite and blast a township at a stroke?" Churchill never ceased to find inspiration from Wells. In the 1930s, Churchill, who himself became one of the most popular writers and syndicated columnists in the world, had the pleasure of inviting Wells to his Chartwell estate for many intellectual dinner conversations; often joining in was Churchill's close friend and lifelong science advisor, physics professor Frederick Lindemann. Churchill also initiated Wells' membership in his politically influential "Other Club."

That private club's membership boasted what would become some of the most outstanding leaders in Britain from 1911 until well after Churchill's death. They included Prime Ministers, senior military officers, economists, writers, artists, and academics with whom Wells would have circulated. Among some of the famous names were Arthur Balfour, Herbert Asquith, Anthony Eden, Margaret Thatcher, Lord Kitchener, Lord Roberts, Rudyard Kipling, and John Maynard Keynes.

H.G. Wells's works also influenced a young Hungarian physicist from Europe named Leo Szilárd. Wells's books were apparently in his mind in 1933 when Szilárd was standing at London's Russell Square intersection across from the historic London Museum, patiently waiting to cross while intently staring at a traffic signal. At that moment, he had an inspiration that seemed to jump into his head. He recounted.

. . . it suddenly occurred to me that if we could find an element which is split by neutrons and which would emit *two* neutrons when it absorbed *one* neutron, such an element, if assembled in sufficiently large mass, could sustain a nuclear chain reaction.[13]

In the 1930s, Churchill was in his political "wilderness years." This was in part because of his opposition to appeasement of Adolf Hitler's National Socialist expansive rule. Churchill and Professor Lindemann, often in company with Wells, would speculate over the possibility of atomic energy during those times. Lindemann affectionately called the "Prof," was also in his own wilderness years of academic life when he befriended a number of curious people, like Leo Szilárd. Lindemann guessed that Szilárd, who had been a close friend of Albert Einstein, was on to something. Lindemann knew Einstein, so he trusted Szilárd and secured a part-time job for him at Oxford. The Prof also got Szilárd's ideas on atomic energy translated into two patents, which Lindemann designated, thanks to Churchill's influence, with a high level of secrecy by the Admiralty.

Szilárd initially became more influenced by Wells's concept of using the power of the atom to aid humanity than in building a weapon. Yet Szilárd, a Jewish emigrant, had seen the danger of Hitler firsthand and became an important proponent of trying to deter Nazi Germany from building a fission weapon. That

---

[12] "Shall We All Commit Suicide?" published September 1924, *Nash's Pall Mall*. His same article appeared later in 1924 as a separate pamphlet (omitting "All" from the title) by the Eilert Printing Company in New York. In 1929, it was adapted within the conclusion of *The Aftermath*, the fourth volume in Churchill's landmark history of the First World War, *The World Crisis*.

[13] The July 1945 Szilard Petition on the Atomic Bomb Memoir by a signer in Oak Ridge, Howard Gest Distinguished Professor Emeritus of Microbiology, Departments of Biology, and History & Philosophy of Science, Indiana University, Bloomington, Indiana, p 4; and William Lanouette, *Genius In The Shadow: A Biography of Leo Szilard, the Man Behind the Bomb, by Bela A. Szilard and Jonas Salk* (New York, NY: Scribner's Sons, 1992); and Richard Rhodes, *The Making of the Atomic Bomb* (New York: Simon & Schuster, 1986), 24. ISBN 978-0-684-81378-3. OCLC 17454791. OL 7721091M.

was a novel idea at the time, but Szilárd was farsighted and proactive.

Einstein had also fled Germany, and he and his old associate Szilárd were keeping in touch. A few years later, Einstein, under Szilárd's urging, pointed out in a letter to Franklin Roosevelt in 1939 that Germany could indeed develop an atomic weapon of great destructive power.[14]

Bertrand Russell was another early author theorizing about the atom. Russell, a noted British mathematician, logician, and philosopher who in the 1950s advocated nuclear disarmament, published an article as early as 1924 titled *The ABC of Atoms,* where he foretold the release of energy through the splitting of atoms. This influenced a famous 1928 Broadway play written by Robert Nichols and Maurice Browne and directed by Rouben Mamoulian titled *Wings Over Europe*. The leading fictional character, Francis Lightfoot, portrayed a young British genius who discovered how to engineer advanced bombs utilizing the power of the atom.[15]

Moving from such influential fiction to reality, James Chadwick discovered the neutron at Cambridge University's Cavendish Laboratory in 1932. That same year at the Cavendish, under Ernest Rutherford's directorship, an Irish physicist, Ernest Walton, working with John Cockcroft, became the first to artificially split the nuclei of a lithium atom. It may be no coincidence that they were also fans of Wells' works.

Also of note, Walton and Cockcroft were among the first to experiment with a particle accelerator, which would eventually lead to more breakthroughs in understanding the atom and its power. The concept that tremendous energy could be released through a fission process slowly developed. Yet, leading scientific elders like Rutherford and even young pioneers in the new field of quantum theory like Robert Oppenheimer thought it would take massive amounts of fissionable material, maybe tons, to create such an energy release. They, therefore, doubted that it could be practically achieved. Rutherford, in fact, called the whole idea "moonshine." To the contrary, Wells, in his science fiction novels, sublimely sensed it to be possible, even inevitable.

Yet physicists, no matter how much they loved escaping with the works of Wells, remained practical. Understandably, enthusiasm for research on an "atomic bomb" was thus cool by the early 30s—even though the great American comic strip, Hollywood serials, and radio series "Buck Rogers" and "Flash Gordon" talked of "atomic" weapons as if they already existed.

---

[14] Hungarian physicists Edward Teller and Eugene Wigner, also readers of Wells, had consulted on that letter, which Szilárd wrote and Einstein then edited and signed and addressed to President Roosevelt on August 2, 1939. The Einstein-Szilárd letter urged Roosevelt to start a US program. With help from Great Britain, the Manhattan Project slowly started to take form. History does not give enough credit to the British in the long effort to build a nuclear bomb. It is ironic because Wells, as an Englishman, first conceived of such a power in his science fiction novels.

[15] Charles A. Carpenter, *A Dramatic Extravaganza' of the Projected Atomic Age: Wings Over Europe* (1928), Modern Drama 35, no. 4 (1992).

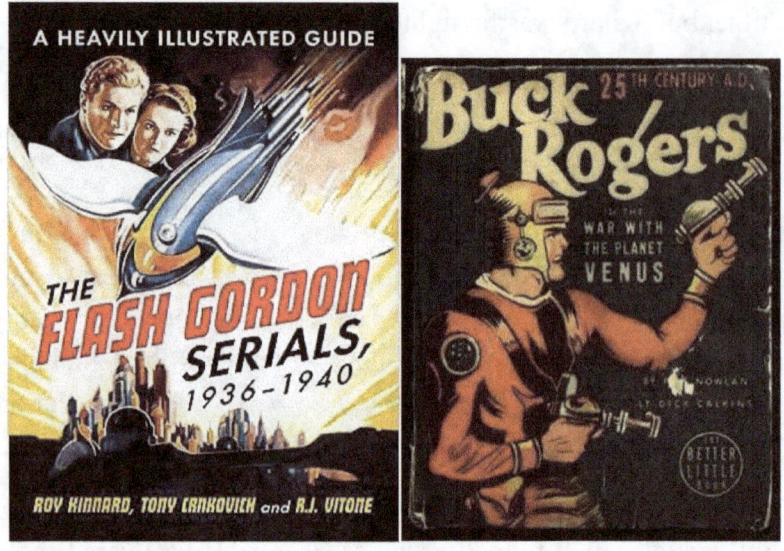

Public record images of the early sci-fi serials.
https://mcfarlandbooks.com/wp-content/uploads/978-0-7864-6615-
3.jp, file:///C:/Users/Michael/OneDrive%20-
%20National%20Atomic%20Testing%20Museum/Desktop/Picture2.jp

In Germany in December of 1938, Otto Hahn, with his pupil Fritz Strassmann, demonstrated that uranium atoms could be split when bombarded with neutrons. Nevertheless, it took pioneering nuclear theorist Lise Meitner's calculations to prove Hahn and Strassmann correct. By the spring of 1940, two German refugee physicists, Otto Frisch (nephew of Lise Meitner) and Rudolf Peierls, made a tremendous breakthrough. Working at Birmingham University, Frisch and Peierls authored two papers that theorized that if the rare Uranium-235 isotope could be separated, a practical weapon would result based on the principle of a fissionable chain reaction.

What made their theory so startling stemmed from a calculation that only a few kilograms, rather than tons, of material, would be needed—just as Wells had envisioned. A workable "super bomb" with the force of a thousand tons of TNT now seemed feasible. Not an unattractive idea for Britain, in the wake of France's fall at the beginning of the Second World War, was soon to be facing Nazi Germany alone in a battle for the very survival of their island nation. As a result of that potentially important concept outlined in what is now remembered as the Frisch-Peierls Memorandum in March of 1940, the British government formed the "Maud Committee."

That group facilitated research into fission and, in doing so, gathered together some of the leading British scientists of the day. By March 1941, the Maud Committee concluded that the theory proposed by Frisch and Peierls could realistically lead to a weapon. They'd formalized it into a report (see Appendix IV) and shared it with the United States.[16] It is no surprise that all these people had been readers of Wells at one time

---

[16] Otto Peierls would later be knighted by the British for his contributions to the Maud Committee. By March 1941, the Maud Committee had made a report called the Maud Report. Their research actually culminated in two reports, "Use of Uranium for a Bomb" and "Use of Uranium as a Source of Power." The aim was to analyze the feasibility and necessity of an atomic bomb for the war effort. Accordingly, the British created a nuclear weapons project code-named Tube Alloys. The Maud Report (see appendix IV) was shared with the United States before Pearl Harbor and became an embryonic step in the Manhattan Project by

or another.

Public record image of H.G. Wells, https://4.bp.blogspot.com/-bhorxkvWwgE/T3zpSx Hihfl/AAAAAAAAVSs /FtQZM90h8Nw/s1600/ h+g+wells.jpg.

Britain then began exchanging nuclear information with the US. In October of 1941, President Roosevelt approved a project to examine the feasibility of an "atomic bomb." After America entered into World War Two following the Pearl Harbor attack, the military formalized the Manhattan Project, and in cooperation with Britain, a race began to build the bomb before Germany might be able to. Britain worked closely with America on that project under the code name "Tube Alloys."

H.G. Wells lived to learn of the dropping of atomic bombs on Hiroshima and Nagasaki in the summer of 1945. He was often asked about the uncanny accuracy of his predictions concerning the use of aircraft in war and atomic bombs. Before he died, almost exactly one year later, Wells, with an affable yet eclectic personality to the end, often repeated a famous phrase from his 1908 book *The War in the Air*, "I told you so. You damned fools."[17]

1942.

[17] Wells apparently asked that this phrase become his epitaph and be used on his gravestone. His request was not carried out because he was cremated.

CHAPTER TWO
THE WOMAN BEHIND THE BOMB

Public record image of Lise Meitner, circa 1906,
https://study.com/academy/lesson/lise-meitner-biography-discovery.

Lise Meitner always regarded her youth as her center. During her long life, she would never cease to be comforted and sustained by those early memories. Born into a prominent Jewish family in Austria in 1878, Lise indeed had a privileged and nurtured childhood. Inspired by her father, who served as one of the first Jewish lawyers in Vienna, she became determined from an early age to make her own mark. Her father and mentor did not practice their ancestors' religion, and she became known as a very liberal and free-thinking individual. Emulating him, Lise's fascination for learning never ceased as she developed a very independent personality with a particular love of music. By age eight, she was keeping a notebook of what she interpreted as a wondrous and mysterious natural world around her.

Being her father's daughter, Lise excelled in elementary and middle school. Unfortunately, as a woman, society prevented her keen interest in math and science from being tolerated outside her home. Nor was she allowed to attend high school; the fairer sex simply was not encouraged to seek a higher education. Lise's parents nevertheless made a private education possible for her so that she could pursue the growing passion she had developed for physics. Her father purchased any books she requested. Her parents strongly supported all of their eight children, yet Lise matched every honor and every achievement of her siblings.

In 1901, Lise petitioned to take a "Matura" examination to prove her self-taught equivalency and eligibility for schooling at a university. She passed that challenging testing for higher learning and became one of the first women to study physics at the University of Vienna. There she had the opportunity to study under the renowned theoretical physicist Ludwig Boltzmann. A living legend, Boltzmann took Lise under his wing, and because of his charismatic teaching style and encouragement, in 1905, she went on to become the second woman to earn a doctoral degree in physics at the University of Vienna. Breaking other boundaries, she converted to Christianity in 1908.

As the 20th century progressed, Lise continued to be encouraged by her parents and became one of the first women to attend the Friedrich-Wilhelms-Universität in Berlin. It would be an understatement to say this proved a bold move because, in Lise's own words, "[W]omen were not tolerated in German science." In fact, her continuing education only became possible with the assistance of the most prominent German physicist, Max Planck.

By that time, Ludwig Boltzmann had died. Planck had been asked to fill his seat in Vienna but insisted on staying in Berlin. With her former scientific mentor gone, Lise similarly recognized that Planck was the only one from whom she could expand her knowledge of modern physics. Planck allowed her to attend his lectures in Berlin. However, as far as her studies within the university, she could only work in backroom workshops or cellars, as women were not allowed in the main spaces of the facilities.

Such a situation was even more of a challenge for Lise as she was still very young and extremely shy in any public forum. Nevertheless, she soon proved her abilities and became Planck's assistant. She also worked with another of Planck's associates, Otto Hahn. Hahn concentrated on chemistry and had a laboratory in the university's back areas, making it easier for her. Working closely with Hahn, they jointly discovered several new isotopes. In 1909, she authored two important papers on beta-radiation. She and Hahn then discovered "radioactive recoil," which deals with radioactive decay.

The success of their work centered on the very fact that they were so different, Hahn as a chemist and Lise as a physicist. These were then considered divergent disciplines. Before that time, chemistry garnered the focus of most research because of its practical industrial applications. Theoretical physics, on the other hand, had no appreciated place, yet it would eventually come to inspire great discoveries as the 20th century progressed. Physics would validate and explain why certain chemical processes occurred.

In 1912, the newly founded Kaiser-Wilhelm Institute opened in Berlin-Dahlem, and Hahn and Meitner gravitated to this Mecca of scientific research. Unfortunately, being discriminated against as a woman scientist, Lise had to work without a salary in Hahn's department of radiochemistry. Although, by 1913, she had attained a paid position. All abruptly ended as war came in 1914. Most of German academia focused on the war in one form or another. Hahn would work with the chemical weapons unit headed by Fritz Haber, where he volunteered for dangerous trials in which he personally tested gas masks.

Fritz Haber, a close colleague of Albert Einstein, had previously created the Haber-Bosch process, which synthesized ammonia from nitrogen and hydrogen gas to make nitrogen fertilizers. The innovation literally saved Europe's growing population, which had exhausted its farmland's soil after centuries of agriculture. However, Haber's enthusiasm and talent in helping his beloved Germany then turned to waging war with deadly chemical agents derived from his previous research. He created the age of "weapons of mass destruction." In later years, Hahn would have great problems coming to grips with such work. Hahn could never forget the inhumanity of their chemical toxins as science became so powerfully employed in warfare.[18]

The war similarly changed Lise. She volunteered as a nurse, handling X-ray equipment for the Austrian army. She served very near the Eastern Front and saw massive casualties. Such an experience added to her growing maturity. The war put in motion a long process that would give women across Europe a greater

---

[18] Thomas Hager, The Alchemy of Air: A Jewish Genius, a Doomed Tycoon, and the Scientific Discovery that Fed the World but Fueled the Rise of Hitler, 1st ed. (New York, NY: Harmony Books, 2008).

voice, slowly gaining more respect. By war's end, Lise had returned to the Institute, where she and Hahn discovered the first long-lived isotope of the element protactinium by 1919. This discovery won her the Leibniz Medal from the Berlin Academy of Sciences. The award earned her the privilege of her own physics section at the Kaiser-Wilhelm Institute for Chemistry.

Lise Meitner continued to excel and became more integrated into her circle of associates. In 1922, she discovered the "Auger Effect," which deals with the emission of electrons with signature energies. (The effect was named for Pierre Victor Auger, a French scientist who independently discovered the phenomenon in 1923.) By 1926, at the University of Berlin, Lise became the first woman in Germany to attain a full professorship in physics. She soon headed the physics department of the Kaiser-Wilhelm Institute for Chemistry[19].

By then, Hahn was the Director of the Kaiser-Wilhelm Institute. Together, they started research into what are now known as transuranic elements. These are elements with an atomic number greater than 92 which eventually led to the discovery of nuclear fission. In those post-World War One years, the academic environment in which Lise worked was unparalleled in history. She developed close friendships with Albert Einstein and Leo Szilárd. Lise taught a seminar on nuclear physics and chemistry with Szilárd.

She then began to correspond with James Chadwick. Chadwick worked at the premier English research center, the Cavendish Laboratory at Cambridge University. There, in 1932, Irish physicist Ernest Walton, working with John Cockcroft, became the first to artificially split the nuclei of a lithium atom. They did this using an early form of a particle accelerator. Just weeks before that momentous experiment, James Chadwick identified the neutron at Cambridge. Chadwick and Lise Meitner were among a select group who were beginning to understand the mysteries of nuclear physics.

In England, Chadwick was not the preeminent physicist. That distinction went to Ernest Rutherford. Rutherford, as noted in the previous chapter, did not promote nuclear physics as a potential energy source because he calculated it would take so much exotic material to facilitate a release of energy that the whole concept, he thought, was nonsense or, as he said, "moonshine."[20]

Of course, other groups were studying the atom and its potential. In France, the leader became Jean Frédéric Joliot-Curie, and in Italy, Enrico Fermi. Meitner and Hahn were at the forefront in Germany. Today, history looks back on all of this as the race for the atomic bomb. However, at the time, this was not a clear goal by any means. The thought of a nuclear weapon, although gradually becoming a theoretical possibility, was not what any of the scientists wanted. Their passion was abstract research at that point.

As also previously mentioned, the single exception to this was Leo Szilárd. He did not favor using science to make a weapon of great destruction, but he was clearly the first to be entranced by that possibility. As early as 1933, this young Hungarian physicist had come up with the novel idea that it might be possible to

---

[19] Ruth Lewin Sime, *Lise Meitner* (Berkeley, CA: University of California Press, 1996), p. 138.

[20] *The* (London) *Times*, September 12, 1933, coverage on page 7 detailing Ernest Rutherford's speech to The British Association titled *Breaking Down the Atom, Transformation of Elements*. He stated in a section called *Hope Of Transforming Any Atom*: ". . . we could not expect to obtain energy in this way. It was a poor and inefficient way of producing energy . . . anyone who looked for a source of power in the transformation of the atoms was talking moonshine."

build an "atomic bomb."

At the time, Szilárd had recently been an academic at the University of Berlin, where he earned his Ph.D. in physics. But he, like many scientists of Jewish descent, saw the need to leave Germany while they still could of their own free will. When Adolf Hitler came to power in 1933, Lise also found herself in a precarious situation. Despite her conversion to Christianity, Lise's Jewish heritage forced her dismissal from the head of the physics department of the Kaiser-Wilhelm Institute for Chemistry. She later commented on how naive all the intellectuals had been. She stated, "I cannot believe my adopted homeland of Germany was capable of that."[21]

The same dangerous situation held true for those like Szilárd, Einstein, and Meitner's nephew Otto Frisch. Frisch, as this book often emphasizes, would later have a significant impact on the atomic bomb project in England and then on the Manhattan Project. Of course, it is so important to be reminded that one of many reasons Germany would later fall in defeat during the Second World War was specifically because its ideology became so intolerant of human values.

Lise wanted to stay in Germany to try to outlast what she saw as a temporary and reactionary trend. She had one advantage in that she still held her Austrian citizenship, which protected her for a while. Max Planck, President of the Kaiser-Wilhelm Institute, and Otto Hahn, Director of the Institute, also protected her the best they could. For a time, she remained employed. Then in March 1938, Anschluss came. After that point, her Austrian passport ceased to serve as an official document, while her Jewish heritage made applying for German papers a potential death sentence. It became a ridiculous, although seriously dangerous, situation. To some extent, she still refused to believe Germany could have become so repressive.

Hahn then forced Meitner to see reality and convinced her she had to flee. Assisted by Hahn and Dutch physicists Dirk Coster and Adriaan Fokker, she fled to the Netherlands with only the clothes she wore, being forced to abandon all her worldly possessions except for a diamond ring Hahn had given her to barter her way through the border. She did not have to sell the ring, but at the age of 60, she had only ten marks in her purse and no idea what to do. With a lifetime of accumulated wealth totally wiped out, Lise felt devastated. Finding work as a woman in the academic world, let alone nuclear physics research proved extremely difficult. Finally, she found employment in the Manne Siegbahn Laboratory in Stockholm despite extreme prejudice against women in that organization. Although finding refuge in Sweden did prove an opportunity of sorts, it allowed her to establish a working relationship with Niels Bohr in nearby Denmark. Besides being that country's most distinguished citizen and a worldwide scientific superstar, Bohr was the top expert in theoretical physics.

Both Lise Meitner and her nephew Otto Frisch attended a noted lecture by Bohr in Copenhagen at the Niels Bohr Institute in November of 1938. Then, in December, back in Germany, Hahn demonstrated that uranium atoms could be split when bombarded with neutrons with his pupil Fritz Strassmann. Nevertheless, it took Lise's calculations through correspondence to prove Hahn and Strassmann correct.

Ironically, Hahn had no problem corresponding with Lise because Germany had a very efficient

---

[21] Lise Meitner gave this quote in a letter to her friend Eva von Bahr-Bergius in 1939, after she had fled from Nazi Germany to Sweden.

overnight airmail service with Stockholm. Hahn commented years later that Lise was the only one who really understood the physics of his chemistry and, therefore, the physical significance or explanation for the processes in his laboratory. Frisch later said his aunt had these brilliant moments when she made the connection between the experiments and theory. Lise realized by 1938 that the trans-uranium elements she and Hahn had studied for so many years were actually fission fragments. She then started to interpret some of the physical dynamics of fission. Soon, she and Frisch calculated that the potential release of energy could be a thousand times greater than any chemical reaction.[22]

In continued consultation and correspondence with Lise, Hahn, and Strassmann, worked on a second publication after a hurried first article in *Naturwissenschaften*, a peer-reviewed scientific journal now called *The Science of Nature*. In that second article in February of 1939, Hahn and Strassmann used, for the first time, the name *Uranspaltung* or uranium fission. However, the actual term "fission" came from Lise and her nephew Frisch, who published a letter in *Nature, International Journal of Science* in February of 1939:

. . . It seems therefore possible that the uranium nucleus has only small stability of form, and may, after neutron capture, divide itself into two nuclei of roughly equal size (the precise ratio of sizes depending on finer structural features and perhaps partly on chance). These two nuclei will repel each other and should gain a total kinetic energy of c. 200 Mev., as calculated from nuclear radius and charge. This amount of energy may actually be expected to be available from the difference in packing fraction between uranium and the elements in the middle of the periodic system. The whole 'fission' process' can thus be described in an essentially classical way, without having to consider quantum-mechanical 'tunnel effects', which would actually be extremely small, on account of the large masses involved. After division, the high neutron/proton ratio. . . .

Lise Meitner, Physical Institute, Academy of Sciences, Stockholm.

O.R. Frisch, Institute of Theoretical Physics, University, Copenhagen.[23]

Word of the experiment spread from Meitner to Szilárd and through him to Einstein. Einstein and Szilárd wrote to Franklin Roosevelt in August 1939, highlighting their fear that Nazi Germany might develop a fission weapon. The letter had originally been intended for the King of Belgium, whom Einstein knew through the Royal Family. Einstein wanted to alert the King to the importance of his country's large stockpiles of uranium ore.

However, just one month before Hitler invaded Poland, they deemed the situation so serious that they realized the letter should instead go directly to the US President. Unfortunately, the letter did not reach Roosevelt's attention until October. Many by then realized that Germany had the know-how and scientists to start a nuclear program, which they feared would be fully supported by Hitler.

---

[22] Otto Frisch, *What Little I Remember* (Cambridge University Press, 1979), ISBN 0-521-40583-1. OCLC 861058137.

[23] Otto Frisch first detected evidence of fission fragments on January 13, 1939. The *Nature* article did not appear until February, at which time Jean Frédéric Joliot-Curie published similar conclusions from identical experimentation in Comptes Rendus. Jean Frédéric Joliot-Curie was a French physicist and husband **of** Irène Joliot-Curie. They were jointly awarded the Nobel Prize in Chemistry in 1935 for their discovery of the principles of radioactivity or more specifically, how to induce or produce artificial radioactivity.

The famous letter urged Roosevelt to start a US program. Britain, which had just begun breaking into the German secret Enigma codes, learned that Germany had indeed had several such nuclear research projects underway by 1940. Roosevelt thus authorized the formation of the "Uranium Committee" in the Bureau of Standards, the only government entity other than the US Navy remotely connected to scientific work. Then, in the spring of 1940, Lise's nephew, Frisch, who had made his way to England along with German refugee Rudolf Peierls, came up with his astounding discovery.[24] Frisch and Peierls, working at Birmingham University, authored two papers theorizing that "if" the rare isotope uranium-235 could be separated from the more common element uranium-238, a weapon would result based on the principle of a fissionable chain reaction. They, in fact, by then standardized the word "fission."

As already discussed, what made their theory so startling was that it calculated that only a few kilograms, rather than tons, of this material would be needed. This was a eureka moment. The realization is certainly attributed to Meitner's influence on Frisch and Peierls.

Finally, Ernest Rutherford's "moonshine" argument was put to rest. A practical "super bomb" with the force of at least a thousand tons of TNT now seemed feasible. Meanwhile, in Germany, the calculations over-estimated the amount of enriched material they thought it would take for a bomb, and the project lost its priority status. Of course, no one outside Germany could yet know that or the fact that Hahn and Strassmann had refused to work on any nuclear project, although they did remain in Germany.

Other renowned German scientists who stayed, like Werner Heisenberg, were known to be working on a bomb project. Heisenberg, in fact, led the project. Yet Germany's best scientists had already left. Since 1933, eleven Nobel prize-winning physicists and four chemists had fled, including Hans Bethe, Felix Bloch, Max Born, Albert Einstein, James Franck, Heinrich Gerhard Kuhn, Peter Debye, Dennis Gabor, Fritz Haber, Gerhard Herzberg, Victor Hess, George de Hevesy, Erwin Schrodinger, Otto Stern, and Eugene Wigner.

So many key scientists feared Hitler could master an atomic weapon that it influenced leaders like Churchill and Roosevelt to push for an all-out endeavor. As previously discussed, that initially became the formation in Britain of what was known as the Maud Committee, which sanctioned the theory proposed by Frisch and Peierls. First, however, they needed to isolate and understand the fissile material. Although most scientists still felt a fission bomb would be immensely difficult to create, Frisch and Peierls realized that natural uranium (U-238) could be subjected to an isotopic separation to produce a fissile material called U-235. That material would be extremely challenging to produce, but the British by that point, understood the way forward in which to make an atomic bomb a reality.

Thus, England initially took the lead in a race for the bomb, largely thanks to Lise's pioneering influence on others. Lise had been in Denmark visiting Niels Bore when Germany invaded it on April 9, 1940. After discussions with Bohr, she continued to send telegrams to Frisch and other scientific leaders. Some felt the formation of the Maud Committee was named after one of Lise's communications with British physicist Owen Richardson. She had written that Bohr and his family were unharmed following the invasion and

---

[24] Lise Meitner had also intended to go to England but waited too long and missed her opportunity. In July 1939, she was in Cambridge when John Cockcroft offered her a position at the Cavendish Laboratory. After returning to Stockholm, she delayed too long in organizing her move and got trapped by the exploding war in Europe. Fortunately, her nephew Otto Frisch made it, and with her influence, he helped prove an atomic bomb to be a practical concept.

ended her communication with an enigmatic message. It read, "... please inform Cockcroft and Maud Ray Kent," supposedly meaning "radium being seized." Actually, Lise was simply referring to a woman by that name. Her telegram, however, was reviewed by many physicists who were becoming concerned that Germany must be pursuing an atomic bomb and thus collecting vital materials. German scientists did utilize the cyclotron at Bohr's Institute for Theoretical Physics in Copenhagen and the cyclotron in Joliot's laboratory in Paris.[25]

Lise found herself trapped in Stockholm, and Sweden remained neutral throughout the war. Her location and neutral status did afford her the means to maintain correspondence with many colleagues in Germany and around Europe, by which time she had learned small bits and pieces of various work in nuclear physics and Werner Heisenberg's attempts to build a nuclear reactor for Germany. Heisenberg had traveled to Denmark to visit Niels Bohr in Copenhagen in September of 1941, which Lisa and others were to hear stories about.[26] She passed her various items of interesting information onto Frisch and then further tidbits to Bohr after he escaped Denmark in 1943 and fled to England and then America, where he visited the Manhattan Project Y laboratory in Los Alamos, New Mexico. So, whether intentional or not, Lise served as a recourse for the Allied war effort.

Frisch stated long after the war that by 1943, the Allies had wanted to get Lise to Los Alamos but that she had refused their attempts and any offers involving war work.[27] This was very atypical among scientists within the Allied camps as nearly all were driven by fear of Nazi Germany. Those who knew about the race to build the bomb saw it as just that, a race to get the bomb before Hitler could. Yet Lise, said by those who knew her, simply could not work on a weapon of war.

By 1943, Britain concentrated all its own efforts, then known as the "Tube Alloys" project, toward

---

[25] In early 1929, Leo Szilárd filed patent applications for the concept of what became the cyclotron. Ernest Lawrence built the first cyclotron at the University of California Berkeley by 1930. Cyclotrons were the most powerful particle accelerator technology until the 1950s. The cyclotron played a crucial role in the development of the atomic bomb. The cyclotron functions as a particle accelerator and can produce fissile isotopes of uranium or plutonium known as uranium-235 and plutonium-239 which became critical to sustaining nuclear chain reactions.

[26] My son and I have always been fascinated with this story of a mysterious meeting between Niels Bohr and Werner Heisenberg in Copenhagen in September of 1941, just a year after Denmark fell to the Nazis. The two men were then the preeminent physicists of the day; however, Bohr had become a victim of Hitler's aggression while Heisenberg worked for the Germans. Both men were longtime friends and at heart remained so even though they were then on opposite sides of a growing world war. Bohr would eventually escape to the Allies and find his way to Los Alamos to tell his story of the meeting with the man who was then, for all apparent perceptions, in charge of Hitler's atomic bomb project. Thus, that meeting in Copenhagen has tremendous historical significance. The problem for scholars is that Bohr and Heisenberg, who survived the war, tell very different stories of what they said to each other so many years ago in Copenhagen. Over the decades, their accounts kept changing. Yet it remains an extremely important historical event to define because the Allies then needed to know how far the Germans were along in building an atomic bomb and if Heisenberg was going to deliver that bomb or purposely confuse Hitler and Albert Speer into thinking it an impractical venture. We still do not know the whole story nor exactly why Heisenberg failed to develop the bomb. After the war, Heisenberg, mainly in his own defense, claimed he was only working on developing a nuclear reactor and not a weapon.

[27] Otto R. Frisch, *Lise Meitner, 1878-1968, Biographical Memoirs of Fellows of the Royal Society* (London, 1970), p. 414.; *and* Otto

R. Frisch, "*Lise Meitner, Nuclear Pioneer,*" *New Scientist* (November 4, 1978), pp. 426-428.

supporting the Manhattan Project. This would not, however, involve Lise Meitner, who refused any involvement in making a weapon based on nuclear fission. She stated, "I will have nothing to do with a bomb!"[28] At the end of the war, Lise shunned the atomic bomb and resisted any connection to it and her work. She also denounced her colleagues who remained in Germany. She wrote a letter in June of 1945 expressing her anger to Hahn, which, although very revealing, was claimed to have never been received by her long-time colleague:[29]

You all worked for Nazi Germany. And you did not even try passive resistance. Granted, to absolve your conscience you helped some oppressed person here and there, but millions of innocent human beings were murdered and there was no protest. Here in neutral Sweden, long before the end of the war, there was discussion of what should be done with German scholars once the war is over. What then must the English and Americans be thinking? I and many others are of the opinion that the one path for you would be to deliver an open statement that you are aware that through your passivity you share responsibility for what has happened, and that you have the need to work for what can be done to make amends. But many think it is too late for that. These people say that first you betrayed your friends, then your men and your children in that you let them stake their lives on a criminal war – and finally that you betrayed Germany itself, because when the war was already quite hopeless, you never once spoke out against the meaningless destruction of Germany. That sounds pitiless but nevertheless I believe that the reason I write this to you is true friendship. In the last few days one had heard of the unbelievably gruesome things in the concentration camps; it overwhelms everything one previously feared. When I heard on English radio a very detailed report by the English and Americans about Belsen and Buchenwald, I began to cry out loud and lay awake all night. And if you had seen those people who were brought here from the camps. One should take a man like Heisenberg and millions like him, and force them to look at these camps and the martyred people. The way he turned up in Denmark in 1941 is unforgettable.[30]

After the war, Lise chose to concentrate on peaceful uses of nuclear energy. In Sweden, she worked at Siegbahn's Nobel Institute for Physics, the Swedish Defense Research Establishment, and the Royal Institute of Technology in Stockholm. There, she had a laboratory and participated in research on "R1, " Sweden's first nuclear reactor.

She became a Swedish citizen when a position was created for her at the University College of Stockholm. Finally, in 1947, she earned the salary of a professor at the Swedish Council for Atomic Research. Completely overlooking Lise's contributions to the discovery of fission, the Royal Swedish Academy of Sciences awarded Otto Hahn the Nobel Prize in Chemistry for the discovery of nuclear fission. Only later could Lise finally bring herself to revisit Germany, where she renewed her six-decade-old friendship with Hahn.

---

[28] Ibid.

[29] Otto Hahn claimed he never received the letter, but Lisa may have had some disbelief about his claim.

[30] Letter from Lise Meitner to Otto Hahn, June 1945; Ruth Lewin Sime, *Lise Meitner* (Berkeley California: University of California Press, 1996), p. 310.

In 1960, she relocated to England. By 1966, in an effort to partly rectify her exclusion from a Nobel Prize, the US Department of Energy jointly awarded Hahn, Meitner, and Strassmann the US Fermi Prize. In 1968, Lise Meitner and Hahn died within months of each other, both at the age of 89. A final fitting gesture to her life came from her nephew, Otto Frisch. Frisch provided the epitaph on her gravestone, "Lise Meitner: a physicist who never lost her humanity."

Lise Meitner (1878–1968)
© Anne Meitner, Malcom Farrer-Brown,
Tony Brown (lottemeitnergraf.com).
https://www.mpg.de/11721986/Lise-Meitner.

CHAPTER THREE
THE MAN BEHIND THE BOMB

**Public record image of Leo Szilárd,
https://quotesgram.com/leo-szilard-quotes-
atomic-bomb/.**

As early as 1933, a young Hungarian physicist named Leo Szilárd had come up with the novel idea that it might be possible to build an "atomic bomb." As already examined, his vision was not unlike that of the popular science fiction stories of H.G. Wells. At the time, Szilárd had recently been an academic at the University of Berlin, where he earned his Ph.D. in physics. Because of his Jewish heritage, he wisely fled from Germany, just as he had originally fled from Hungary years earlier, from fascists and communist radicals. Arriving in England that year, a handful of businessmen provided him with minor support in his pioneering passion for atomic research, most of which took place with makeshift equipment in modest hotel flats. While in England, he helped the Academic Assistance Council to aid other refugees from Hitler's Germany.[31]

By 1933, the 35-year-old Szilárd had acquired an interesting background. He originally came from a privileged childhood filled with governesses and private tutors. His father had been an accomplished civil engineer. Szilárd also studied that profession, as did his brother, who became a noted engineer himself. So, although he eventually became a physicist, Szilárd had a unique understanding of not just the theory of how things worked but how actually to put theory into practice with his own two hands. He also understood that at the time, physics, in general, was not considered that vital of a discipline as it later became during WWII and after. The "heroes," in fact, were still the chemists who, during WWI, produced poison gas.

After surviving the horrors of the First World War as a soldier in the Austro-Hungarian Army and then a bout of the Spanish flu, Szilárd eventually became a student under Max Planck and Albert Einstein in Germany. He and Einstein developed a good friendship. At one point, they even collaborated on ways to improve home refrigeration technology and patented a number of designs. In post-World War One

---

[31] The Academic Assistance Council (AAC) was founded in April 1933 by William Beveridge. To this day, various incarnations of that group and its original core mission continue to assist academics fleeing from repressive governments.

Germany, ways of making ends meet became a necessity even for theoretically gifted minds like Einstein and Szilárd.

Also in the real world, the theory of a nuclear chain reaction slowly came into being by the 1930s. It conceived of one neutron splitting an atom, producing two neutrons that split two atoms, which would produce four neutrons that split four atoms, and so on. In Szilárd's mind, if this chain reaction could be triggered all at one time, a huge release of energy might be possible. The next problem was to determine what element could release the initial two neutrons after splitting its atom. Beryllium was first considered long before uranium seemed a logical choice. A critical mass of energy had to be reached, the part in which no one could yet determine if it was even possible or practical. Practical became the keyword. Most physicists thought that if a chain reaction could be attained, it would require many tons of fissile material. Such considerations, as this book has already examined, meant it would be impossible to wield a weapon and even be questionable as a practical energy source.

Various bits and pieces of nuclear research like this had been evolving for a long time. The neutron itself had only recently been understood. As already detailed and must, for the sake of further understanding, repeat, an Irish physicist, Ernest Walton, working with John Cockcroft, became the first to artificially split the nuclei of a lithium atom. They did this using an early form of a particle accelerator at the Cavendish Laboratory at Cambridge University in 1932. Just weeks before that momentous experiment, James Chadwick identified the neutron at Cambridge. That discovery certainly percolated in Szilárd's head when he had his epiphany of a chain reaction while standing under a London street signal in 1933. The concept that tremendous energy could be released through a fission process continued to develop in a piecemeal fashion.

Also, as has been discussed but must be further stressed, a highly respected New Zealand-born physicist from Cavendish Laboratory, Ernest Rutherford, became vocal on that point. He stated that a nuclear energy release would never be practical, even if it could be proven possible. Known as the father of nuclear physics, he had discovered the concept of radioactive half-life. As a result, his skepticism dissuaded serious attention to the concept of nuclear energy, which Rutherford called "moonshine" science. This upset Szilárd and countered that "experts only tell you what can be done, not what could be done."[32] Of course, Rutherford should not be judged too harshly because Robert Oppenheimer, a much noted but younger physicist at the time, had similar doubts.

This is where H.G. Wells made one of many key and memorable appearances on the historical stage as we saw in chapter one. Szilárd met Wells face to face while in London and became convinced that the possibility of a nuclear power source was too important to ignore. He hoped, however, that splitting the atom would not mean a weapon but a way to propel mankind to the stars. Modern scholars view Wells as only a vague inspiration to 20th century visionaries, but for people like Szilárd and even Winston Churchill, Wells became a real-life player where science fiction became real science. Nevertheless, enthusiasm for research on an "atomic bomb" remained cool when Szilárd approached the British Army in the fall of 1933 about a patent for his concept. This would include concepts for a nuclear reactor meant for peaceful purposes. He, in fact, developed a patent based on the chain reaction principle in collaboration with physicist Enrico Fermi in 1934. Also, in 1934, Szilárd, with physicist T.A. Chalmers, discovered a way to chemically isolate

---

[32] William Lanouette, *Genius in the Shadows: A Biography of Leo Szilard: The Man Behind The Bomb* (New York: Skyhorse Publishing, 1992), pp. 131–132.

radioactive isotopes in a laboratory. Szilárd was nominated for a Nobel Prize; however, he had virtually no chance for the award as almost no one in the establishment took him seriously after talking about such radical concepts as chain reactions.

Szilárd was not as concerned about his own popularity or even exploiting the business applications of nuclear theory. He did, however, seek funds for research because he feared that Hitler's Germany might gain a lead, which was the point he wanted the experts to focus on. Ironically most scientists and governments, including Nazi Germany, did not yet feel the idea of atomic weapons to be practical. Despite this, Szilárd persisted and, in 1936, secured another British patent involving atomic energy principles. Szilárd next approached the General Electric company of Britain and confidently told them that his patents would lead to a new energy source, making coal and oil obsolete. Of course, much of the modern economy revolved around coal and oil, so such statements generally branded him as a lunatic. That is, until one day, he met Winston Churchill's longtime friend and science advisor, Frederick Lindemann.

Churchill was not yet in power, but in the 1930s, Churchill and Lindemann, often in company with Wells, speculated over long dinners and late-night cocktail socials at the Other Club and Churchill's country estate Chartwell about the possibility of atomic energy. Lindemann, as we have seen, was perceptive enough to realize that Szilárd was not talking nonsense. He secured him a part-time job at Oxford. Lindemann also had the influence to get his two patents designated with a high level of secrecy by the Admiralty. Thus, Szilárd's patents were useless for commercial purposes because they became highly classified. Later, when the US Army offered to pay Szilárd for his patented ideas, he refused payment. Instead, he asked only to be reimbursed for the expenses he had incurred. Szilárd never pursued monetary gain; instead, he sought official recognition and funding to conduct cutting-edge research.

It is worth noting that Szilárd finally received recognition from some of the right people in the 1930s, even if those people were not yet the key players. Lindemann went on to be Churchill's supreme war-time scientific advisor and leader of Britain's Second World War science endeavors under the cabinet office of Paymaster General. He was an equivalent to Vannevar Bush in the US, who later became the behind-the-scenes overlord of the US atomic bomb project. If Britain had not taken an initial interest in nuclear research under Churchill, it is unlikely the United States would have ever ended up with a formal project like the Manhattan Project, or at least not in time to make a difference.

As the war came closer, Szilárd was not the only scientist on the early trail of atomic research. As has already been covered in earlier chapters, in December of 1938, Berlin University chemist Otto Hahn, with his pupil Fritz Strassmann, in consultation with Lise Meitner, proved that uranium atoms could be split when bombarded with neutrons. French Nobel Prize-winner Jean Frédéric Joliot-Curie proved, only weeks later, that he could split a uranium atom and, in doing so, demonstrated that the split atom would release more than one neutron, thereby substantiating Szilard's theory that a chain reaction was possible. As world war loomed in 1939, the larger scientific community still felt the idea of an atomic bomb to be impractical, if not impossible. However, many did start to sympathize with Szilárd's concerns over Germany. More and more leaders in science and government feared what might happen if the Nazis committed to an all-out project toward atomic research. By that point, the idea of an atomic super weapon was gaining increased attention as Joliot-Curie and French physicists and engineers drafted a patent for a theoretic uranium bomb as well as a working nuclear reactor.

When Szilárd left Britain for America in 1938, he initially teamed up with his friend, Italian physicist Enrico Fermi, at Columbia University. There, they confirmed Joliot's neutron experiments. Soon, they would envision a design for what would later become the world's first working reactor. At the same time, deep inside the Third Reich, the Ministry of Education became interested in the possibility of building a uranium fission reactor. The Nazi ministry ordered theoretical physicists to form a research project called the "Uranium Club." The USSR Academy of Sciences then set up an early project looking into the "Uranium Problem." These were very important early projects that were not given wide support by their own countries.

Then, as has been covered, the foundation shook for those still skeptical of the importance of atomic research. In the spring of 1940, an Austrian and German-born British physicist, Otto Frisch, nephew of Lise Meitner and Rudolf Peierls, made an astounding discovery. Frisch and Peierls, working at Birmingham University, authored two papers theorizing that if the rare isotope uranium-235 could be separated from the more common U-238, a weapon based on a moderate amount of enriched material would result—based on the principle of a fissionable chain reaction. Next came the previously mentioned Frisch-Peierls Memorandum, as the British government formed the "Maud Committee." That organization facilitated research into fission. Yet the committee, with Peierls' help, soon realized that such a weapon would require a large-scale enrichment program for the separation of U-235. Therein lay Britain's catch-22. The costs and logistics of producing highly enriched uranium were then beyond their ability. Britain's national resources were already stretched to the breaking point in the war with Hitler. Yet, Hitler was the very problem.

Einstein and Szilárd pointed out in their famous letter to Franklin Roosevelt in August 1939 that the fear centered on the nightmare scenario that Nazi Germany could build a fission weapon. In response, President Roosevelt authorized the formation of the "Uranium Committee."

The first meeting of the Uranium Committee in the late fall of 1939 was attended by Teller, Wigner, and Szilárd. All three were looked upon at the time with great suspicion because of their foreign accents and heritage; however, they were primarily the only experts in America familiar with the physics of what an atomic bomb might look like. Certain Army officials advocated restricting their access to atomic secret information, little realizing that those three individuals were largely the origin of everything then considered secret in nuclear physics.

That also proved a period in which British Prime Minister Winston Churchill successfully solicited more and more aid and close cooperation from Roosevelt. The cash-strapped Britain had little to offer in return except scientific know-how, which Churchill quietly and subtly used as a bargaining chip. Aside from nuclear theory, Britain was far in the lead in areas like radar, anti-submarine warfare, ballistics research, and aircraft engine technology.

In the nuclear field, they had another unique insight. The Cavendish Laboratories predicted in 1940 that a new artificial element called "Element 94" could be created by bombarding uranium-238 with neutrons. Simultaneously, the Berkeley Radiation Laboratory discovered under physicist Edwin McMillan that a byproduct of the operations of nuclear reactors was that same new Element 94. Element 94, soon called plutonium by McMillan, was subsequently first produced by chemist Glenn Seaborg in early 1941 on the cyclotron at the University of California Berkeley. Plutonium would eventually make large-scale production of nuclear weapons possible. Ernest Lawrence had built the first cyclotron at the University of California Berkeley by 1930. Since then, a close colleague and instructor of his at Berkley, Robert Oppenheimer, had been a leader in theoretical physics. By 1940, Oppenheimer was seen making sketches on his classroom blackboard of what he envisioned an "atomic bomb" might look like.

In March, Einstein sent a second letter to Washington. In it, he pointed out that since the outbreak of war in Europe, Germany had made moves to secure uranium stockpiles. This, he stressed, was a very troubling sign. In October of 1941, the American Manhattan Project formally began. Britain worked closely with the US on this project under the code name "Tube Alloys." Sir Wallace Akers led that British team overseeing the Directorate of Tube Alloys in England. Initially, with America's continued inability to enter the war as a combatant, the British had planned to design a plant to produce enriched uranium on their own despite the great costs associated. More significantly, the Maud Committee's continued research and formal 1941 report (see Appendix IV) motivated Roosevelt to step up the Uranium Committee's pace with Vannevar Bush's help. After Pearl Harbor on December 7, 1941, and coupled with the US entry into the war, a greater American commitment followed.

The Manhattan Project, which was a US Army endeavor, started in June of 1942 to try to get a practical nuclear project underway. Later, under the Quebec Agreement in 1943, Britain abandoned its own Tube Alloys effort, and via an agreement between Churchill and Roosevelt, Britain became a partner, albeit a junior partner, in the Manhattan Project. Churchill, in fact, understood the immense economic burden that a nuclear project posed and willingly looked to the United States to make the tremendous investment required.

Szilárd remained an important player during those early days. By January of 1942, his talents were being used at the Metallurgical Laboratory in Chicago. On December 2, 1942, he took part with Enrico Fermi in the first artificial self-sustaining nuclear chain reaction under the viewing stands of Stagg Field. Fermi and Szilárd would eventually be granted the official patent for devising a nuclear chain reaction. Soon, he became a research associate and later the chief physicist at the Metallurgical Laboratory.

Manhattan Project Military Director Brigadier General Leslie R. Groves Jr. tried to dismiss Szilárd because of his European background and the implied security concerns. Szilárd had retained his German passport and was technically considered an enemy alien. Groves also developed an intense personal dislike of the free-thinking Szilárd and ordered the FBI to track his every move. Fortunately, the Secretary of War, Henry L. Stimson, who was aware of his impressive background and close association with Professor Lindemann, intervened, and Szilárd survived.

Science Photo Library, Illustration of the first man-made self-sustaining nuclear chain reaction under the viewing stands of Stagg Field, https://ichef.bbci.co.uk/news/800/cpsprodpb/D016/production/_84007235_chainreaction_cut.jpg.

After that, partially because of Groves's continued prejudice, Leo Szilárd became a minor figure in the atomic bomb story, but his early warnings of Germany were prophetic. True, they never came to pass, but they could have. That unthinkable possibility was enough to warrant the tremendous Allied effort to build the bomb. Certainly, there were German scientists of note, like Werner Heisenberg, who stayed in Germany and whose scientific reputation soared far above that of his American counterpart, Robert Oppenheimer.

German efforts never succeeded, yet in 1940, no one could have guessed that. Fortunately for the Western powers, Germany had lost many of its scientific all-stars due to Hitler's anti-Semitic policies. Another fact that is not often given adequate attention concerns the mindset of the German physicists who remained in Germany. It is a small but critical observation. The group who worked to try to deliver an atomic bomb to Hitler have often been seen as hesitant in their efforts to give the Nazi leader such a weapon. That, in fact, may not have been the case at all. Accounts actually show an intensive effort in the earlier days of the war. However, one major miscalculation was made by the elite physicists. It was not much about mathematical miscalculation but about human miscalculation.[33]

Unlike their counterparts in America who eventually came together in the Manhattan Project, the Germans under Werner Heisenberg developed a prejudice toward the value of integrating engineers into the inner circle. In the American project, the engineers were utilized to their fullest by Manhattan Project physicists. That became a significant success of the project under the coordination of Robert Oppenheimer. Manhattan Project engineers came to outnumber the physicists by ten to one, working all the while in a close teamwork atmosphere.

This was not so in Germany, where Heisenberg's physicists insisted on doing much of the engineering work themselves, which proved an extremely unproductive use of their time. More importantly, unlike individuals like Szilárd, who had practical engineering experience, Heisenberg's men were not as able in using their hands as their minds. It is a small but critical detail first pointed out by authors Peter Pringle and James Spigelman in 1981 and still overlooked by mainstream historians.[34]

Another critical mistake the Germans made was the fact that they never realized that the commercial-grade graphite they were using to moderate the uranium reactions in their first reactor contained impurities made up of boron. Boron absorbed neutrons and retarded the reaction process.

They instead moved to heavy water for the process of which they could never get enough. Szilárd learned very early on that commercial-grade graphite contained boron and worked with manufacturers to develop 99 percent pure graphite. That is what made the first self-sustaining nuclear reactor in Chicago a success. It is not an exaggeration to say that if the Germans would have had Leo Szilárd, they may have gotten the bomb! Noted writer and researcher Jonathan F. Keller has additional insights on Germany's failure to win the race to make an atomic bomb:

The most nightmarish of World War II alternative history scenarios is the one in which Nazi

---

[33] Werner Heisenberg and the Germans made a calculation mistake regarding the practicality of a U-235 fission device. They thought a bomb required far too much U-235 to make it a practical weapon. After the war, Heisenberg suggested that the project failed because he deliberately sabotaged it. Maybe to save his own reputation, he intimated that he knew the proper critical mass for U-235 in 1942, but deliberately misled Nazi officials. The full truth may never be known.

[34] Peter Pringle and James Spigelman, *The Nuclear Barons* (New York: Holt, Rinehart and Winston, 1981).

Germany acquires atomic weapons. In fact, by the spring of 1945, when America's massive nuclear program was reaching its culmination, the Nazi atomic program consisted of one experimental reactor in a cave in southern Germany, operated by scientists who lacked a clear conception of how to build an atomic weapon. . . In the late 1930s, the most famous physicist in Germany (Einstein having left Germany for New Jersey) was Werner Heisenberg. Heisenberg was internationally renowned for his work in quantum mechanics and the Uncertainty Principle that usually bore his name. In 1932, Heisenberg was awarded the Nobel Prize for Physics for his work on the Uncertainty Principle. . . While not a card-carrying Nazi, Heisenberg was a loyal and patriotic German. . . By 1941, the Germans were operating two experimental reactor projects, but German success had in fact been limited. Heisenberg's team in particular made certain engineering decisions that put the German program almost immediately at risk. Very basically, a nuclear reactor operates by inducing a chain reaction in masses of Uranium 238 within the reactor. To initiate a reaction, the flow of neutrons around the radioactive isotope must be moderated by another substance, such as graphite or deuterium (heavy water). The Germans chose to use heavy water, which is rare in nature and difficult to manufacture. In 1940, the Germans captured a heavy water plant in Vermok, a Norwegian town 100 miles north of Oslo. British intelligence had learned the basic outline of the German reactor project and realized that the Norwegian heavy water supply was a weak link. . . Despite the continuing attacks on the heavy water supply line, by 1941 German scientists had come to several broad theoretical conclusions that mirrored American conceptions of how to build an atomic device: (1) an enriched uranium fission device, (2) a plutonium-based fission device, or (3) a "reactor bomb." While the United States would build successful atomic reactors and both uranium and plutonium bombs by the end of the war, the German scientists never approached a working conception for actual production of a successful atomic machine.[35]

As the war progressed, Szilárd became ambivalent about the bomb project. This became all the truer when Germany surrendered by May of 1945, and the decision was made to continue the race for a bomb and use it against the Japanese. By July of 1945, he had authored the "Szilárd petition." In it, he urged US President Harry S. Truman to inform Japan of the atomic bomb and allow Japan to either accept or refuse the terms of surrender before the bomb was used. Seventy scientists and employees at the Chicago Metallurgical Lab and the Oak Ridge project site signed his petition, but it never made it to President Truman in time to be considered, nor was it made public until 1961.

July 17, 1945

A PETITION TO THE PRESIDENT OF THE UNITED STATES

Discoveries of which the people of the United States are not aware may affect the welfare of this nation in the near future. The liberation of atomic power which has been achieved places atomic bombs in the hands of the Army. It places in your hands, as Commander-in-Chief, the fateful decision whether or not to sanction the use of such bombs in the present phase of the war against Japan.

We, the undersigned scientists, have been working in the field of atomic power. Until recently, we

---

[35] Jonathan F. Keller, *Why The Nazi Atomic Bomb Never Happened*, https://warfarehistorynetwork.com/why-the-nazi-atomic-bomb- never-happened/; and https://www.mdpi.com/2673-4362/2/1/2/htm.

have had to fear that the United States might be attacked by atomic bombs during this war and that her only defense might lie in a counterattack by the same means. Today, with the defeat of Germany, this danger is averted and we feel impelled to say what follows:

The war has to be brought speedily to a successful conclusion and attacks by atomic bombs may very well be an effective method of warfare. We feel, however, that such attacks on Japan could not be justified, at least not unless the terms which will be imposed after the war on Japan were made public in detail and Japan were given an opportunity to surrender.

If such public announcement gave assurance to the Japanese that they could look forward to a life devoted to peaceful pursuits in their homeland and if Japan still refused to surrender our nation might then, in certain circumstances, find itself forced to resort to the use of atomic bombs. Such a step, however, ought not to be made at any time without seriously considering the moral responsibilities which are involved.

The development of atomic power will provide the nations with new means of destruction. The atomic bombs at our disposal represent only the first step in this direction, and there is almost no limit to the destructive power which will become available in the course of their future development. Thus a nation which sets the precedent of using these newly liberated forces of nature for purposes of destruction may have to bear the responsibility of opening the door to an era of devastation on an unimaginable scale.

If after this war a situation is allowed to develop in the world which permits rival powers to be in uncontrolled possession of these new means of destruction, the cities of the United States as well as the cities of other nations will be in continuous danger of sudden annihilation. All the resources of the United States, moral and material, may have to be mobilized to prevent the advent of such a world situation. Its prevention is at present the solemn responsibility of the United States — singled out by virtue of her lead in the field of atomic power.

The added material strength which this lead gives to the United States brings with it the obligation of restraint and if we were to violate this obligation our moral position would be weakened in the eyes of the world and in our own eyes. It would then be more difficult for us to live up to our responsibility of bringing the unloosened forces of destruction under control.

In view of the foregoing, we, the undersigned, respectfully petition: first, that you exercise your power as Commander-in-Chief, to rule that the United States shall not resort to the use of atomic bombs in this war unless the terms which will be imposed upon Japan have been made public in detail and Japan knowing these terms has refused to surrender; second, that in such an event the question whether or not to use atomic bombs be decided by you in light of the considerations presented in this petition as well as all the other moral responsibilities which are involved.

Leo Szilard and 69 co-signers

Signers listed in alphabetical order, with position identifications added:

1.DAVID S. ANTHONY, Associate Chemist

2.LARNED B. ASPREY, Junior Chemist, S.E.D.

3.WALTER BARTKY, Assistant Director

4.AUSTIN M. BRUES, Director, Biology Division

5.MARY BURKE, Research Assistant

6.ALBERT CAHN, JR., Junior Physicist

7.GEORGE R. CARLSON, Research Assistant-Physics

8.KENNETH STEWART COLE, Principal Bio-Physicist

9.ETHALINE HARTGE CORTELYOU, Junior Chemist

10.JOHN CRAWFORD, Physicist

11.MARY M. DAILEY, Research Assistant

12.MIRIAM P. FINKEL, Associate Biologist

13.FRANK G. FOOTE, Metallurgist

14.HORACE OWEN FRANCE, Associate Biologist

15.MARK S. FRED, Research Associate-Chemistry

16.SHERMAN FRIED, Chemist

17.FRANCIS LEE FRIEDMAN, Physicist

18.MELVIN S. FRIEDMAN, Associate Chemist

19.MILDRED C. GINSBERG, Computer

20.NORMAN GOLDSTEIN, Junior Physicist

21.SHEFFIELD GORDON, Associate Chemist

22.WALTER J. GRUNDHAUSER, Research Assistant

23.CHARLES W. HAGEN, Research Assistant

24.DAVID B. HALL, position not identified

25.DAVID L. HILL, Associate Physicist, Argonne

26.JOHN PERRY HOWE, JR., Associate Division Director, Chemistry

27.EARL K. HYDE, Associate Chemist

1.JASPER B. JEFFRIES, Junior Physicist, Junior Chemist

2.WILLIAM KARUSH, Associate Physicist

3.TRUMAN P. KOHMAN, Chemist-Research

4.HERBERT E. KUBITSCHEK, Junior Physicist

5. ALEXANDER LANGSDORF, JR., Research Associate

6. RALPH E. LAPP, Assistant to Division Director

7. LAWRENCE B. MAGNUSSON, Junior Chemist

8. ROBERT JOSEPH MAURER, Physicist

9. NORMAN FREDERICK MODINE, Research Assistant

10. GEORGE S. MONK, Physicist

11. ROBERT JAMES MOON, Physicist

12. MARIETTA CATHERINE MOORE, Technician

13. ROBERT SANDERSON MULLIKEN, Coordinator of Information

14. J. J. NICKSON, [Medical Doctor, Biology Division]

15. WILLIAM PENROD NORRIS, Associate Biochemist

16. PAUL RADELL O'CONNOR, Junior Chemist

17. LEO ARTHUR OHLINGER, Senior Engineer

18. ALFRED PFANSTIEHL, Junior Physicist

19. ROBERT LEROY PLATZMAN, Chemist

20. C. LADD PROSSER, Biologist

21. ROBERT LAMBURN PURBRICK, Junior Physicist

22. WILFRID RALL, Research Assistant-Physics

23. MARGARET H. RAND, Research Assistant, Health Section

24. WILLIAM RUBINSON, Chemist

25. B. ROSWELL RUSSELL, position not identified

26. GEORGE ALAN SACHER, Associate Biologist

27. FRANCIS R. SHONKA, Physicist

28. 54. ERIC L. SIMMONS, Associate Biologist, Health Group

29. JOHN A. SIMPSON, JR., Physicist

30. ELLIS P. STEINBERG, Junior Chemist

31. D. C. STEWART, S/SGT S.E.D.

32. GEORGE SVIHLA, position not identified [Health Group]

33. MARGUERITE N. SWIFT, Associate Physiologist, Health Group

34. **LEO SZILARD, Chief Physicist**

35. **RALPH E. TELFORD, position not identified**

36. **JOSEPH D. TERESI, Associate Chemist**

37. **ALBERT WATTENBERG, Physicist**

38. **KATHARINE WAY, Research Assistant**

39. **EDGAR FRANCIS WESTRUM, JR., Chemist**

40. **EUGENE PAUL WIGNER, Physicist**

41. **ERNEST J. WILKINS, JR., Associate Physicist**

42. **HOYLANDE YOUNG, Senior Chemist**

43. **WILLIAM F. H. ZACHARIASEN, Consultant**[36]

Szilárd had co-authored a report with Glen Seaborg that preceded that petition in which they cautioned over the post-war reputation of the United States if Japan was not provided with a proper warning of the power of the atomic bomb. In short, they warned that the nation could lose its moral authority.[37] Szilárd and others continued to insist that scientists have more input into the use of atomic weapons. Predictably, General Groves continued his distrust of Szilárd and efforts to persecute him.

By the war's end and afterward, Szilárd, like Robert Oppenheimer, feared a future nuclear arms race and the development of ever more destructive nuclear weapons. Szilárd warned of the development of increasingly larger thermonuclear (fusion) bombs that could also be infused or "slated" [38] with radioactive cobalt-60. Although the concept has never been tested, radioactive cobalt-60 in a hydrogen bomb could cause fusion-generated neutron activation and enhance the gamma dosage, resulting in hazardous fallout and leaving much of the world uninhabitable.[39] His talks on that subject likely led to that theme being used in numerous nuclear war science fiction novels of the age.

------

[36] US National Archives, Record Group 77, Records of the Chief of Engineers, Manhattan Engineer District, Harrison- Bundy File, folder #76. Note: The spelling of the first names of Wilfrid Rall and Katharine Way was corrected on July 10, 2015. Source note: The position identifications for the signers are based primarily on two undated lists, both titled "July 17, 1945," in the same file as the petition in the National Archives. From internal evidence, one probably was prepared in late 1945 and the other in late 1946. Signers were categorized as either "Important" or "Not Important," and dates of termination from project employment were listed in many cases. It seems reasonable to conclude that the lists were prepared and used for the purpose of administrative retaliation against the petition signers.

[37] Lawrence Badash, *American Physicists, Nuclear Weapons in World War II, and Social Responsibility* (Physics in Perspective, 2005),

[38] The term "salted" refers to the manufacturing process that involves the incorporation of radioactive cobalt-60 and from the expression "to salt the earth," meaning to render an area unusable.

[39] Carey Sublette, *Types of Nuclear Weapons – Cobalt Bombs and Other Salted Bombs*, Nuclear Weapons Archive Frequently Asked Questions, May 1, 1998, https://nuclearweaponarchive.org/Nwfaq/Nfaq5-1.html (accessed January 7, 2020).

Szilárd certainly wanted nothing more to do with the creation of nuclear weapons and went into the field of biology, which he saw at that time as a still very uncharted territory for an innovative mind like his. In 1961, Szilárd authored a book of short stories called *The Voice of the Dolphins*.[40] In that work, he dealt with the moral and ethical issues raised by the Atomic Age, as well as his own role in the development of atomic energy.[41] Szilárd retained tremendous admiration among both sides of the nuclear scientific community. The legendary Edward Teller, said Leo Szilárd is the "most ingenious person I ever met." While Szilárd may have initially been the man behind the bomb, once Hitler had been defeated, he devoted the rest of his life to peace.[42]

---

[40] Leo Szilárd, *The Voice of the Dolphins and Other Stories* (New York: Simon and Schuster, 1961).

[41] *The Voice of the Dolphins* derives its title from a television program by the same name where a group of American and Russian scientists conspire to save the world, https://www.goodreads.com/book/show/45997458-the-voice-of-dolphins-and-other-stories.

[42] The nuclear age actually saved Szilárd's life. Radiation treatments in the early 1960s cured him of bladder cancer.

CHAPTER FOUR
MUNICH BY JAMES HALL

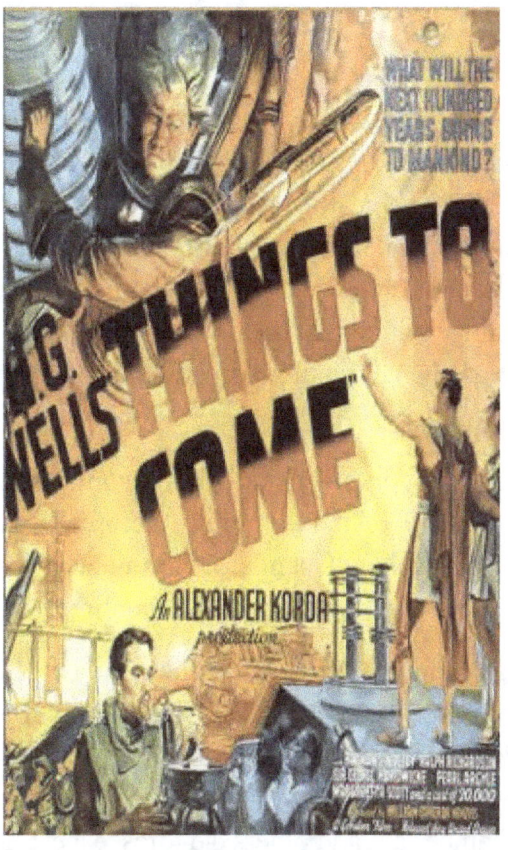

Things to Come movie poster, 1936, produced by Alexander Korda and
written by H.G. Wells,
https://cinescopia.com/wp-
content/uploads/2016/06/150705145313_hg_wells_poster.jpg; and
https://ichef.bbci.co.uk/news/320/cpsprodpb/5F5A/production/_840
01442_poster.jpg.

Eight decades ago, European powers sacrificed an independent country to Hitler's Nazi aggression out of fear of generating a wider war. That agreement negotiated in Munich, Germany, to violate the sovereignty of the Sudetenland of Czechoslovakia, did prevent a European war for one critical year. Yet, when war came, it soon led to a worldwide conflict. However, in that time from October 1938 to September 1939, Britain made significant advancements in new technologies, including radar and monoplane fighter design, that would eventually turn the tide against Germany.

British Prime Minister Neville Chamberlain is the most identified figure in what became known as the Munich or Sudeten Crisis. He is remembered today for a then-popular policy of appeasing Hitler and Mussolini—as those fascist leaders attempted to rebuild and expand their way out of a worldwide depression. It is hard to impress upon people today how popular appeasement really became in the 1930s. That was true, particularly among those in France, Britain, and America who were horrified by the great loss of life that ensued during World War One. France had its former ability to field large-scale armies crippled because so many of its youth had died in the trenches. Britain also, quite literally, lost an entire generation of people who never lived to have families or serve the business of the Empire. The population loss was only aggravated by the pandemic of 1918-1920. The United States had low casualties, but its loss came in the form of a diminishing desire ever to get involved in European affairs again. The US increasingly became isolationist.

Chamberlain, of course, became criticized by history for being naïve about Hitler's intentions. Post World War II appeasement was blamed for letting the situation escalate into a broader war. Undoubtedly, the example of appeasement led to an overzealous paranoia by both the USA and USSR during the many decades of the Cold War. Certainly, Russia's recent move against Ukraine has rekindled everyone's fears of letting appeasement lead to a widespread war. Yet, in the day, no one had that mindset, which only came later thanks to the hindsight of history. Initially, Winston Churchill became one of only a handful of people during the 1930s questioning appeasement, creating political tensions between him and Chamberlain, who belonged to the same Conservative party.

The Churchill-Chamberlain rivalry continued into the Second World War, during which Churchill served under Chamberlain as First Lord of the Admiralty before he succeeded him as Prime Minister. Although often at loggerheads, Churchill kept Chamberlain in the top position in his own government and praised Chamberlain as a man of peace and honor up until his untimely death in late 1940.

Neville Chamberlain was that. Contrary to popular thought, Discoveries of private correspondence between Chamberlain and his two sisters reveal that he did not actually trust Hitler's promises of "no further territorial demands."[43] Chamberlain, instead, was determined never to lose another generation to war, as had happened during the First World War. Britain, along with France, thus desperately wanted to avoid conflict with Germany or Italy. Chamberlain and others did foresee the dangers of a coming war but, in short, wanted to cut a deal to stop it at any cost.

Europe did not want war for any reason, although it became increasingly apprehensive over Germany's ever-growing numbers of sleek modern bombers rolling off Nazi assembly lines. The English and French were engrossed by the 1936 film *Things to Come*, which was produced by Winston Churchill's close friend, Alexander Korda[44]. The futuristic movie was written by H.G. Wells, who, as we have explained, was another longtime associate of Churchill. *Things to Come* imprinted images in everyone's mind of a bombing of London and European capitals.

Stanley Baldwin, a former PM and associate of both Chamberlain and Churchill, famously stated that in a modern war, "[T]he bomber will always get through."[45] Baldwin warned of that as early as 1932. It

---

[43] Neville Chamberlain to Hilda Chamberlain; and Book review of *Burying Caesar* by James Hall, National Atomic Testing Museum webpage, https://nationalatomictestingmuseum.org/2022/07/13/book-review-burying-caesar/; and Graham Stewart, *Burying Caesar, The Churchill-Chamberlain Rivalry* (Woodstock, New York: The Overlook Press, 2001).

[44] Churchill had contributed screenwriting to a number of Korda productions.

[45] The phrase "The bomber will always get through" is attributed to Stanley Baldwin, a British statesman who served as Prime Minister of the United Kingdom three times between 1923 and 1937. The phrase was not stated verbatim, but Baldwin did express similar sentiments in a speech to the House of Commons on November 10, 1932, during a debate on air defense. In his speech, he said: "I think it is also good for the man in the street to realize that there is no power on earth that can protect him from being

proved true during the London Blitz of 1940-41. On May 10, 1941, one year after Churchill assumed leadership as PM, the gallery of the House of Commons was destroyed by a German bomb. The bombing saw both the Commons chamber and the roof of Westminster Hall set ablaze. In a post-bombing tour of the Palace of Westminster, Churchill observed in silence the devastation. Perhaps he was recalling to himself his many speeches in that historic chamber over the last decade, where he repeatedly warned of the growing threat of Hitler's burgeoning Luftwaffe. Leaders in Churchill's own Conservative party under both Baldwin and Chamberlain scoffed at his notions, and for many years, Churchill was a lone but prophetic voice of apprehension of Adolf Hitler. Yet once war came and surprised so many with its blitzkrieg intensity, he never once used the phrase "I told you so." He simply was too great a gentleman for that.

Before and after photos of the May 10-11 bombing of the Palace of Westminster, which destroyed the House of Common's gallery. The adjacent House of Lords section fared better when a bomb also struck, but it passed through the floor of the Chamber without exploding. Eventually, the historic galleries were meticulously restored during the course of a decade of reconstruction. Getty Images, https://www.gettyimages.co.uk/detail/news-photo/government-england-pic-circa- 1930s-the-houses-of-parliament-news-photo/79043241.

One of the best ways to study the cause of the Second World War and Hitler's attempted enslavement of Europe is to read the three-volume biography of Winston Churchill by William Manchester. The books tell the story of such past wars and reflect on historical events that very much remind us of current times. Insightfully, these books are, therefore, not just about the famous life of Winston Churchill but also about the changing times he lived in and how those years shaped the world we now live in.

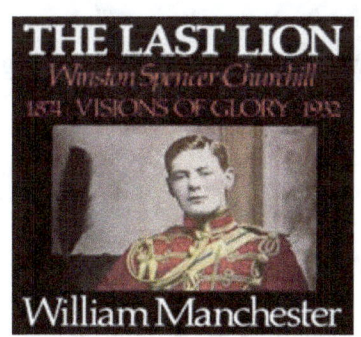

Public record image, cover of first of three volumes, Manchester biography of Winston Churchill.

The greatest lesson in Manchester's works is how easily the world found itself engulfed in a world war, not once but twice. We are all familiar with Churchill's 1940 darkest hour as Prime Minister, which led to Britain's finest hour. However, he had seen the world go to war in 1914 when he played a similar center-stage

---

bombed. Whatever people may tell him, the bomber will always get through." The exact wording varies slightly in different sources, although this statement captures the essence of what Baldwin was trying to convey regarding the potential threat of aerial bombing.

role as First Lord of The Admiralty. Volume One of "The Last Lion" tells the first story of nations stumbling into war.

No one could have then imagined what a modern war would look like; however, they did have foreshadowing hints. At the end of the 19th century, Churchill participated in what is considered to be the last great cavalry charge in British military history while serving as part of the Egyptian Expeditionary Force leading up to the battle of Omdurman. There, he saw the coming effectiveness of modern weapons used, albeit still from horseback, against primitive firearms and swords.

On September 1, 1898, the day before the Battle of Omdurman, Lieutenant Winston Churchill of the Queen's 4th Hussars attached himself as a war correspondent to the 21st Lancers. They scouted villages on the west bank of the Nile opposite Khartoum and the site of General Gordon's legendary defeat thirteen years earlier. A noted cavalry charge ensued, and Major General Kitchener vowed to avenge and restore imperial order as the 20th century dawned. Public record image, https://www.triexz.com/wp-content/uploads/2016/10/M- Omdurman-1-4C-Jun11-crop-2000x1125.jpg.

Churchill next witnessed what modern German Krupp artillery and machine guns could do to what he had considered a modern British army when he served as a *Morning Post* correspondent in the Boer War in 1899. In that conflict, events oddly but fortuitously smiled on Churchill when he was captured and then escaped from the Boers. That escapade, which made newspaper headlines around the world for weeks, not only made him an international figure but catapulted him into a 60-year political career that matched none other in history. By 1911, when Churchill began his first term as First Lord of the Admiralty, the greatest weapons in the world were then the great Dreadnoughts. On the eve of war in 1914, Churchill had not only overseen, with Grand Admiral Jackie Fisher, the construction of the most advanced of the new largely oil-fueled "Super Dreadnoughts" but also commanded two-thirds of the world's arsenal of those mighty steal castles. He played center stage in what became the first great arms race.

Kaiser William II's Germany possessed the bulk of the remainder of mega, coal-fueled weapons of the day. If a great war ever came, it was then assumed such a conflict would be decided by a battle of Armageddon between these beasts, which had proved highly effective as a deterrent to any such war. That battle, Jutland, eventually happened. The results, however, were indecisive. World War One instead became one of stagnant immobility despite Churchill's creation of armored cars and the first tanks, as well as the first use of aircraft from ships. Yet it was the machine gun and tens of thousands of artillery pieces that caused the greatest carnage.

As soldiers chewed barbwire, Churchill tried to force the Dardanelles with his older dreadnoughts to relieve Russia via the Black Sea and keep her in the war against Germany. It proved a failure that almost destroyed him. Forced out of the cabinet, he chose to go to the trenches in France, leading a brigade as a major and then a battalion as a lieutenant colonel. His 700 men loyally followed him after they saw him routinely walk the frontlines alone—often on the outside side of the barbed wire defenses. They became even more enthusiastic when he shared his portable bathtub with everyone for a "war on lice." Churchill proved quite a novelty in that war of mud and horror.[46]

The outbreak of that First World War had surprised even Churchill, who never really thought the great European powers would be so careless as to endanger their own growing economies and progressive social programs by falling into war. Momentous events had simply outweighed man's ability to control them. War would prove far more expensive than peace. Volume two of Manchester's Churchill biography tells a similar story, illustrating how, prior to the outbreak of the Second World War, Neville Chamberlain and most Western leaders thought that Adolf Hitler, although clearly trying to reconstitute former German lands, would never be so careless as to start a general war and thereby ruin his great achievements in rebuilding Germany.

Churchill was a lone figure who could see the dangers of a then widely popular policy of appeasement. Yet we tend to forget that Churchill, although holding almost every major political office over the years, was not only the most successful political figure in history but also, at many times, the most controversial. He eventually proved himself right most of the time, but when he was wrong, it became disastrous. Still, he remained, in his own words, "a glow worm." The 1930s was one of those many periods when he was a lone wolf in the wilderness. If not for his prophetic warning of Hitler's rise, he may have been forgotten in history. Not until the eleventh hour did a majority appreciate his stand and allow him to assume political power again; by then almost too late. Many still wanted to make a deal with Hitler even after Dunkirk. So assuredly, if not for this one brash, stubborn, often ill-tempered, but seasoned figure, England may have given in to a Nazi-ruled Europe almost two years before America entered the war. Deposing of National Socialism would have been impossible after that point, and we would likely still be dealing with it to this day.[47]

---

[46] Suggested reading of William Manchester, *The Last Lion: Winston Spencer Churchill*, 3 vols. (New York: Little, Brown, 1983-1988), vol 1-3. Also highly recommended is the book *Appeasement*, Tim Bouverie, *Appeasement* (London: Vintage, 2019.)

[47] Volume three showed how Churchill had yet another distinguished, although controversial, political career, returning as Prime Minister in the early 1950s when Britain became its own nuclear power. Of most significant note, throughout Churchill's long life, he was, above all else, a prolific and highly paid writer, publishing 58 books and 1,100 articles. I am also a writer and hope to be like him, expressing a strong vision with a passionate artistic style about past, present, and future trends. Churchill even saw acclaim as a landscape painter and brick mason. Churchill exemplified a true Renaissance man of the ages, the likes of which we will not see again! Books like William Manchester's three-volume study of Churchill and his works clearly pose questions as to whether the Western powers are again becoming too complacent with potential threats or, on the other hand, overreacting. Edmund Burke stated: "The only thing necessary for the triumph of evil is for good men to do nothing." My father read Manchester's books and Churchill's works when he was young. Now, I repeat the education.

CHAPTER FIVE
THE DAY THE WORLD CHANGED FOREVER

Robert Oppenheimer, by *Science Source*,
https://fineartamerica.com/profiles/photo-researchers-inc/shop/canvas+prints/robert+oppenheimer.

The first nuclear test in history occurred on July 16, 1945, in the desolate region of New Mexico known as *Jornada del Muerto*. The early Spanish explorers coined this phrase, which means "route of the dead." Our world soon changed forever after that experiment in the wilderness. The leader of the scientific project to build an atomic bomb, Robert Oppenheimer, had a premonition of that change. A highly learned man in many disciplines, poetry moved him to select a codename for this momentous operation, inspired by the John Donne poem, *Holy Sonnet XIV: Batter My Heart, Three-Personed God*. "Trinity," as Oppenheimer interpreted it, became the term ever-linked with that first nuclear test. The poem went:

Batter my heart, three-personed God; for you as yet but knock, breathe, shine, and seek to mend; That I may rise and stand, o'erthrow me, and bend Your force to break, blow, burn, and make me new. I, like an usurped town, to another due, Labor to admit you, but O, to no end; Reason, your viceroy in me, me should defend, but is captived, and proves weak or untrue. yet dearly I love you, and would be loved fain, But am betrothed unto your enemy. Divorce me, untie or break that knot again; Take me to you, imprison me, for I, except you enthrall me, never shall be free, Nor even chaste, except you ravish me.[48]

John Donne called upon God in his poem to envelop him in spiritual majesty. The atomic test that

---

[48] John Donne, "Holy Sonnet XIV: Bater My Heart, Three Personed God," in *Norton Anthology of English* Literature, edited by Stephen Greenblatt et al, 10 ed. Vol. B (New York: Norton, 2018), p.1137.

Oppenheimer and his collection of physicists and engineers worked so hard to achieve created just such an image. Great planning went into this first-of-its-kind explosion. The revolutionary device was placed on top of a 100-foot steel tower for the test. This elevation was designed to keep as much of its tremendous fireball off the ground as possible to decrease the amount of dust stirred by the radioactive explosion and thus limit the spreading of radionucleotides through the debris. In later years of nuclear testing, the towers would be much higher and thus more efficient, but this symbolized the first attempt. However, another primary reason for the elevation focused on the evaluation of an airburst explosion. The idea gained favor that the best use of atomic bombs over Japanese cities (the war with Germany by then over) would be to explode them at least 1,000 feet or more in the air. The theory was that an elevated explosion would maximize the destructive power of the bombs as opposed to a conventional bomb that had to hit its target point-blank.

The Trinity bomb used plutonium in contrast to enriched uranium, which was in much shorter supply. In fact, the enrichment program of uranium and the production of the manufactured element plutonium as a byproduct of uranium reactors became as complex of a project as building the actual bomb itself. It proved a race against time, but enough enriched plutonium was produced for three bombs and enough enriched uranium for one bomb by mid-1945. The uranium core bomb resulted in an extremely simple and elegant design. To vastly oversimplify it, this involved smashing one sub-critical mass of enriched uranium into another sub-critical mass of enriched uranium in a gun barrel-like chamber to make it become a critical mass. (Critical mass simply means enough fissile material to start/and/or maintain a nuclear chain reaction, which in this case was 60 kilograms of material that was at least 80 percent pure, known as uranium-235.) They knew this formula would work, and the bomb design called Little Boy soon destroyed Hiroshima. Little Boy's components were already in preparation for transport to the Pacific as the Trinity test readied. The plutonium bomb, however, had far more challenges.

The problem with the plutonium bomb design was that it could not simply collide with another piece of itself to attain a chain reaction. Enriched plutonium possesses spontaneous fission characteristics, leading to a high rate of neutron emission. Utilizing a gun assembly method would prematurely detonate this material due to the abundance of neutrons, thereby preventing a controlled critical-mass explosion. Achieving a controlled critical release with plutonium necessitates using an implosion method to compress it uniformly and simultaneously. This task poses significant engineering challenges.

Nevertheless, this method had advantages because it required only one-tenth the fissile material used in a U-235 gun-type design. The resulting Fat Man bomb, eventually used on Nagasaki, required a mere 6 kilograms of plutonium. The first plutonium bomb was basically a mockup for what became the core of the Fat Man design, and that is what they tested at Trinity. The test was required because the process of developing conically-shaped explosive lenses to implode the plutonium into a critical mass was such a radical idea that no one could guarantee it would work. Trinity, of course, did work so that the second plutonium bomb would be used on Nagasaki on August 9, 1945, just three days after the Hiroshima uranium bomb, Little Boy, was dropped. A third plutonium Fat Man bomb sat on standby to be sent to the Pacific, but thankfully, America did not have to exhaust its first nuclear arsenal before Japan surrendered. Putting these complicated bomb designs into a small and light enough casing for an airplane to deliver them was also a tremendous engineering challenge. The word engineering is stressed! The great success of the Manhattan Project is due to the fact that Oppenheimer emphasized using engineers. That was a key mistake the Germans made in their bomb project—not utilizing skilled engineers in tandem with physicists.[49]

The Trinity bomb would explode with a force of 21 kilotons of TNT, yet before the test detonation took place, no one knew its scope. That yield figure thus became a critical data point. Prior to this, no one could precisely calculate how plutonium would react. Edward Teller, the future father of the hydrogen bomb, once openly questioned if the A-bomb could fuse nitrogen atoms in the Earth's atmosphere and ignite it. Teller estimated the Trinity yield could be as high as 45,000 tons. Oppenheimer, on the other hand, estimated only 300 tons. He also had a bet it would not work at all. Enrico Fermi, the inventor of the first nuclear reactor with Leo Szilárd, offered to take wagers from his fellow scientists on whether or not the bomb would merely destroy New Mexico or the entire world. This infuriated the military leader of the Manhattan Project, General Leslie Groves, who had little patience for such conjecture.

When General Groves arrived at the Trinity site that Sunday night, rain poured from the sky. Other officials who had just arrived were similarly concerned about the weather. These distinguished observers included Vannevar Bush, James Conant, Ernest Lawrence, Thomas Farrell, and James Chadwick.

Oppenheimer and Groves obsessed over the weather and realized a delay from the scheduled 4:00 a.m. zero hour was unavoidable. Understandably, nerves were frayed. Groves's main concern centered on what he would tell his superiors if the test failed. The new President, Harry Truman, was about to negotiate with the Russians at the Potsdam Conference, and the stakes were very high. How would Japan be defeated? The casualty figures for an invasion of mainland Japan were estimated at a million American fatalities as well as twice that many for Japan. Everyone felt a sense of duty to bring the war to a final end. So, the Manhattan Project group was interested in watching their work succeed. Some dissension arose over using the bomb on Japan in light of Germany having lost the nuclear race, but none of that diminished anyone's patriotism nor scientific curiosity.

The observation area for the key groups consisted of two makeshift bunkers, each located 10,000 yards from ground zero. As the weather reports remained bleak, Edward Teller grated everyone's nerves by applying multiple layers of sunscreen, advocating that it would help shield from the radiological effects. To ease the tension, Oppenheimer bet $10 against George Kistiakowsky's entire month's pay that the bomb would not work at all. That did not make Groves feel any better.

They missed the 4 AM test time because of persistent foul weather but hoped for a 5 AM or 5:30 AM. Groves got so frustrated with the meteorologists that he demanded they give him a time when the weather would improve and then insisted they sign their forecast. He also contacted the governor of New Mexico, warning that martial law would be required if there was an unforeseen disaster.

As it looked like a window in the weather would prevail, key personnel moved to different locations with the idea that observing from multiple areas would circumvent some catastrophic accidents and prevent all the important observers from being killed at once. Groves accordingly went further back to base camp with Bush and Conant. Oppenheimer remained at a forward bunker. Then, a final nerve-racking incident

---

[49] In his school days, Oppenheimer was considered a miserable failure in applied science because, in the laboratory, he proved himself helpless to work with his hands. On the other hand, he became a brilliant theoretical physicist, and that led to his fame and perhaps his own appreciation for the value of engineers and scientists who, with their skilled hands, can apply concepts to physical reality.

occurred that became almost surreal to the tense atmosphere. The intercom system they were using for communication inexplicably picked up music on a local radio station.

Finally, the Trinity test detonated at 5:29 AM on Monday, July 16. The blast wave vaporized the test tower, sending a shock wave strong enough to knock some observers off their feet. The sand at the base of the tower turned into a green glass and is now known as trinitite. Some rare forms of trinitite samples have iron embedded in them, which represents the remains of the vaporized tower. Many wore welders' glasses, and through those dark prisms, they saw a light that previously had only been observed in the celestial images of a star.

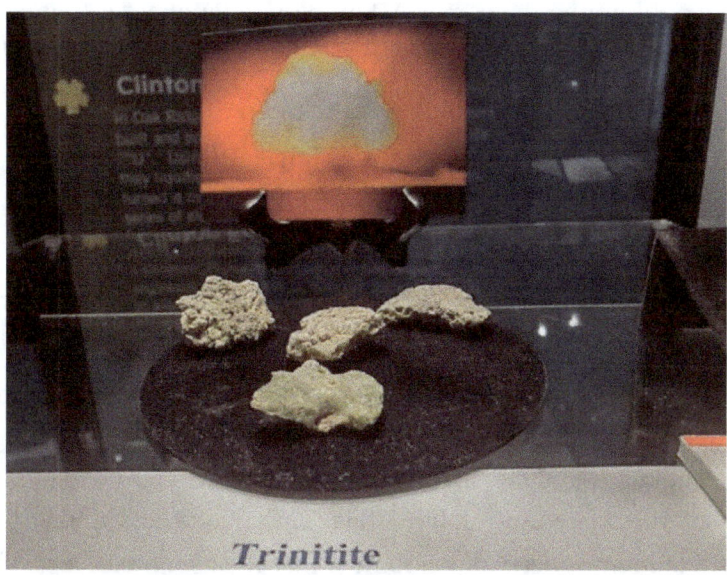

Trinitite

Trinitite, also known as atomsite or "Alamogordo glass," represents the glassy residue left on the desert floor after the plutonium- based Trinity nuclear bomb test on July 16, 1945, near Alamogordo, New Mexico. The glass is primarily composed of arkosic sand or quartz grains that were melted by the atomic blast. It is usually a light green, although color can vary. It is mildly radioactive but safe to handle. Photo belongs to authors who curated this pictured trinitite exhibit.

The overriding recollections focused on the intense wind and radiated heat waves produced by the bomb. These were felt on everyone's skin during that damp, cool morning. As the first man-made nuclear explosion took place, a great mushroom cloud rose above the blast, giving an iconic signature to what has ever since been known as the Atomic Age.

The official government radiological report on Trinity published in 1982 stated, "A column of smoke and debris rose as high as 15,000 feet before drifting eastward. Offsite monitoring teams in an area northeast of ground zero encountered gamma readings ranging from 1.5 to 15 R/h (Roentgen per hour) two to four hours after the detonation."[50]

In his reflections, Oppenheimer added a great emotional and artistic bend to Trinity. He stated that the

---

[50] Manhattan Project exhibit, "The Day the World Changed," National Atomic Testing Museum; and Defense Nuclear Agency, publication NNA 6006F, November 23, 1982.

explosion reminded him of a line from the Hindu holy text called the *Bhagavad Gita*. He later stated:

I remembered the line from the Hindu scripture, the Bhagavad Gita. Vishnu is trying to persuade the Prince that he should do his duty and to impress him, takes on his multi-armed form and says, "Now, I am become Death, the destroyer of worlds.[51]

No one recalls Oppenheimer commenting on that at the time. Physicist Frank Oppenheimer, who stood right beside his brother, only remembers "Oppie" (as everyone called him) saying simply: "It worked." Shortly after the war, Oppenheimer did publicly comment on that moment at Trinity:

"We knew the world would not be the same. A few people laughed, a few people cried. Most people were silent… We thought of the legend of Prometheus, of that deep sense of guilt in man's new powers, that reflects his recognition of evil, and his long knowledge of it. We knew it was a new world. . .[52]"

As the Trinity test unfolded, the sure-fire Little Boy bomb components were already being loaded aboard the *USS Indianapolis* in San Francisco Bay for transport to an airfield on Tinian in the Mariana Islands supporting B-29 operations. The war with Japan ended with the atomic bombing of two key Japanese cities. America did not have to risk a million estimated US casualties. Nor did Japan lose as many lives as had been estimated in a conventional invasion of the mainland which no longer had to take place.

Of course, many scholars have argued that the Japanese would have had to eventually sue for peace. Yet, the fact remains that the plans for the invasion of Japan were already in motion by August 1945 and scheduled for that fall. The invasion would have taken place, as would have the continued mass fire bombings of cities, which had already proved far more destructive than both atomic bombings. Oppenheimer, in fact, commented at the time of the bomb's early development that it was more of a psychological weapon than a strategically useful device, considering the limited number of nuclear cores available.

Yet, the bigger significance of Trinity is the war that did not follow the Second World War. The concept is as controversial as the decision to use the bomb on Japan. However, many do subscribe to this theory that nuclear weapons, even with current problems of proliferation, have created an actual deterrent to another world war. Admittedly, while nuclear weapons have not prevented conventional wars, and although they do hold us hostage to our own devilish invention, we have yet had to face another world war since the dawn of the atomic age.

Others argue that nuclear weapons have only made war a civilization-ending proposition. It is worth paraphrasing Albert Einstein, a self-admitted pacifist who had initially lobbied President Roosevelt for an atomic bomb project out of fear that Nazi Germany might get it first. Einstein later stated, "The splitting of the atom changed everything except the way we think." Certainly, as Oppenheimer described almost 75 years ago, there is no doubt that Trinity was the day the world changed forever.[53]

---

[51] J. Robert Oppenheimer, *"The Decision to Drop the Bomb,"* in *Conversations with Oppenheimer*, ed. Alice Kimball Smith and Charles Weiner (Hanover, NH: University Press of New England, 1995), p. 47.

[52] J. Robert Oppenheimer, interview by Stephane Groueff, June 4, 1965, in Voices of the Manhattan Project, interview transcript,

[53] For suggested further reading about the Manhattan Project and the creation of the first nuclear weapons and their immediate aftermath, read: *Project Y: The Los Alamos Story*, by David Hawkins and Edith C. Truslow (Los Angeles: Tomash

The only well-exposed color photo of the Trinity detonation was taken by Jack W. Aeby. He captured the historic shot of the first nuclear explosion with a Perfex 33 camera with a 35mm lens set at a shutter speed of 1/100 with a f4 stop, using Anscochrome color movie film stock. Image copy from authors' collection.

Publisgers,1983).

CHAPTER SIX
LORD OF THE UNDERWORLD

**Public record image oil on canvas
painting "Orpheus in the Underworld"
by Henryk Hector Siemiradzki.**

Plutonium is symbolic of this classical painting by Hector Siemiradzki, having been named after the planet Pluto, which was derived from the mythical Roman god of the underworld. Pluto served as a giver of gold and silver from the subterranean world of mysterious metals. The earlier Greeks used the term Plouton, and earlier than that, Hades—meaning lord of the underworld of exotic metals, a very appropriate title for one of the world's most mysterious manmade elements. Known originally only as Element 94, Enrico Fermi at the University of Rome had theorized of such a material as early as 1934. It was, however, not until December 14, 1940, that this substance was first isolated by Glenn Seaborg, Joseph Kennedy, Edwin McMillan, and Arthur Wahl. They were using Ernest Lawrence's new invention, the cyclotron. Based at the University of California, Berkeley Radiation Laboratory, that team bombarded uranium-238 with deuteron.

In doing so, they created or synthesized neptunium-238, which decayed with beta emissions with a half-life of a little over two days. So, it basically decayed to form a new, heavier element with an atomic number of 94, which had a much greater half-life. McMillan named the first element neptunium after Neptune. Pluto, of course, was the next celestial body in line. By February 23, 1941, they chemically identified and confirmed element 94, which took on the name plutonium by then. A year earlier, colleagues warned Seaborg that he could lose his reputation for experimenting with such an exotic concept.

At the time, Seaborg did not feel he had a reputation to lose; by this time, he had stumbled upon a real bombshell. In March, Seaborg submitted a paper on plutonium to the journal *Physical Review*, but it soon became evident that plutonium, in a particular form called plutonium-239, could theoretically facilitate a fission reaction.

Then research projects in England and America quickly formed to investigate Leo Szilárd's long-held belief that a fission reaction could trigger a bomb of immense destructive force, prompting the swift withdrawal of the paper. The kinetic energy from the resulting chain reaction would be far stronger than any chemical reaction could ever be. A new-found urgency and caution revolved around the fear that Hitler's Nazi Germany may unlock the secret first. The discovery of plutonium became a guarded secret, along with work on its parent element, uranium, because only such elements possessed properties suitable for making a fission reaction possible. Of course, no one yet had a clue how to do that. So, uranium and plutonium research work were key parts of several embryonic experiments that would all come together in the later Manhattan Project's goal to create an atomic bomb.

The initial focus centered on how to take standard uranium 238 and refine out the more fissile isotope U-235. This was extremely difficult to do, so the prospect of additional fissile material like plutonium-239 coming from this process became intriguing. The British at the time were much further ahead in atomic research due to the urgency of Hitler's invasion of Europe, which had not yet brought America into the war.

Their research was much more centralized at the Cavendish Laboratory in Cambridge. In 1940, physicists at the Cavendish Laboratory, similar to the Seaborg group in California, predicted that they could create a new man-made element by bombarding uranium-238 with neutrons. Egon Bretscher and Norman Feather at Cavendish realized that a slow neutron reactor fueled with uranium could potentially produce element 94 (plutonium) as a by-product, even though it had to be laboriously chemically separated.

They calculated that this manufactured element would be chemically different from uranium but had the advantage of being more fissile per kilogram and somewhat easier to produce in quantities than enriching uranium 235 from 238. This simultaneous discovery of plutonium by Britain and America made an atomic bomb seem a practical achievement. Thus, atomic research received a top priority because Western scientists felt it a realistic goal to produce the actual fissile material that would be required to make such a fission bomb.[54]

Meanwhile in Germany, those working on atomic theory with Werner Heisenberg allegedly stumbled in their calculations by overestimating the amount of enriched material they thought it would take to make a bomb. Deemed not of immediate war- winning potential, the atomic projects lost their priority status because they did not seem practical to the Nazi hierarchy. Of course, no one could know that in the West. Fearing Hitler would soon master the bomb, key leaders like Winston Churchill and then Franklin Roosevelt pushed for an all-out endeavor. A great deal of research had to first go into transuranic elements—elements with an atomic number greater than 92. The rare elements needed for nuclear fission are starting with uranium and this odd derivative material called plutonium. Uranium-238 is what you basically find in natural forms and is mined in various areas of the world. It comes out of the ground as a rock and is then often milled into a powder called yellow cake. U-238, in its natural form, cannot sustain a fission chain reaction. It has to be separated into a more unstable or fissionable isotope called U-235, representing only 0.7 percent of natural uranium-238.

Plutonium is made from uranium enrichment, but it proved a complicated process to master. To produce plutonium, you must create a self-sustaining fission reactor and use the spent fuel, which has to be chemically separated to develop a fissile plutonium-239 product. This Pu-239 only represents one percent of the residual waste of the reactor, which is highly toxic and radioactive. Of course, physicists and engineers

[54] Conversation with retired Los Alamos National Laboratory Director Siegfried S. Hecker, June 2015, National Atomic Testing Museum.

first had to make a uranium reactor.

Two legendary scientists—Enrico Fermi and Leo Szilárd—figured that process out at the Metallurgical Laboratory at the University of Chicago, which became the focus of this research for the newly formed Manhattan Project. On August 18, 1942, scientists conducted a significant experiment where they successfully isolated and measured a minimal amount of plutonium—just one microgram. This tiny amount of plutonium was officially identified as plutonium-239. The experiment reinforced the idea that plutonium would become an ideal material to facilitate a fast chain reaction, which would be needed for this still rather nebulous idea of building an atomic bomb.

Earlier on December 2, 1942, Fermi, with Szilárd at the Metallurgical Laboratory, mastered the first artificial self-sustaining nuclear chain reaction under the viewing stands of Stagg Field. They achieved the chain reaction in a graphite and uranium pile known as CP-1. There is a fascinating story related to the graphite design of the reactor. Szilárd learned very early on that commercial-grade graphite contained boron. That was not good because boron absorbs neutrons and retards the reaction process. So Szilárd worked with manufacturers to develop 99 percent pure graphite. His inspiration made the first self-sustaining nuclear reactor in Chicago a success.

As stated in earlier chapters, the Germans, at that same time, also experimented with a graphite reactor but never fully appreciated the boron problem with the graphite they were using. They instead moved to heavy water for the process of which they could never get enough. With great endeavor, the British kept sabotaging the Axis production of heavy water in Norway because the Allies understood why the Germans were so interested in that hard-to-produce substance. It truly is no exaggeration to say that if the Nazis would have refrained from expelling scientists of Jewish ancestry like Szilárd in the 1930s, they may have perfected the bomb!

CP-1 only served as a learning experiment. It was not designed to be a nuclear power-producing reactor because it produced less than a watt of energy. Although, it did prove a chain reaction could be created. It also confirmed that a slow neutron reactor could produce usable amounts of plutonium as a by-product when uranium atoms absorb neutrons. The isotopes can make plutonium-238, 239, 249, 241, and 242. Those are all fissionable because the atom's nucleus can be split apart by a neutron.

Plutonium-239 has a high radioactivity with a half-life of 24,100 years. Plutonium, in other safer forms, emits alpha particles which have a short half-life, so they can be used as a battery-like power source called plutonium-328. Radioisotope thermoelectric generators and radioisotope heaters in satellites are examples. However, in nuclear weapons research, the more fissile forms like Pu-239 emit beta particles and gamma-rays which have a very long half-life, making those forms toxic. Plutonium is not only dangerous because it is radioactive but because, like all heavy metals, it is chemically toxic.

When plutonium particles decay, they can eventually transform back to uranium or neptunium, which remains radioactive to different degrees. These facts became evident by November 1943, at which time the Metallurgical Laboratory finally produced enough plutonium in the form of a metal to actually be seen with the naked eye. This may not sound like a great accomplishment, but it proved, just as in all of the other stair-step experiments, that an atomic bomb would be an attainable goal. By late 1943, physicists had understood that when neutrons hit a plutonium-239 atom, a neutron would be released, which would hit another neighboring plutonium atom, and soon a chain reaction of tremendous speed, releasing a similar amount of energy, would occur. Next came the challenge of how to mass-produce enough plutonium to make at least a few bombs by the war's end. Not only would a massive infrastructure have to be created, but production facilities would have to be designed, along with blueprints for the enrichment equipment, which did not yet even exist. Then, everything would have to be constructed in time to have an impact on the allied war effort.

Despite this, a plutonium production site quickly emerged in Hanford, Washington, while uranium enrichment facilities had been built in Oak Ridge, Tennessee. The real success of the Manhattan Project was the ten-to-one proportion of engineers to physicists and their excellent record of cooperation. That was another reason Germany failed in their atomic project because Nazis-directed nuclear physicists never valued their engineers as much as the American project did under Robert Oppenheimer's masterful leadership.[55]

It became a race against time, but enough plutonium was produced for three bombs and enough uranium for one bomb by mid-1945. The core or "pit" of plutonium was about the size of an orange, just as Winston Churchill had envisioned in his 1924 *Pall Mall Gazette* article, detailed in chapter one. The first plutonium bomb went to a site codenamed Trinity near Socorro and Alamogordo, New Mexico, for the first nuclear test on July 16, 1945.

In the 75 years of the nuclear Cold War and nuclear peace, almost 500 tons of plutonium have been produced worldwide. What most people do not understand is that the American nuclear stockpile still depends on plutonium. Yet, America, unlike Russia, China, and other nuclear powers, no longer builds new nuclear weapons nor manufactures new plutonium. Thus, we no longer have a plutonium production capacity or a large-scale ability to enrich uranium. This is of concern because our nuclear weapons are aging along with their plutonium pits. Now, this may be changing. Since the Russian invasion of Ukraine, the US has recognized its growing dependence on nuclear fuels, which are largely exported from abroad, and that includes reliance on Russian exports of uranium. With the increase in sanctions and rising tensions with Russia, Congress is urging reconstituting our ability to domestic enrich uranium on an industrial scale.[56] Oilprice.com reports the following:

For a long time, the US has been heavily reliant on Russian uranium, and imported about 14 percent of its uranium and 28 percent of all enrichment services from Russia in 2021 while the figures for the European Union were 20 percent and 26 percent for imports and enrichment services, respectively. And, there seems to be no end in sight despite Ukrainian President Volodymyr Zelenskiy last year calling on the US and the international community to ban Russian uranium imports following the Russian shelling near Ukraine's Zaporizhzhya power plant. US companies are sending more than $1 billion each year to Russia's state-owned nuclear agency, Rosatom, and imported another $411.5 million in enriched uranium in the first quarter of 2023 alone.[57]

---

[55] Conversation with retired Los Alamos National Laboratory Director Siegfried S. Hecker, June 2015, National Atomic Testing Museum.

[56] *The New York Times,* June 14, 2023, https://www.businessinsider.com/

[57] Https://oilprice.com/Alternative-Energy/Nuclear-Power/US-Takes-Bold-Step-To-Break-Free-From-Russian-Uranium-Hold.html.

Because the US no longer produces plutonium, the weapons laboratories must recycle and literally reshape it by hand, reconditioning pits when needed. Plutonium, as we have seen, is not a natural element. It is a very unusual substance that still holds many mysteries. One expert on plutonium recently stated that "there is a reason God did not make plutonium."[58] The point here is that plutonium has only been around for the past 80 years, and physicists are still unclear on what happens as the strange and unpredictable substance ages. One fact that is known is that plutonium produces helium bubbles and, over time, dilutes and changes the actual composition of the enriched compound, which today serves as the triggering mechanism in every American nuclear weapon.

Today, the United has what is called a stockpile stewardship program that addresses our aging nuclear stockpile. Since 1992, it has observed (but not ratified) a moratorium on underground nuclear testing while significantly decreasing the total nuclear arsenal. That important achievement has allowed the US to pursue a variety of nonproliferation and disarmament goals. It has also required the National Nuclear Security Administration (NNSA) and its weapons laboratories to replace the functions of nuclear tests with a science-based Stockpile Stewardship Program to ensure the safety, security, and effectiveness of the stockpile. Much of that work takes place at the Nevada National Security Site (NNSS), formerly known as the Nevada Test Site, where most nuclear tests took place from 1951 to 1992. Today, of course, we cannot explode a nuclear weapon in any kind of test. Although the modern-day NNSS, still affectionately called the "Test Site," continues to serve as an ideal place to do experiments that are called subcritical, or in other words, not producing a chain reaction. Subcritical tests continue routinely to this day.

The national weapons laboratories are instrumental as well. Los Alamos National Laboratory and Lawrence Livermore National Laboratory, with the assistance of the NNSS operation contractor Mission Support and Test Services (MSTS), conduct subcritical experiments (hydronuclear) at an underground facility called U1a. I have toured that site several times, and it is an extremely impressive operation.[59] This is much as in the days of nuclear testing when the Test Site in Nevada proved the ideal place for the national laboratories to do their work. Other experiments, called equation-of-state experiments, also occur at a test site facility called the Joint Actinide Shock Physics Experimental Research Facility (JASPER). These stockpile stewardship programs combine nonnuclear experiments, highly accurate physics modeling, and improved computational power to simulate and predict nuclear weapon performance across a wide range of conditions and scenarios. This predictive power affords scientists and engineers the necessary tools to assess the stockpile, maintain its performance, continuously improve safety, respond to technological surprises, and support future treaties. The primary mission of the modern-day Test Site is to help ensure that the nation's nuclear weapons stockpile remains safe, secure, and reliable. The Test Site has been handling weapons-grade plutonium since 1951, and it is the only place in our country where such testing can be conducted. The NNSS has described all of this modern-day stockpile stewardship initiative in one very comprehensive statement:

Since the United States no longer conducts full-scale nuclear tests – the US voluntarily ended underground nuclear testing in 1992 – Stockpile scientists and engineers now obtain data from

---

[58] Conversation with Ambassador Linton Brooks, former Under Secretary for Nuclear Security and Administrator of the National Nuclear Security Administration, July 2019, National Atomic Testing Museum.

[59] Nevada National Security Site: https://nnss.gov/mission/stockpile-stewardship-program/u1a-complex/.

breakthrough scientific experiments, engineering audits and analysis, high-tech computer simulations, and world- class diagnostic measurement systems. To keep existing warheads reliable, secure, and safe, every aspect of a weapon's performance is meticulously studied so the national laboratories can predict not only what will happen during an explosion (i.e., measurements within billionths of a second), but also measure what will happen to a device as it changes and ages over time, as the nuclear arsenal is now more than 50 years old.[60]

That covers the science of plutonium, but the actual logistics of maintaining our nuclear stockpile and its plutonium needs is another story. An *Associated Press* article from September 24, 2023, titled "Inside the Delicate Art of Maintaining America's Aging Nuclear Weapons," gives some of the best information available to date:

The Associated Press was granted rare access to key parts of the highly classified nuclear supply chain and got to watch technicians and engineers tackle the difficult job of maintaining an aging nuclear arsenal. Those workers are about to get a lot busier. The US will spend more than $750 billion over the next 10 years replacing almost every component of its nuclear defenses, including new stealth bombers, submarines and ground-based intercontinental ballistic missiles in the country's most ambitious nuclear weapons effort since the Manhattan Project.

It's been almost eight decades since a nuclear weapon has been fired in war. But military leaders warn that such peace may not last. They say the US has entered an uneasy era of global threats that includes a nuclear weapons buildup by China and Russia's repeat threats to use a nuclear bomb in Ukraine. They say that America's aged weapons need to be replaced to ensure they work.

"What we want to do is preserve our way of life without fighting major wars," said Marvin Adams, director of weapons programs for the Department of Energy. "Nothing in our toolbox really works to deter aggressors unless we have that foundation of the nuclear deterrent."

By treaty the US maintains 1,550 active nuclear warheads, and the government plans to modernize them all. At the same time, technicians, scientists and military missile crews must ensure the older weapons keep running until the new ones are installed.

The project is so ambitious that watchdogs warn that the government may not meet its goals. The program has also drawn criticism from non-proliferation advocates and experts who say the current arsenal, though timeworn, is sufficient to meet US needs. Upgrading it will also be expensive, they say.

"They are going to have extreme difficulty meeting these deadlines," said Daryl Kimball, executive director of the Arms Control Association, a non- partisan group focused on nuclear and conventional weapons control. "And the costs are going to go up."

He cautioned that the sweeping upgrades could also have the undesired effect of pushing Russia and China to improve and expand their arsenals.

---

[60] Nevada National Security Site: https://www.nnss.gov/pages/programs/StockpileStewardship.html.

Where it begins

The core of every nuclear warhead is a hollow, globe-shaped plutonium pit made by engineers at the Energy Department's lab in Los Alamos, New Mexico, birthplace of the atom bomb. Many of the current pits in use come from the 1970s and 80s. That can be problematic, because there's a lot about plutonium's aging process that scientists still don't understand.

The key radioactive atom in the plutonium pit has a half life of 24,000 years, which is the amount of time it would take roughly half of the radioactive atoms present to decay. That would suggest the weapons should be viable for years to come. But the plutonium decay is still enough to cause concern that it could affect how a pit explodes.

President George H.W. Bush signed an order in the 1990s banning underground nuclear tests, and the US has not detonated pits to update data on their degradation since. When the last tests were performed, they provided data on pits that were at most about two decades old. That generation of pits is now pushing past 50.

Bob Webster, deputy director of weapons at Los Alamos, said scientists have relied on computer models to determine how well such old pits might work, but "everything we're doing is extrapolating," he said.

That uncertainty has pushed the department to restart pit production. The US no longer produces man-made plutonium. Instead, old plutonium is essentially refurbished into new pits.

This task takes place inside PF-4, a highly classified building at Los Alamos that's surrounded by layers of armed guards, heavy steel doors and radiation monitors. Inside, workers handle the plutonium inside steel glove boxes, which allow them to clean and process the plutonium without being exposed to deadly radiation.

In the final production steps, a lone employee in the vault takes the almost- completed pit into both of her gloved hands and shapes it into its final form. "Things have to fit a certain way, and everything is by touch, by feel," said the Los Alamos employee, who the AP has agreed not to name because she is one of only a handful of people in the US, and the only female, who performs this sensitive task.

For about the last 10 years technicians have been practicing on "test" pits that aren't ready for the stockpile. The US is planning to fully recycle its first weapon-ready pit next year — and quickly increase annual production to as many as 80 new pits. The painstaking and hazardous work has led a government watchdog to express doubts about whether the US government can meet that goal.

"The United States has not regularly manufactured plutonium pits since 1989," the Government Accountability Office noted in a January 2023 report, adding that the Energy Department's National Nuclear Security Administration has provided "limited assurance that it would be able to produce sufficient numbers of pits."

Webster has been at Los Alamos since Ronald Reagan was president. He could have retired years ago, but has remained to shepherd the first new plutonium pits through to production. The lab is starting to feel a bit like it did in the 1980s, during the Cold War, he said. Los Alamos scientists are having intense discussions about weapon design — how much each can weigh, its explosive punch, how far it must travel.

"We need our nation to be back making pits," Webster said. "We just have to be able to do that."[61]

One primary observation that stands out in maintaining plutonium, which exists for the primary purpose of maintaining a nuclear arsenal, is the associated costs. The amounts are hard to even rationalize. It goes without saying that in a world of nuclear weapons, the only recognized way to prevent one nation from using such a weapon is the principle of deterrence. So the idea is that we as a nation have to maintain nuclear weapons, even if they hold us hostage to the very technology of mass destruction we created so many years ago. Maybe deterrence is a valid concept, and maybe not, yet the idea of dismantling one's nuclear arsenal is simply never going to happen. I, personally, along with many, would nevertheless like to see total disarmament, but it will never happen, just as wars and conflict will never cease to exist. It is rather surreal to think just how much wealth we could be devoting to education and healthcare. Instead, we use that treasure to maintain a weapon of mass destruction that could potentially destroy the very society we labor to improve.

Later in the book, we will examine the interesting history of the heyday of nuclear testing and provide more insights into the nuclear arsenal and how it developed. To me, the final word on plutonium is an ironic one. The slowly decaying form of plutonium-238, with a half-life of 87 years, safely powered the *New Horizons* space probe to this strange element's namesake, Pluto, in July of 2015. Plutonium and Pluto finally met.

Artist's conception of *New Horizons* at Pluto. Image credit: NASA.

---

[61] AP staff article, *Inside the Delicate Art of Maintaining America's Aging Nuclear Weapons*, Associated Press, September 24, 2023, https://www.voanews.com/a/inside-the-delicate-art-of-maintaining-america-s-aging-nuclear-weapons/7276214.html.

CHAPTER SEVEN
THE BIKINI

The image is reproduced with permission from Geoffrey Beaumont's family.

A few years ago, I had the great fortune to meet Geoffrey Beaumont, the son of famed Californian Impressionist painter Arthur Beaumont. Geoffrey has devoted a lifetime to educating the public on his late father's noted works. His generous collaboration led to three art exhibits focused on a period when Arthur Beaumont created stunning watercolors of the first postwar atomic tests around Bikini Atoll in the Marshall Islands during Operation Crossroads.[62]

Conducted in the summer of 1946, Crossroads involved two detonations. These were the Able and Baker shots, one an airdropped "Fat Man" type of plutonium fission device as used in the Nagasaki bombing, and the second a submerged test of an identical bomb. Both detonations had a yield of 21 kilotons each. A planned third test was canceled. Crossroads determined the effects of nuclear weapons on a fleet of aging Second World War naval vessels.

These were historically significant ships. The legendary battleship USS *Nevada* served as the target ship, and amazingly, it went on to survive both atomic detonations. Other famous ships anchored in Bikini lagoon included the Japanese Pearl Harbor Flagship *Nagato*, the German battleship *Bismarck's* companion *Prince Eugene*, and the first of the 1930s heavy aircraft carrier series, the USS *Saratoga*.

The United States Navy utilized the artistic talents of watercolor painter Arthur Beaumont (and Beaumont's friend, oil painter Captain Charles Bittinger) to capture the historic scenes of those first postwar nuclear tests. Aside from Beaumont's impressive artistic accomplishments, which included commissions by three US Presidents, he also served as the official artist for the United States Navy for almost fifty years. Beaumont witnessed many historical events in his Navy association, ranging from World War Two to

---

[62] The Beaumont art exhibits I oversaw as the Executive Director of the Smithsonian-affiliated National Atomic Testing Museum were thanks to Geoffrey Beaumont, son of the noted artist Arthur Beaumont, who worked with Admiral Samuel Cox, Commander of the Naval History and Heritage Command in Washington Naval Yard to bring part of this noted collection to our museum. I also thank Head Curator Gale Munro of the National Museum of the United States Navy.

Antarctica explorations, including Operation Crossroads. His vibrant watercolors of those spectacular nuclear tests are now famous thanks to publications and lectures by Geoffrey Beaumont.

Aside from the vivid watercolors of the ships taking part in nuclear tests, Arthur Beaumont painted life on the island of Bikini. He frequently captured scenes of the officer's club, where a few dozen nurse lieutenants spent their off hours. One of those historic paintings documents the first example of a two-piece bathing suit, which the nurses crafted themselves in order to cope with the hot and humid island paradise.

These were true handmade inspirations—there then being no such thing yet as a manufactured two-piece bathing suit. Arthur Beaumont took quite a liking to these colorful outfits worn by the female lieutenants. He is pictured here documenting the stylish phenomenon with a nurse standing behind him, who also appeared in the painting on the next page. That image is a zoomed-in portion of the larger painting displayed at the beginning of this chapter, titled *Bikini Village*. At that time, the artist in him decided that this creative clothing design needed a name. He christened it the Bikini.[63]

Zoomed in detail of chapter cover photo, *Bikini Village*, 1946, watercolor by Arthur Beaumont, depicting scenic tropical Bikini Island during atomic tests of Operation Crossroads with nurses wearing first true bikinis. Loaned for the exhibit from the Naval History and Heritage Command in Washington, DC. Reprinted with generous permission by Geoffrey Beaumont.

No one knows that story today because in that exact same period, on the opposite side of the world, a

---

[63] Conversation with Geoffrey Beaumont, August 2019, National Atomic Testing Museum.

famous clothing designer got the same idea. Radio news stories of the nuclear tests on Bikini Atoll were broadcast around the world all summer long. The barrage of news coverage reached Paris, where it was hot that summer. Listening to his radio in his workshop, the nuclear tales inspired noted Parisian swimsuit designer Louis Réard to coin his latest two-piece summer swimsuit creation, the Bikini. He, of course, got the credit, not Arthur Beaumont nor the nurses who hand-made the first true bikini.

The results of the first post-war nuclear tests were disappointing. One of the most beautiful islands in the Pacific remains contaminated to this day. The original population of 167 Bikini islanders was never able to return to their beloved homeland for any period of time. Less significant but regrettable is the fact that numerous historic and iconic ships were sunk or irradiated. The tests came at a time when there was still hope by people like Robert Oppenheimer that the new United Nations might be able to take charge of all nuclear energy and keep it confined to peaceful uses. Unfortunately, Crossroads convinced the USSR that nuclear weapons would become part of the United States's permanent arsenal.

Many Scientists like Niels Bohr and Oppenheimer had hoped that nuclear weapons would prove so intimidating and dangerous that they would make war obsolete. Unfortunately, the opposite happened as the weapons simply became too tempting not to possess, making war much more potentially destructive. The United States did create its own Atomic Energy Commission by the end of that summer, and thanks to President Truman's founding intent, nuclear energy has ever since been placed in the hands of civilian and not military control. Unfortunately, as the years went by, atomic weapons evolved into thermonuclear weapons with yields a thousand times the size of the Hiroshima and Nagasaki bombs and the bombs used at Crossroads. A costly and dangerous arms race followed. Thankfully, no nuclear weapons have been used in post-WWII warfare. Their existence, in theory, has deterred another world war, or at least that has remained true to the date of this writing.

In the final analysis, Operation Crossroads focused on interservice rivalry and the growing fear of the vast expansion of the USSR following WWII. At the end of the war, the United States Army and its air arm separated with the formal establishment of the United States Air Force in 1947. The United States Navy, which had been severely reduced in size and funding with the conclusion of the Pacific war, became concerned with accusations by Air Force advocates that atomic weapons made the air arm the dominant military service and primarily made surface ships obsolete. The Navy hoped to disprove that with the Crossroads tests.

Indeed, both the airdrop and submerged detonations at Crossroads surprised everyone. The majority of ships survived. Yes, a fleet could fight a nuclear war. The problem, however, slowly became apparent that a ship within a mile's range of a nuclear air blast would be subjected to such significant gamma radiation that its crew would be fatally incapacitated within three hours. The submerged Baker test proved even more concerning because the bomb vaporized a portion of the coral lagoon. The rising water vapor and debris carried plutonium contamination over the ships and the island. The fallout-ridden ships proved almost impossible to decontaminate, making it problematic even to reboard. Alpha and beta contamination also infiltrated the air systems and made occupation of the vessels all the more hazardous.[64] Most remaining ships were left to sink or intentionally sunk. Los Alamos scientists warned the Navy of this possible outcome

---

[64] Chapter Twenty-Five will explain alpha, beta, and gamma radiation.

and recommended against the test, but the military ignored them.[65] A third and final test was subsequently canceled.

Crossroads did provide a certain legacy. Aside from the magnificent images produced by Arthur Beaumont, today, the lagoon of Bikini serves as an underwater park preserving some of the most amazing warships ever built. Although still contaminated, experienced divers can and do safely visit this ocean museum yearly and will likely do so for centuries. Following are a few select images of Arthur Beaumont's magnificent watercolors from the Operation Crossroads tests:

Watercolor on Paper by Arthur Beaumont; 1946, depicting the Japanese battleship *Nagato* after the Baker blast.

Watercolor on Paper; by Arthur Beaumont; 1946,
depicting underwater Baker Bomb Test.
(From the *Jonathan Art Foundation*, this Beaumont watercolor once hung in the *National Gallery*.)

---

[65] Conversations with and work in assisting in presentations by Nelson Cochran, former Nevada Test Site engineer and nuclear testing engineer and historian.

These pencil sketches were never exhibited, but they were produced by Arthur Beaumont during the Crossroads tests in preparation for his watercolor paintings. The author thanks Geoffrey Beaumont for sharing these images from his personal collection.

Watercolor on paper by Arthur Beaumont 1946, depicting the Sinking of the Saratoga. (Although the Navy mounted an extraordinary effort to save the USS *Saratoga* (CV-3), the warship was too contaminated to conduct extensive damage control procedures, and she eventually sank. One of the legends of the Pacific war, the officers and men of Operation Crossroads, USN Heritage Command, keenly felt her loss.)

Arthur Beaumont (1890-1978)

Arthur Edwin Crabbe was born in Norfolk County, England. He emigrated to the United States and studied art at the University of California's Mark Hopkins School of Art. In 1915, he changed his name to Arthur Beaumont and moved to Los Angeles. Two years later, he opened his first commercial art studio. He continued his studies at the Chouinard Art Institute and the Slade School of Fine Art in London. Then, in Paris, he studied at both the Academie Julian and the Academie Colarossi.

During the Great Depression, Beaumont taught watercolor classes at Long Beach Harbor. In 1933, he was given a commission as a lieutenant in the Navy, with orders to join the fleet on the major endeavors at sea. Some of these included traveling through the Panama Canal and meeting President Roosevelt in New York Harbor.

When World War II started, Beaumont's paintings focused more on military activities, ship paintings, and battle scenes. His paintings of the war were seen in War Bond campaigns and across the front pages of newspapers. Beaumont's work didn't end with the war. He continued sketching and painting onboard scenes on the USS *Iowa*, the USS *Los Angeles*, and the USS *Midway*. In July of 1946, he was designated as the Navy's Official Task Force Artist for Operation Crossroads, the atomic bomb tests at Bikini Atoll.

Beaumont continued painting, pausing during the Korean War due to several injuries to his painting hand. He continued to travel with the Navy, including heading to the Arctic, Antarctica, and Vietnam. (All text is printed with the permission of Geoffrey Beaumont. A more in-depth biography can be found at WWW.NavyArt.com).

CHAPTER EIGHT
INTROSPECTION BY JAMES HALL

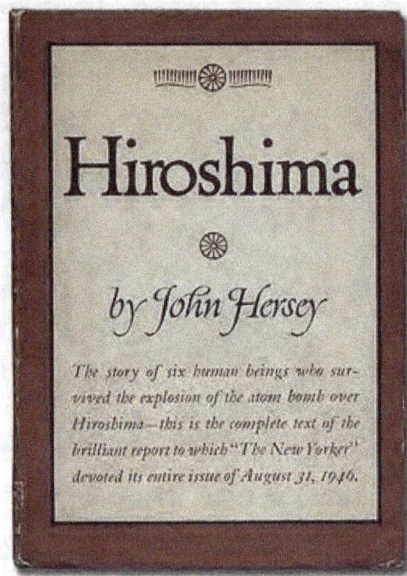

**The John Hersey Book is based on his August 31, 1946,
article in *The New Yorker*.**

It took about a year before the public developed some sort of retrospective of the first use of atomic weapons. The atomic bombing of Japan was initially considered a miraculous war-winning maneuver. To this day, veterans and their surviving stories overwhelmingly look at the A-bombing as a closure to what they considered were lost years for service members and their families during a long period of world conflict. Most also considered it a lifesaving event because so many did not feel they would have survived the war if they would have had to endure a land invasion of Japan.

That image remained the focus of the general public until an astounding article appeared in *The New Yorker* on August 31, 1946. In that seminal feature, John Hersey told the story of six survivors of Hiroshima. The former war correspondent told an account that moved people so much that they printed the article as a book titled *Hiroshima* just a few months later. To this day, the story influences public perceptions of nuclear weapons and even their potential future impact.[66]

Because of Hersey's article, a large segment of the population became more appreciative of the power of science and its endless possibilities. On the one hand, that realization prompted people, especially American youth, to dream of space travel and endless supplies of energy. On the other hand, it caused those who had already lived through two world wars to wonder just how much worse war could get. And it then made them wonder if humanity had finally reached the point where it could cause its own extinction.

---

[66] During the Pacific war, writing for *Life*, Hershey documented a young lieutenant John F. Kennedy's journey through the Solomon Islands.

Accounts in Hersey's story accurately introduced America to the horrors of radiation sickness. The human suffering caused by the survivors of Hiroshima began to resonate with readers who realized that nuclear weapons, if ever acquired by another power, could be used on them as well. So, while the story gave an important historical perspective of a critical turning point in history, its primary impact was making people think about the future instead of the past. At that time, the US still had a nuclear monopoly, and many naively thought that proprietorship would continue for decades. Yet, the idea could never be dispelled after anyone read John Hersey that they were just as human as the citizens of Hiroshima and may one day be just as vulnerable. When I read this story, although I was born fifty-five years after the event, I did not just see Japanese people facing the horrors of a nuclear holocaust. I visualized myself in that position.

A final personal insight concerns the style Hersey used. He became one of the first to tell a non-fictional story using fictional storytelling techniques. That style has since become called "New Journalism." That basically means writing a non-fictional story like a novel. *New Yorker* political commentator and journalist Hendrik Hertzberg wrote that: "Hersey's reporting was so meticulous, his sentences and paragraphs were so clear, calm and restrained, that the horror of the story he had to tell came through all the more chillingly."[67] John Hersey, a long-time *Time-Life* reporter, disappointed the magazine founder Henry Luce when Hersey chose to publish in *The New Yorker*. However, Hersey considered the magazine's format and style, giving him much more freedom. This was especially true since *The New Yorker* devoted the entire issue to Hersey's article.

This is a story that everyone should read, especially in our current times when the nuclear issue is so grave again.

---

[67] Hendrik Hertzberg, *Obituary of John Hersey*, The New Yorker, March 29, 1993, http://www.newyorker.com/archive/199304/05/1993_04_05_111_TNY_CARDS_000365222.

CHAPTER NINE
ATOMIC HEYDAY - PART ONE

Al O'Donnell circa 1946 from authors' collection.

I never had the privilege of knowing Al O'Donnell, but I heard many informative stories from his close associates concerning his years involved in the field of nuclear testing. Mr. O'Donnell is so unique because he not only goes back to the early period of nuclear testing in the Pacific but also to the four decades of nuclear testing at the Nevada Test Site. It is believed he witnessed and participated in more of the 1,054 US nuclear tests than any single individual. To me, his unpublished memoirs, which I have had the good fortune to read, are important because they are part of a bigger story. That story concerns the Cold War and nuclear deterrence.

The tale goes back to March of 1949 when the Berlin Airlift remained in full swing as tensions with the Soviet Union grew into an ever-evolving Cold War. That month, Major General Kenneth Nichols, who had succeeded Manhattan Project-era leader General Leslie Groves, attended a luncheon in Secretary of Defense James Forrestal's office.

In attendance were the three Joint Chiefs and Under Secretary of State Robert Lovett, Secretary of the Air Force Stewart Symington, and Secretary of the Army Kenneth Royall. Dwight Eisenhower also attended the meeting. He had stepped down from Chief of Staff of the Army in 1948 to become president of Columbia University. Still affectionately called "Ike," he may have simply been invited to the meeting as a courtesy. The times were certainly dark, and the Berlin crisis dominated the discussion, so it made sense that a man such as Eisenhower would have been there.

No one questioned his assertiveness as Ike effortlessly and naturally took command of the meeting—perhaps just out of habit. Historians, in fact, later considered him at this time, the first "informal chairman" of the Joint Chiefs of Staff.[68] At one point, Eisenhower asked the big question that was on everyone's mind: if needed, could nuclear weapons be sent to England? Sixty B-29s had just been moved there in a highly publicized manner; however, no one understood that they were not modified to carry nor armed with nuclear weapons. The only B-29s anywhere that could do that were then in the 509th Bomb Group in Roswell, NM.

To Eisenhower's surprise, they told him there were only a handful of nuclear weapons in the stockpile, and they were not really part of a "stockpile." On the contrary, the nuclear arsenal was basically a collection of 56 World War II-era "Fat Man" type plutonium devices. They did not even assemble the would-be bombs, nor did they design the bombs to be permanently assembled. They were, in all reality, a collection of parts held in the custody of the Atomic Energy Commission, and no protocol yet existed for the military to access them. In his wisdom, however, that is exactly how President Truman wished it. He wanted civilian, not military control of atomic energy, which was the very reason for the formation of the Atomic Energy Commission or AEC.[69]

Eisenhower then learned that there were very few scientists who had remained from the Manhattan Project days who were qualified to assemble or arm those nuclear weapons if needed. No military men were yet qualified; however, before General Groves retired, he had initiated a program at Sandia Base in Albuquerque, NM, to train service members to become specialists in the vacuum created by so many Los Alamos scientists leaving the Manhattan Project and going back to university work.

At that time, the Operation Sandstone tests, which followed the Operation Crossroads tests, had been completed the previous summer at Enewetak. There were only a handful of men capable of assembling and arming nuclear devices, and they were at those tests. Many worried about having "all the valuable eggs in one basket."[70] The Sandstone detonations provided the US with a tested practical, mass-producible atomic bomb design called the Mark 4, which would create a true nuclear stockpile by June of 1949 of 169 devices.[71]

---

[68] General Omar Bradley became the first official Chairman of the Joint Chiefs of Staff from 1949 to 1953. The position was established in 1949 as a result of the National Security Act of 1947, which reorganized and unified the US military under civilian control and created the Department of Defense.

[69] The Atomic Energy Commission (AEC) was established in 1946 following the end of World War II. It adopted the responsibility under civilian, not military, leadership for both the development of nuclear weapons and the promotion of peaceful uses of nuclear energy. In 1974, the AEC was abolished, and its functions were divided into two new agencies: the Department of Energy (DOE) and the Nuclear Regulatory Commission (NRC). The DOE focused on energy research, nuclear weapons development, and nuclear waste management, while the NRC was responsible for regulating nuclear power plants and ensuring their safety.

[70] After the Crossroads tests, the following year, on June 27, 1947, President Harry Truman authorized a new test series. The resulting Operation Sandstone was conducted at Enewetak Atoll in the Marshall Islands with three detonations in April and May of 1948. While Crossroads had been weapons effects tests, Sandstone tested the first new atomic bomb designs since World War II.

[71] [7]The Mark 4 bombs had a "levitated" composite uranium and plutonium pit design, called an open pit, which was more easily removable. It was stored separately in a special capsule called a birdcage. The Mark 4 had an explosive yield of up to 31 kilotons.

Eisenhower did not worry as much about the Berlin crisis as others did because he knew the Soviets were not yet logistically ready for war. They had literally stripped all the steel rails from every rail line in East Germany except one and sent the steel east as reparations. The only rail line they had left went from Berlin to Poland in varied gauges. Most roads and bridges were still largely unserviceable from wartime bombing and shelling. Thus, Eisenhower knew the Soviets could not supply or support any offensive west of Poland in 1949. Ike did worry about how the equation of nuclear weapons would play out in future disputes; however, not even he could have then guessed that the Soviets were only six months away from testing their first nuclear weapon, which took place on August 29, 1949, at their Semipalatinsk Test Site in Kazakhstan. After that, at Truman's urging, Ike took an extended leave from the university to become the Supreme Commander of the new North Atlantic Treaty Organization (NATO) in December of 1950.

Al O'Donnell comes into this story as one of those unique individuals who played a vital role in early nuclear testing. He was one of the few men in March of 1949 who could work with nuclear weapons. He came from Boston, Massachusetts—born and raised there since 1922. After Pearl Harbor, he joined the Navy to do his part in the war. He had recently married when the war started, and he and his wife awaited the birth of their first child.

At the baby's birth in 1942, tragedy struck. The newborn died of heart failure. The military shipped O'Donnell out that very day, and they made no allowances for him to stay with his wife in the hospital. The Navy ordered him to the Pacific, which would be the first of many times he had to make sacrifices in service of his country. O'Donnell would not see his wife for more than a year, and he went on to serve all through the Pacific war, including the invasion of Saipan, Tinian, and Okinawa. By late 1945, he had returned to civilian life and found a job in Boston at the Raytheon Manufacturing Company. Raytheon had a contract with the Massachusetts Institute of Technology (MIT).

At Raytheon, three men, who later formed their own company called Edgerton, Germeshausen, and Grier, Inc., were conducting secret work related to the wartime Manhattan Project. These founders of EG&G deserve special mention because Al O'Donnell became a key part of that company. EG&G traces its origin to Dr. Harold "Doc" Edgerton, Kenneth J. Germeshausen, and Herbert E. Grier. These three MIT scientists had partnered on patents and flash photography since the 1930s. When officially incorporated in November 1947, their company assisted the newly formed Atomic Energy. Commission in "design and operation systems that timed, monitored, photographed and triggered nuclear tests."[72] A fourth man by the name of Bernard (Barney) J. O'Keefe entered the picture just before O'Donnell came on board and played an equally crucial part in the story.

Barney O'Keefe became involved with the Manhattan Project during World War II. He worked on firing systems while participating in the actual deployment of the nuclear bombs from Tinian Island that were dropped on Hiroshima and Nagasaki. Prior to that, the trio of Edgerton, Germeshausen, and Grier worked at MIT on night aerial flash photography, which came to be used in both theaters of war for reconnaissance. Simultaneously, Germeshausen worked with the MIT Radiation Laboratory and concentrated on radar work with his study of thyratrons and modulators. Grier worked in the Draper Laboratory, the MIT aviation

---

[72] Authors viewed the original government contract between AEC and EG&G in private files of Peter Zavattaro, the last president of the EG&G corporation. Those files were generously donated to the Smithsonian affiliated National Atomic Testing Museum by Mr. Zavattaro and curated by Executive Director Michael Hall.

instrumentation laboratory known as the Confidential Instrument Development Laboratory. Apparently, that became a connection to the Manhattan Project and Barney O'Keefe.

Prior to that and earlier in the war, Edgerton, Germeshausen, and Grier oversaw the production of electrical discharge systems for aerial stroboscopic photography made in the same Raytheon factory where O'Donnell came to work. Raytheon emerged from the 1920s-era American Appliance Company, which early MIT engineering professor Vannevar Bush helped found. Raytheon led the way in early electronic innovations and patents and became a key defense contractor during World War II, specializing in radar equipment.

At some unknown point, the Manhattan Project took special note of the Edgerton, Germeshausen, and Grier-designed equipment used in aerial stroboscopic photography. Los Alamos deemed it valuable for producing bomb firing sets for the atomic bomb. The exact details of how that happened and how Edgerton, Germeshausen, and Grier got involved are not absolutely clear, but Edgerton's private notes show that he had developed some close associations with defense and military leaders. It also certainly involved Barney O'Keefe. As an engineering and radar specialist in the US Navy, "Ensign O'Keefe" had been transferred during the war to Santa Fe, New Mexico, as part of "Project Y." That was code for the Los Alamos Laboratory established by the Manhattan Project and operated by the University of California during the war.[73]

That is how O'Keefe got involved in the Manhattan Project's early phases. He was then transferred to the Manhattan Project firing group called X-5 in the Explosives Division at Los Alamos because of his radar background, which was thought to be useful in triggering atomic bombs at specific altitudes above a target. The last president of EG&G, Peter Zavattaro, writes in his history on EG&G: "Since the time was so short for the design and production of the firing sets at Los Alamos, the procurement people set out to find an existing factory they could commandeer. They found a Raytheon factory in Boston, MA, assembling night aerial photography systems under Herb Grier's supervision."[74]

Certainly, the godfather administrative figure in the Manhattan Project, Vannevar Bush, could account for knowledge and preference for Raytheon, which he helped create. So, the Raytheon employees who worked on the photography units were redirected to make the firing systems, although they were not privy to what they were doing at the time. Grier and his associates, who had already been connected with the photography work, contributed to the actual design and construction of the equipment now used as firing sets. Under Grier's supervision, engineers used equipment like large electrical capacitors, which were re-purposed for the detonating process. The capacitors required a simultaneous timing system for the explosive lenses that surrounded the core of the plutonium bombs.

Early Raytheon Company logo, Authors' files.

---

[73] Dr. Harold Edgerton's private papers, MIT archives:
https://archivesspace.mit.edu/repositories/2/resources/603/collection_organization.

[74] Peter Zavattaro, *EG&G* (Las Vegas, Nevada: Nevada Test Site Historical Foundation, 2007), pp. 1-2.

How much Herb Greer initially knew about the Manhattan Project is not clear. Another man of note is William F. Lightfoot, who worked specifically on the fuse of the plutonium device. Lightfoot served as manufacturing and test equipment manager for Raytheon. Little detail is known about his role. Al O'Donnell came into this secret Raytheon factory setting in late 1945. O'Donnell stated in his memoirs:

I didn't know it at the time, but Raytheon had a working agreement with MIT's Dr. Edgerton, Germeshausen and Grier. That was a partnership at the time. I said, "OK" and asked, "What am I going to be doing?" They replied, "You will find out in good time." I knew what I was going to be doing but I didn't know whom for. I sort of surmised that whatever I was going to do involved the federal government in some way or other. I thought that it was probably classified as secret. Eventually, I got my "Q" clearance and they told me what the project was. . . "What's going to happen to me now?" I asked. "We're going to send you to MIT in Cambridge, across the Charles River…You are going over there to a laboratory under the control of a Dr. Edgerton."

I never went back to Raytheon. I stayed there with Doc Edgerton. That's where I met the other two members of the partnership – Germeshausen and Grier. We shortened Germeshausen's name down to "Germs;" it was too long to say Germeshausen.[75]

At this point, we need to provide a little more background on EG&G to help understand Al O'Donnell's story. After the war, Barney O'Keefe, a continuingly important historical figure in our story, left the Navy and received an appointment from MIT as a research associate in 1946. O'Keefe, just like O'Donnell, served in Dr. Edgerton's MIT Electrical Engineering department. O'Keefe also served as an associate project manager in Grier's classified MIT research work, to which O'Donnell would also gravitate.

By then, Grier had been contracted to redesign the firing sets that had been originally produced at the makeshift assembly operation at the war-time Raytheon factory. Those early firing sets, as noted, developed from the cumbersome electrical high voltage discharge systems for aerial stroboscopic photography, which sufficed to fire the first atomic weapons—however, a more refined design needed to be developed. Grier was tasked with that, and O'Keefe assisted him in that work. O'Keefe had key qualifications for this because he had not only formally worked on the Manhattan Project but had been the one to oversee the final wiring of the "Fat Man" plutonium bomb on Tinian.

While working at MIT with the original EG&G trio, O'Keefe also pursued private consulting activities. One of those ventures involved a former associate from Los Alamos, Donald Hornig, who had invented a radiation measuring device for infrared spectroscopy that was being built in the Cambridge laboratory. They later formed a partnership and then a corporation called the Radiation Instruments Company. However, O'Keefe continued to work simultaneously at MIT.

O'Keefe and Hornig later sold their company, and O'Keefe went to work for EG&G when it was formed. Nevertheless, the experience O'Keefe initially gained in forming his early company greatly assisted Edgerton, Germeshausen, and Grier when they formed EG&G. From 1945 to 1947, Edgerton, Germeshausen, and Grier handled many classified weapons development projects at MIT. The classified work at MIT actually centered more on Grier. Edgerton liked to focus the majority of his time on electrical

---

[75] Al O'Donnell DRI Oral History Project, University and Community College System of Nevada, 2004, p. 3.

measurement studies, serving as a professor in that department at MIT. Germeshausen wanted to continue his work on radar, which began during the war, focusing on his inventions of the hydrogen thyratron and modulators.

Simultaneously, these men were retaining their interest and apparent entrepreneurial dappling in the field of high-speed commercial photography. Before the war, their invention of a small strobe unit, the Strobotac, was licensed to the General Radio Company. Peter Zavattaro writes in his history about this early period, "The three never had a written agreement, nor was there any specific point at which the partnership was formed. They simply pooled their resources and worked together on whatever consulting task or measurement problem came along." O'Keefe became more and more of a protégé of all these men as time went on in this post-war period, and his innate business sense became a valuable resource for them." This is the environment into which Al O'Donnell fell.

These were logical people to call upon after the war, as nuclear research continued in the wake of the growing Cold War. Logistically, the formation of the Atomic Energy Commission in 1946 became the "government conduit" to engage such brilliant people. This was a critical move because by forming the AEC, President Truman took nuclear weapons out of the hands of the military, and to this day, they remain under civilian authority.

In 1947, Norris Bradbury (successor to World War II Los Alamos Manhattan Project laboratory leader Dr. Robert Oppenheimer) went to MIT to visit the scientific EG&G trio. Bradbury asked the men to build special firing triggers for the upcoming tests to take place in the Pacific. Accepting the job, however, soon overloaded the MIT contracting and procurement system, which Grier had previously set up to handle such sensitive government contracts.

At this time, MIT administrators were actually looking to focus more on aeronautical research and more non-military-related pursuits. So, Grier suggested that he and his partners form a legal corporation to take over their own sensitive and highly specialized work. They decided to move their new firm and its work offsite from MIT. That, at any rate, is one version of the story. The exact details depend on which source you read, and there are a number. O'Donnell described it this way:

We stayed at MIT until late 1946. The president of MIT asked Dr. Edgerton if he would transfer everything pertaining to the US Government out of his laboratory. He said that Dr. Edgerton's work on strobe lights could stay at MIT but that everything else had to go. MIT wanted nothing more to do with wartime or any federal projects. He had already devoted enough of his time and efforts at MIT to the war, so Doc said, "yes, no problem." What they did then was form the corporation from the partnership, and it was going to be known as Edgerton, Germeshausen and Grier, Inc. (EG&G). We went to the Boston area across the Charles River and rented or leased a garage that had an upper and lower level. It was set up so that you went down a ramp to the lower floor and you went up a ramp to the upper floor. The draftsman's office, the library and the executive offices were all on the upper floor. The laboratory and stockroom, where all of the work was done, were on the lower floor. At least, we thought so.

Dr. Edgerton was more interested in his own strobe light work. Dr. Germeshausen was a great scientific physicist who knew what he was doing but really wasn't interested in what Dr. Grier had latched onto. It was Dr. Grier who got EG&G involved in the Manhattan Project. I found out later

that when I was working at Raytheon, I was working on the main control point equipment. It was a matter of my getting acquainted with the sequence timer and the system; how it operated from the control point and how it went out to the substations. I knew what I was doing but I didn't know whom I was doing it for or how it was going to be used.

I finally put the pieces together and realized that it was Grier who spearheaded EG&G's involvement in weapons testing of a nuclear detonation. Grier was given the responsibility by the Manhattan Project to design, develop and execute a timing and firing system. It was not only used in dropping a bomb from an airplane but also exploding it from [testing] towers and balloons. I latched onto him or he latched onto me, one of the two. I was on the design of the system to do the arming and firing of the units that were used at the Nevada Proving Grounds, later to be known as the Nevada Test Site (NTS).[76]

There are many versions of the story. Barney O'Keefe's memoirs state:

In the summer of 1947 I was called to a meeting at MIT with Norris Bradbury, the new Los Alamos director, and Alvin Graves, his test division leader. They were planning a series of weapons development tests, to take place in 1948 at the atoll of Eniwetok, two hundred miles east of Bikini. Since those in our MIT group were the principal designers of the firing sets, we were asked to design a special version for the Eniwetok tests. When we agreed, they asked the next question: Would we design a system of signals to actuate unmanned instrumentation and coordinate them with the firing signal? We said we would, but pointed out that, among other things, we would have to procure several million dollars' worth of underwater signal cable to string around the atoll. Our MIT contract was funded at only $200,000 total. Money was no problem, we were told. At the time, the MIT director of research programs heard about the proposed tenfold expansion of our little project, he balked, saying that MIT was trying to get out of military research, not into it. He suggested to Grier that he and his partners form a corporation to take over the government business and move it off campus... We moved to an old garage on Brookline Avenue in Boston. We were also asked to do a neutron measurement because we were also experts in fast pulse measurements.[77]

Whether this new postwar work in nuclear testing was the only reason for the formation of EG&G as a company has never been clearly documented; however, it is highly likely. Basically, EG&G was formed because MIT simply could no longer facilitate the work. Nor did it seem they wished to. In his unpublished autobiography, Dr. Edgerton claimed the AEC requested that MIT establish a comprehensive test system at Enewetak with Grier in charge. However, according to Edgerton, MIT gave a firm "no" to that idea.[78]

At the time, MIT President Karl Compton was faced with the challenge of reversing the huge wartime direction in military research, which the university had engaged in during World War II. By then, he wanted MIT resources to focus more on pure research. Compton's successor in 1948, James Killian, supported this

[76] Al O'Donnell DRI Oral History Project, University and Community College System of Nevada, 2004, p. 2.

[77] Bernard J. O'Keefe, *Nuclear Hostages* (Boston: Houghton Mifflin Company, 1983), p. 126.

[78] Dr. Harold Edgerton's private papers, MIT archives: https://archivesspace.mit.edu/repositories/2/resources/603/collection_organization.

and started a strict redeployment of research for peace. Edgerton and O'Keefe recorded in their memoirs that the unofficial partnership the three scientific wizards were already enjoying was the convenient inspiration for forming the EG&G corporation. EG&G would take over MIT's unfulfillable commitments to the AEC, and O'Keefe and O'Donnell would go to work for these three founding men.

Al O'Donnell gave invaluable insights on this early period in his accounts. He detailed EG&G's early evolution from literally a garage operation to one of the nation's largest interests. Many remember EG&G as a huge contracting and engineering firm employing thousands of people with subcontracts, associated business partners, and firms. It did become that picture of power and complexity; however, EG&G's beginnings were humble. O'Donnell described how the noted firm started in an old, converted parking garage on Brookline Avenue in Boston in late 1947.

Initially, the three founding scientists had a total of six employees, including Barney O'Keefe. (Apparently, O'Donnell, at this time, was being paid through MIT because he was still technically attached to Dr. Edgerton's lab.) Soon, the company grew to 15 employees, with one security guard who sat at an old wooden desk in the corner of the garage workshop space. Dr. Edgerton, Germeshausen, and Grier remained the critical decision-makers for many years.

O'Donnell documented how, early on, the company's responsibilities expanded from timing and firing systems to alpha measurements, which helped determine the yield of the bomb tests. Then, their specialty in photography came into play. The AEC asked them to record the tests using high-speed photography. O'Donnell recalled in his memoirs that they had a specialist at EG&G named Charlie Wycoff who developed a unique type of film with Eastman Kodak to take pictures that could filter out or correctly expose the intense glare of the atomic flashes. O'Donnell notes that Edgerton headed up the photography, Germeshausen the alpha measurements, and Grier the timing and firing as the Pacific tests started. O'Donnell recounted:

It was a great company to work for. . . I knew whom I was working for and what my tasks were. I also knew that I was going to be traveling and I swore to God that I would never leave my wife again. I was told that everything we were doing was to be shipped overseas to Eniwetok and Bikini in the South Pacific. These were two places that I vowed I'd never return to. I said, "I want no more of that. The memories are bad, too vivid in my mind." To see all of the destruction all over the beach and bodies and everything. I said, "No way, I'm not going back," Then my wife said, "You know something, I'm pregnant. We have a house to pay for, bread and butter must go on the table," I said, "Yes, you're right. Would you mind if I left you for a while?" Anyway, we settled in and I would be gone for a period of two, three months, up to six months, but she was a trooper and she stayed with me.[79]

The impression made by the authors using the original EG&G logo stamp.

Most of the year 1948 kept the young company busy with Operation Sandstone. After Operation Crossroads, Sandstone became the next or second post-war test series. Sandstone tests were carried out at the Pacific Proving Grounds on Enewetak Atoll. In fact, Sandstone and other soon-to-follow far-off Pacific

[79] Al O'Donnell DRI Oral History Project, University and Community College System of Nevada, 2004, p. 3.

equatorial tests so overwhelmed everyone with the vast logistics involved with working thousands of miles from their Boston office that little time had initially gone into organizing a company structure for EG&G.

On Enewetak Atoll, deep in the heart of the Marshall Islands, Grier, Edgerton, and O'Keefe had significant responsibility for the timing and firing of the three tests conducted during a two-month period. They personally oversaw many dry-run tests and checked out the timing and installations of complex cable arrangements right up to the time that the islands were evacuated before each detonation. They then served on the firing parties and operated the equipment that armed and fired the devices. In those early years, it was all EG&G could do to keep up with the Atomic Energy Commission's requests for technical assistance.

Germeshausen and O'Keefe, however, did plan and begin work for expanded commercial business operations, which over the years served EG&G well, although the need never ceased in those early years for a priority to be given to the nuclear weapons tests. O'Keefe claimed it would have been much easier just to build a specialized commercial business, but there was much more to these founding men than the bottom line, something with which Al O'Donnell could always identify.

CHAPTER TEN
ATOMIC HEYDAY - PART TWO

**Image of Baker shot during Operation Crossroads,
courtesy of Cold War veteran Richard G. Carter.**

Although EG&G had not yet been organized as a company prior to Operation Crossroads nor participated in that first post-war 1946 test, the MIT group provides advice. Herb Grier trained Al O'Donnell and others to assist in the timing and firing of the underwater Baker shot for Operation Crossroads. O'Donnell recalled that during the Baker shot, he served in the control room in an observation ship. His job required him to watch an auto sequence timer up to a minute before the detonation. At that point, the firing system moved into automation mode. O'Donnell, however, maintained watch over the control panel until almost the last second, when he moved fast to go above deck to get a glimpse of the historic test. As he raced up the ladder, he grabbed a set of dark goggles just before the moment of detonation. The young technician recounted that nuclear detonation as a monstrous but inspiring sight:

It was huge. The cloud. The mushroom cap. Like watching huge petals unfold on a giant flower. Up and out, the petals curled around and came back down under the bottom of the cap of the mushroom cloud. I watched the column as it started to bend, my eyes went back to the top of the mushroom cloud where ice was starting to form. The ice fell off and started to float down. Then it all disappeared into the fireball. Watching your first nuclear bomb go off is not something you ever forget. I forgot, however, to hold on to the rail. When the shock wave came it picked me up and threw me ten feet against the bulkhead. I thought to myself, you damn fool![80]

By 1948, O'Donnell could already call himself a veteran when EG&G as a group became involved in Operation Sandstone. O'Donnell recounted:

EG&G was charged with designing the equipment to explode the bombs, record their yield, and photograph the results. It was a logistical nightmare. You had to set up housing facilities, you had to

---

[80] K.J. Evanslas, *Alfred O'Donnell*, (Las Vegas, Nevada) Vegas Review-Journal, February 7, 1999.

move equipment out there, you had to move not just scientists, but construction people. As for the scientists and technicians, the extensive traveling was tiring and disrupted their home lives. With all these (South Pacific) operations, every time I came home, nine months later, we had another child.[81]

Barney O'Keefe's memoirs provide a very comprehensive background and first-hand insight into EG&G's role in this early period in which Al O'Donnell participated:

The Atomic Energy Commission inherited a confused and deteriorating atomic establishment. Things had pretty much ground to a halt after the Japanese surrender, not only because of personnel departures, but also because of lack of direction from the top... Although the very continuance of the laboratory was in doubt for many months, Norris Bradbury, the new director, began to forge a strong organization. A reserve Navy commander during the war, he understood both the military and civilian points of view. Recognizing that the laboratory would be shorthanded for years, he placed a good deal of confidence in his contractors, such as our group at MIT... The stockpile, if you could call it that, was woefully small. New designs promised to double the efficiency of the implosion, effectively doubling the size of the stockpile at one stroke. Bradbury began planning for a series of tests on the new designs... One series of tests [Operation Crossroads] had been conducted by the military in the summer of 1946... The tests were poorly conceived and inexpertly executed... On the first test, an error in the operation of the radio-controlled timing system prevented crucial instrumentation, such as high-speed cameras, from operating until fifteen seconds after the device had detonated; much valuable data were lost. The next series of tests, Operation Sandstone, was planned to be of an entirely different character. Because of the contamination at Bikini, they were moved to the atoll of Eniwetok, a slightly smaller coral configuration two hundred miles to the west. They were developmental in nature, planned to test the new designs, as opposed to the Crossroads effects tests.

The primary job for our company was to turn on instrumentation after the area had been evacuated and, for measuring devices with short operating times, to start them in the final seconds between the remote arming and the firing of the device. This was the task for Grier and me. Edgerton headed a small group responsible for an optical technique to measure the multiplication of neutrons. Germeshausen stayed home to "mind the store."

We had the responsibility to operate instrumentation automatically during "dry runs" to check out timing installations after the islands had been evacuated, to serve on the arming party with the test director, and to operate the equipment that armed and fired the devices. A nuclear weapons test is a massive undertaking. Thousands of people and hundreds of ships, boats, motor vehicles, and aircraft were organized under the military command of scientific task groups. All equipment, to the last nail and screw, was assembled months before the test and transported thousands of miles to these remote Pacific islands. Towers, hundreds of feet high, were built as platforms for the devices to approximate air burst conditions and minimize ground contamination. Additional towers were constructed to house photographic operations in distant islands, to keep the field of view of the cameras above the salt spray of the lagoon. Heavy concrete bunkers covered with sand were needed to protect instrumentation from blast and radioactivity. Miles of underwater cable were laid to

---

[81] Ibid

connect the various instruments to the timing and firing system. Aircraft for atmospheric measurements and cloud-following aircraft to sample the characteristics of the mushroom cloud and track its direction of travel were based on the larger atoll of Kwajalein, two hundred miles away.

Although I had been working on nuclear devices for years, I had never seen one explode. Because we had to watch the control panel until the last second in the event of a malfunction, we could not wear welders' glasses to protect our eyes; we kept our backs to the windows and our eyes glued to the panels. The flash of light coming in the small window reflecting from the gray metal panel boards was blinding as the image of the meters froze my vision for seconds, then gradually turned to motion as the lights flashed crazily on and off and meters bent their needles against their stop posts from the force of the electromagnetic pulse traveling down the submerged cables with the speed of light. The pulse was so powerful that one of our engineers, halfway around the world in Boston and knowing the scheduled time of the explosion, was able to detect it with a makeshift antenna and an oscilloscope, the world's first detection measurement.[82]

That detection equipment allowed America to confirm the first Soviet test of a nuclear weapon a year later. Soon, the war in Korea created even more urgency and subsequent concerns for securing a proper nuclear defense stockpile. At that time, America's atomic weapons cache was still small. Put simply, maintaining and designing a nuclear arsenal meant the actual testing of nuclear devices. There was then no other way to develop and maintain nuclear weapons without testing.

Soon, the decision was made to develop a thermonuclear or H-bomb to destroy strategic targets. Parallel to this, an initiative started to design smaller, "tactical" or "theater based," nuclear weapons for use on the battlefield. Thus, this changing political climate led to a more favorable environment for supporting nuclear testing, which soon allowed the AEC to establish a state-side nuclear testing facility.

		Operation Crossroads				
2	Able	06/30/46	Bikini	Airdrop	Weapons Effects	21 kt
3	Baker	07/24/46	Bikini	Underwater	Weapons Effects	21 kt
		Operation Sandstone				
4	X-ray	04/14/48	Enewetak	Tower	Weapons Related	37 kt
5	Yoke	04/30/48	Enewetak	Tower	Weapons Related	49 kt
6	Zebra	05/14/48	Enewetak	Tower	Weapons Related	18 kt
		Operation Ranger				
7	Able	01/27/51	NTS	Airdrop	Weapons Related	1 kt
8	Baker	01/28/51	NTS	Airdrop	Weapons Related	8 kt
9	Easy	02/01/51	NTS	Airdrop	Weapons Related	1 kt
10	Baker-2	02/02/51	NTS	Airdrop	Weapons Related	8 kt
11	Fox	02/06/51	NTS	Airdrop	Weapons Related	22 kt
		Operation Greenhouse				
12	Dog	04/07/51	Enewetak	Tower	Weapons Related	81 kt
13	Easy	04/20/51	Enewetak	Tower	Weapons Related	47 kt
14	George First thermonuclear test explosion	05/08/51	Enewetak	Tower	Weapons Related	225 kt
15	Item- First test of the boosting principle	05/24/51	Enewetak	Tower	Weapons Related	45.5 kt

Department of Energy list detailing early post-war atomic nuclear tests, continued on page 92, https://www.osti.gov/opennet/manhattan-project-history/publications/DOENuclearTests.pdf.

In 1951, the Nevada Test Site (originally called Nevada Proving Grounds) began facilitating nuclear tests for the AEC. The emerging expertise of EG&G in photography became critical to documenting the

---

[82] Bernard J. O'Keefe, *Nuclear Hostages* (Boston: Houghton Mifflin Company, 1983), pp. 134-137.

tests and measuring yield data. Analysis at that early point was largely dependent on high-speed photography using Rapatronic Cameras. EG&G had pioneered and patented that equipment. Grier and his associates would also remain key figures in the actual triggering of the atomic weapons. In fact, before the Test Site even opened, key people from Los Alamos were on a plane to Boston in late 1950 to consult with Edgerton, Germeshausen, and Grier, as well as O'Keefe, about the first five tests at the Nevada Test Site to be called Operation Ranger. O'Keefe wrote in his memoirs, detailing the unfolding events that brought EG&G and then Al O'Donnell to the Nevada desert:

In June 1950, the Korean War broke out. A country willing to go to war to save South Korea was in no mood to stop the development of the hydrogen bomb. The decision changed the direction of activities at our little company. Bradbury put Los Alamos on a six-day week; all the contractors followed. Our plans to balance out the government work with commercial activities had to be scrapped, at least for the time being, while we grappled with plans for the upcoming test program, now advanced to the early months of 1951. Because of the shortage of people and housing at Los Alamos, more work was contracted out.

In addition to our timing and firing work, we were asked to do the technical photography on the upcoming tests, of which high-speed photographs of the rate of growth of the fireball were most important. The rate at which the fireball grew in the first thousandths of a second depended on the yield or magnitude of the explosion. Measurement of that rate of growth with high- speed cameras was the primary measurement of yield; thus, it was one of the most important tests performed. . .

Two series of events brought it about. One was the development of small low- yield devices. New implosion techniques and better detonators had made possible much smaller diameter devices than the five-foot-wide Fat Man, devices that could be carried by a light bomber or fighter planes, or possibly in an artillery piece. Better knowledge of plutonium characteristics and better designs of tamper, which enclosed the plutonium and reflected back escaping neutrons, had made possible designs of lower yields than the Nagasaki weapon, some as low as the equivalent of one kiloton of TNT. At the same time an interservice dispute had arisen in the military. Atomic strategy, as it had developed at the turn of the decade, called for massive air strikes with high- yield weapons carried out by Air Force bombers. This left no nuclear task for the Army and Navy, who wanted a piece of the action, and fast.

A war was going on in Korea; the Army and Navy requested small, low-yield nuclear explosives that could be delivered by carrier planes or fired from an artillery piece, so-called tactical weapons. The Defense Department agreed and laid a requirement on the Atomic Energy Commission for development and testing of such devices.

But the AEC and its laboratories were deeply involved with preparations for the upcoming Pacific test. . . The Defense Department and the AEC together asked the President for authority to establish a continental test base, which would not have the logistic problems of the Pacific atolls.

The President acquiesced; [Lieutenant General Elwood] Quesada was directed to locate a site. An Air Force pilot, Quesada knew just the spot. He had flown many times from the Nellis Air Force Base in North Las Vegas, Nevada, to the Las Vegas Gunnery Range, a huge tract of government-owned land a hundred or so miles to the northwest. It was bleak, uninhabited desert with dry lake beds

surrounded by mountains, which would cut down the blast effects. The site, 1350 square miles in area, was turned over to the AEC to become the Continental Test Range. In September another crash program was authorized to carry out the first tests before the Greenhouse operation. As soon as the program, Operation Ranger, was approved, Quesada and Alvin Graves flew to Boston to visit us to describe the program. There would be five shots, with yields ranging from one to twenty kilotons, air-dropped over a dry lake in the new test site, with no firing requirement. They asked Grier and me to coordinate the timing signals and do just enough high-speed photography to determine the yield.

We had no time to design new equipment, borrowing instruments scheduled for the Pacific. For photographic stations we bought secondhand trucks, installed the cameras, checked them out in our converted garage, and had them driven to Nevada.[83]

From that point on, a long association began for the fledgling company at the Nevada Test Site. During the first Operation Ranger tests in January and February of 1951, five bombs were dropped from B-50s flying out of Albuquerque. The first Ranger test devices were Mark 4 bomb designs, which were the successor to the WWII Fat Man bomb designs, which became classified as the Mark 3. The last shot was a test of a new bomb design called the Mark 6.

Germeshausen personally supervised the photography. Soon, as more testing continued after Operation Ranger, nuclear devices were largely fired off from towers on the Test Site just as they had been at the original Trinity test in New Mexico. Grier and O'Keefe were hands-on and boots on the ground with the timing work for those tower detonations. It is said that with few exceptions through the history of US nuclear testing, EG&G remained the "voice" for the countdowns, with Herb Grier himself pushing the button to detonate the shots.

Meanwhile, Grier and O'Keefe had been occupied with the first preliminary thermonuclear-related tests in Operation Greenhouse and then later Operation Ivy, both far back out in the Pacific, in which Al O'Donnell participated as well. Although, the company soon realized it needed more dedicated help at the Test Site. So, in 1952, Grier asked O'Donnell to organize a permanent Las Vegas office. This may have been a good choice because, as Peter Zavattaro recalls, O'Donnell's strength was in administration. He contributed his share on the technical side, and like many early EG&G employees, he was responsible for an extensive assortment of duties that involved more and more administrative matters. It seemed there was no task that some of the pioneer EG&Gers did not have to do at one time or another. O'Donnell's duties soon included recruiting people for the Las Vegas office. Becoming a jack of many trades, O'Donnell recalls that period in Las Vegasb:

We got on the phone and started calling military bases in New Mexico; made trips to San Francisco, doing interviews. We were hiring people on the phone. The first to join EG&G's new office were six radio technicians employed by Reynolds Electrical Engineering Co. By the end of 1952, we had recruited a staff of nearly 50 people in the area. By 1968, there were more than 8,000.

I think it is notable that O'Donnell took the responsibility at this time for hiring the first African American employee at EG&G. He then hired the first disabled person, who had a background that more than

---

[83] Ibid., pp. 148-149.

qualified him for the job. EG&G continued to grow rapidly, although the original players remained hands-on for a number of years. In 1952, EG&G participated in eight tests at the Nevada Test Site in Operation Tumbler-Snapper, and then the first thermonuclear test and the largest fission test in Operation Ivy on Enewetak. O'Donnell recalls:

EG&G grew and grew, as did the experiments overseas. We were involved at the NTS, Eniwetok and Bikini. As it would turn out, we traveled between the Nevada Test Site, Eniwetok and Bikini more frequently than anticipated. EG&G was responsible for those three major areas. Therefore, groups like Los Alamos, Lawrence Livermore, the Navy, the Army, Sandia Lab and numerous other individuals interested in weapons testing were counting on EG&G for their experienced technical people. They wanted us to fire it, take pictures, calculate yield and prepare for the next event. . . That's when the doors of EG&G opened here [Las Vegas] and we found ourselves sort of out in left field. Who in the name of God ever heard of a company called Edgerton, Germeshausen and Grier? We went to the bank to open an account in the company name and when they asked the name of the company, we had to spell it out – all 25 letters. We did have our struggles with the name because . . . if you weren't in the gaming industry, what were you doing here? [84]

In 1953, Grier, O'Keefe, and O'Donnell were involved in 11 tests or "shots" at the Nevada Test Site. The bombs they tested were not really finished weapon designs so were, therefore, not called bombs. They were called "devices." Around that time, the men had an extremely startling experience. At a tower shot, right at the moment of a planned detonation, the firing system failed. The party consisted of Jack Clark, the test director; John Wieneke, a Los Alamos physicist; Grier and O'Keefe, and Al O'Donnell. O'Donnell recounted that experiencing a countdown to zero proved nerve-wracking enough. Reaching the anticipated moment of detonation and having nothing happen was unprecedented, as well as outright frightening. Instrumentation was still very basic then, so there was no way to remotely diagnose the bomb and determine why it did not go off. Also, there was no way to remotely disarm it without physically examining the device itself. O'Donnell recalled that Grier and O'Keefe flipped a coin to see who would make the dangerous trip out to the test tower. O'Keefe ended up going with John Wieneke. They took two Jeeps, just in case one broke down. Associates told them to keep their goggles ready and the Jeep's visor down in case the bomb suddenly went off. O'Keefe sarcastically recounted that after a short while, they decided that if the bomb chose to go off, it really would not make much difference if the visor was down or not. Upon arriving at ground zero, they realized the elevator to the 200-foot tower had been dismantled prior to the test. (That was a security precaution because of fears of Soviet espionage.) Thus, the only access to the bomb at the top of the tower was by a metal ladder attached to the structure. O'Keefe made the climb with Wieneke.[8685] (O'Keefe claimed this episode occurred in 1953; however, it was likely the Operation Snapper, Fox shot on the early morning of May 20, 1952, involving an 11-kiloton device in Area 4. That is the troublesome shot O'Donnell recalled in his accounts.)

It was finally successfully detonated on the 25th at 11 kilotons. Al O'Donnell and Barney O'Keefe had another extremely unique experience in those early years that involved the first and only test of the famous Atomic Cannon. That account is detailed by O'Donnell, who recalled the May 25, 1953, Grable test in

---

[84] Al O'Donnell DRI Oral History Project, University and Community College System of Nevada, 2004, p. 6.

[85] Ibid., p. 31.

Operation Upshot-Knothole that involved a 15-kiloton device in Area 5: O'Donnell recalled the test:

The cannon shot called Grable was an interesting event. The army came and wanted to shoot but they didn't want to be interrupted by anybody. . . Dr. Clark, the deputy test director for NTS, requested two observers at the spot from which the cannon was going to be fired. They agreed to Barney O'Keefe and myself . . . There were about 25 army personnel involved with the cannon shot. When this thing was shot, it was remarkable.

They counted 5,4,3,2,1, pulled a lanyard and the cannon fired. Everyone jumped into the six-foot deep trench and the unit went zooming 15 miles across Frenchman Flat and exploded on target. . . It was the only time the atomic cannon was ever fired.[86]

Al O'Donnell contributed a long and productive career to the Nevada Test Site as an EG&G employee. His personal reminisces are so valuable because they provide a great deal of information about the early days of nuclear testing. He explained the Ranger tests were convenient and necessary "properties tests" in preparation for the first "thermonuclear fundamental tests" scheduled for Operation Greenhouse that spring out in the Pacific.

As more tests were scheduled at the Nevada Test Site, it became clear that a full-time core of workers was needed, in contrast with the previous practice of conducting "campaigns" where specialists were temporarily drawn from different areas to facilitate a test operation. While EG&G continued to serve as a key contractor for timing, firing, and photography, the El Paso, Texas, firm of Reynolds Electrical and Engineering Company (REECo) became another key contractor at the Site. They complemented the growing EG&G workforce, which swelled to more than 50. REECo, however, soon grew to thousands of workers as they became responsible for building and maintaining much of the infrastructure, which had to be built from scratch.

REECo had been a logical choice, just as EG&G had been, because Lou Reynolds, the founder of REECo, subcontracted with Robert E. McKee, who had been the prime contractor at Los Alamos from the early days of the Manhattan Project.

A later REECo president and general manager, Harold D. Cummings, once recalled that "the joint venture with the Robert E. McKee and another, Brown and Olds mechanical and industrial contractor, proved logical for REECo because those early groups already had employees from Santa Fe and Los Alamos who had security clearances." Cummings himself originally worked for Brown and Olds at Los Alamos.

The Haddock Engineering Company became another prime contractor in 1951, which played a role in building infrastructure at the Test Site. (In fact, REECo had a hiatus from the Test Site from March 1951 to January 1952.) Al O'Donnell recalls that Haddock personnel were initially responsible for constructing some of the first buildings at the entrance of the Test Site, which became known as "Camp Mercury," (later just Mercury) a valuable support center that serves the modern-day successor, the Nevada National Security Site, to this day, still known as Mercury.

Haddock also built underground concrete bunkers and testing towers, although they encountered major

---

[86] Bernard J. O'Keefe, *Nuclear Hostages* (Boston: Houghton Mifflin Company, 1983), pp. 156-157.

cost overruns in that work. The overruns may have led to Haddock's exit and a re-acquaintance with REECo for continued construction work. So the contract with Haddock was canceled, and REECo was given responsibility for all the maintenance, starting in 1952.

In those pioneering days, Al O'Donnell recalled that REECo became tasked with a great deal of administrative duties, making a viable facility out of "Camp" Mercury. On January 7, 1952, Frank Rogers came to Las Vegas to become the Deputy General Manager of REECo. The Mercury administrative complex had just taken shape with offices and sleeping dormitories. It was far from elegant and consisted mainly of plywood paneled structures. Oil heaters with a 55-gallon oil drum alongside them were the only source of heat, and bathroom facilities were in separate huts. Existing power and communication systems built by Haddock would not cope with the growing infrastructure. Drilling for a closer water source was also a priority, and it was contracted out to S.R. McKinney and Sons of Las Vegas. Communications became critical, so mobile radios were installed in many vehicles. As the tests ran into more tests and then more tests, additional infrastructure became required.

Roads, phone lines, electrical plants, and wiring followed. The challenges were immense. Harold Cunningham, a close associate of Frank Rogers and a later REECo General Manager, stressed in his memoirs the immense complexity of those tasks.[87] Al O'Donnell recalled there were many other associated contractors, as well, in those early days.

These included Holmes and Narver of Los Angeles, who also did work at the Pacific Proving Grounds. Architecture and engineering assignments often went to Sukas Nasin Company of New York. Foster and McHarg Company of Riverside, California, worked on roads with Dodge Construction Company of Fallon, Nevada, and Vinell Company, Inc. of Alhambra, California, built test towers. McNeil Contracting Company worked on the initial construction of the first control point area, with Lembke-Clough and King of Albuquerque working on the initial phases of Mercury. Despite such expertise, it took time to build facilities. O'Donnell always told the amusing story that his first night at the Mercury encampment was spent on a cot placed on a concrete slab out in the open air.

Initially, like most EG&G workers, O'Donnell commuted back and forth from Boston for the Nevada tests and was often assigned to frequent trips to the Pacific. EG&G was comparatively small in those early days but vital to Pacific and Nevada tests in which O'Donnell became a participant. O'Donnell recalled the next series of Nevada tests, christened Operation Buster-Jangle, in November of 1951:

Without EG&G getting to their stations there would not be a test. Towers were built to facilitate tests because airdrops were not deemed the most efficient way to do a carefully monitored nuclear experiment. We would leave Camp Mercury to go to the forward area and to the control point (CP-1). There, they had bunk facilities downstairs in the basement. Personnel were allowed to sleep there if they were on arming and firing parties because of the distance, but most came back to Camp Mercury. Los Alamos set up their administrative, scientific and engineering types from New Mexico [in Mercury] where EG&G brought out its trailers too. . . All of the equipment was in the forward area and we would be headquartered up at the control point. There weren't enough vehicles to satisfy everyone's needs, so Dr. Greer and Barney O'Keefe picked up a couple of Land Rovers and drove

---

[87] Ha*rold Cunningham*, private notes, author's collection.

them from Los Angeles to Camp Mercury. In those days, the highway between Los Angeles and Las Vegas was nothing more than a two-lane road. The road leading out of Las Vegas to Camp Mercury was, also, nothing more than a two-lane road, which was referred to as the "widow maker." Dr. Greer and Barney took their chances, but they wanted to do the job without having the hassle of getting vehicles for their personnel every morning. [O'Donnell, as most historiography, refers to Herb Grier as "Dr." however this was only an honorary, but well-deserved, distinction.] The vehicles were used, primarily, for the timing and firing personnel so that they could roam the forward areas and distribute signals from the control point as needed. They also had to work out of the distribution station and, therefore, every single participant in each event had to be given a timing signal ranging anywhere between 60 minutes down to five seconds. Consequently, we had to cover every bit of territory possible on Yucca and Frenchman Flat.[88]

Operation Buster-Jangle differed from Ranger in that the Department of Defense had a greater presence because these seven tests involved effects measurements. The tests also involved troop movements cooperating with field maneuvers, codenamed the "Desert Rock Exercises" supported by an Army tent camp called Desert Rock build near Camp Mercury. Weapons development tests were also a part of the series. Diagnostic experiments on the tests were staffed by Los Alamos in coordination with EG&G and some teams from Sandia.

Operation Tumbler-Snapper came next. Its duration was from April to June 1952, totaling eight tests. By then, EG&G knew they were in for a long series of operations. At that time, Grier asked O'Donnell to organize a permanent office in Las Vegas. That cumbersome task was detailed in the prior chapter and is what brought O'Donnell to permanent residency in Las Vegas. Meanwhile, testing continued. The timetable in those hot Cold War days proved urgent as the Korean War raged. The Soviets and, eventually, the Chinese would have a thermonuclear device. Despite the misgivings of Manhattan Project veterans like Robert Oppenheimer, the broader consensus became that America needed to get there first. Many issues in the development of a practical and deliverable thermonuclear device had to be worked out, and those tests would have to take place in the Pacific. However, theoretical work could be applied in smaller tests in Nevada and economize much of the work needed for the higher yield shots to come at the Pacific Proving Grounds.

Al O'Donnell recalled that Operation Tumbler-Snapper filled that role with Snapper serving the developmental or "weapon concept" test, while the Tumbler shots were geared to weapon effects. Tumbler consisted of three airdrops, Snapper one airdrop, and four tower shots. Tumbler comprised all airdrop tests because it explored the relationship between the height-of-burst and the "overpressure" at the ground. O'Donnell recalled that the third test of Tumbler, called the "Charlie" shot of 31 kilotons, became the first in the United States to be made available to the press, or "open," as they called it. In fact, 300 press members attended, 60 civil defense observers, two governors, and other assorted VIPs attended. O'Donnell recalled:

The first nuclear testing observers' post was established a few miles north of Camp Mercury overlooking Frenchman Flat. Wooden benches were placed for those who were going to observe the shots that were to take place at Frenchman Flat. Today, there are a number of weather beaten, dried

---

[88] Al O'Donnell DRI Oral History Project, University and Community College System of Nevada, 2004, pp. 5-6.

out planks that have been here since the early 1950s, and this is where the invited dignitaries observed the early shots. . . The first media group was invited to observe in 1952. One test was televised to the American public from this location. The front row of benches was reserved for dignitaries, such as navy captains, army general. Walter Cronkite was here. I was never there because I was at the control point, where the test director and EG&G personnel were involved with the detonation of the unit.[89]

O'Donnell's early Nevada Test Site accounts are invaluable because he lived this history. It is interesting, however, that O'Donnell's favorite stories centered on the history of Las Vegas. He often commented on the "Atomic" atmosphere of early 1950s Las Vegas. He recounted how some of the mushroom cloud blasts could be seen from downtown casino rooftops. He explained that most tests took place before dawn, so a dark sky could aid the photography as background. Many hotel and casino patrons stayed up to the wee hours of the early morning to see the atomic flashes in the pre-dawn sky. O'Donnell felt he had been on the ground floor of not just nuclear testing but the birth of modern Las Vegas itself. He admitted that it proved a culture shock for both him and his family, who had lived most of their lives in Boston.

Nevertheless, he and his wife and children came to love Las Vegas. They, in fact, helped play their own small part in its growth as a modern American city. His family became active in school functions, sports, and their local church. O'Donnell recounted the Las Vegas experience of the early days of nuclear testing:

I don't think there was a vendor in this town in the early 1950s that didn't have something to do with the Test Site at one time or another, either directly or indirectly. Bankers, shoemakers, everybody supported the NTS and there was no friction. Nevada stood out because it was clearly a community effort. To my knowledge, there was no objection from anybody. There were no objections from the gambling industry, which could have put up the biggest fight, with comments such as, "You stop that blasting up there because we want tourists to come and gamble." We never, ever experienced that. They were very supportive of everything. Even the little mom and pop shops were cooperative. If you needed something machined or a pipe fitted, they were right there—no waiting. They realized that with the opening of the Test Site, there were going to be many millions of dollars poured into the community. They were very, very supportive and I commend them for that. They were a credit to the whole operation. Nevada never got the recognition . . . We proved that you could do it [nuclear testing] on [the North American] continent, although we did a lot of damage downtown on Fremont Street. I think that was on the Backer shot. The shock wave went out and over the mountain range that contains Frenchman Flat and the shock wave arrived downtown. The downtown people were warned that a shock wave might arrive and sway the buildings. Oh God, we blew out plate glass windows in Sears, J.C. Penney's and all those stores on Fremont Street. We did some damage that day, but I swear to God, I don't think I can recall one complaint.[90]

---

[89] Ibid

[90] K.J. Evanslas, *Alfred O'Donnell,* Vegas Review-Journal, Feb. 7, 1999. O'Donnell loved to tell the story of "Miss Atomic Bomb." This atomic-inspired beauty contest or, rather, promotion stunt began in May of 1952 with Las Vegas dancer Candyce King. She became "Miss Atomic Blast." The second beauty contest coincided with the Upshot-Knothole series and featured pageant winner Miss Paula Harris, who rode on a Las Vegas Chamber of Commerce float. That motion picture-themed parade depicted the film "The Atomic City," which earned her the first official use of the name "Miss A-Bomb." The contest endured

Operation Buster

16	Able Radioactivity not detected offsite	10/22/1951	LANL	NTS	Area 7
17	Baker	10/28/1951	LANL	NTS	Area 7
18	Charlie	10/30/1951	LANL	NTS	Area 7
19	Dog	11/01/1951	LANL	NTS	Area 7
20	Easy	11/05/1951	LANL	NTS	Area 7
B₂	Operation Jangle				
	Sugar	11/19/1951	LANUDoD	NTS	Area 9
	Uncle First underground test at the Nevada Test Site	11/29/1951	LANL/DoD	NTS	Area 10

Operation Tumbler-Snapper

23	Able	04/01/1952	LANL/DoD	NTS	Area 5
24	Baker	04/15/1952	LANUDoD	NTS	Area 7
25	Charlie	04/22/1952	LANL/DoD	NTS	Area 7
26	Dog	05/01/1952	LANL	NTS	Area 7
27	Easy	05/07/1952	LANL	NTS	Area 1
28	Fox	05/25/1952	LANL	NTS	Area 4
29	Georae	06/01/1952	LANL	NTS	Area 3
30	How	06/05/1952	LANL	NTS	Area 2

Operation lvi

31	Mike Experimental thermonuclear device	10/31/1952	LANL	Enewetak	- - - -
32	King Largest fission device	11/15/1952	LANL	Enewetak	- - - -

until 1957. The Las Vegas News Bureau retains the rights to the most famous photograph of that last beauty to win the title in that year, Copa showgirl Lee A. Merlin from the Sands Hotel. Her image is immortalized with a cotton mushroom cloud attached to her swimsuit.

Operation Upshot-Knothole

33	Annie	03/17/1953	LANL	NTS	Area 3
34	Nancy	03/24/1953	LANL	NTS	Area 4
35	Ruth	03/31/1953	LLNL	NTS	Area 7
36	Dixie	04/06/1953	LANL	NTS	Area 7
37	Ray	04/11/1953	LLNL	NTS	Area 4
38	Badaer	04/18/1953	LANL	NTS	Area 2
39	Simon	04/25/1953	LLNL	NTS	Area 1
40	Encore	05/08/1953	LANUDoD	NTS	Area 5
41	Harry	05/19/1953	LANL	NTS	Area 3
42	Grable Fired from 280mm gun	05/25/1953	LANL	NTS	Area 5
43	Climax	06/04/1953	LANL	NTS	Area 7

Operation Castle

44	Bravo Experimental thermonuclear device Highest yield nuclear test	02/28/1954	LANL	Bikini Island	- - - -
45	Romeo	03/26/1954	LANL	Bikini Island	- - - -
46	Koon	04/06/1954	LLNL	Bikini Island	- - - -
47	Union	04/25/1954	LANL	Bikini Island	- - - -
48	Yankee	05/04/1954	LANL	Bikini Island	- - - -
49	Nectar	05/13/1954	LANL	Enewetak	- - - -

Operation Teapot

Wasp		02/18/1955	LANL	NTS	Area 7
	Moth	02/22/1955	LANL	NTS	Area 3
	Tesla	03/01/1955	LLNL	NTS	Area 9
53	Turk	03/07/1955	LLNL	NTS	Area 2
54	Hornet	03/12/1955	LANL	NTS	Area 3
55	Bee	03/22/1955	LANL	NTS	Area 7
56	Ess	03/23/1955	LANL	NTS	Area 10
57	Apple-1	03/29/1955	LANL	NTS	Area4
58	Wasp Prime	03/29/1955	LANL	NTS	Area 7
59	HA (High Altitude) Named "HA" for "high altitude" in reference to its intended detonation at an altitude of 40,000 feet	04/06/1955	LANL	NTS	Area 1
60	Post	04/09/1955	LLNL	NTS	Area 9
61	MET (Military Effects Test)	04/15/1955	LANUDoD	NTS	Area 5
62	Apple-2	05/05/1955	LANL	NTS	Area 1
63	Zucchini	05/15/1955	LANL	NTS	Area 7

Operation Wigwam

64	Wigwam North 29 degrees, West 126 degrees	05/14/1955	LANUDoD	Pacific	- - - -

Operation Project 56

65	Project 56 No. 1	11/01/1955	LANL	NTS	Area 11a
66	Project 56 No. 2 Plutonium dispersal	11/03/1955	LANL	NTS	Area 11b
67	Project 56 No. 3 Plutonium dispersal	11/05/1955	LANL	NTS	Area 11c
68	Project 56 No. 4 Plutonium dispersal	01/18/1956	LANL	NTS	Area 11d

Operation Redwing

69	Lacrosse	05/04/1956	LANL	Enewetak	- - - -
70	Cherokee First airdrop by U.S. of a thermonuclear weapon	05/20/1956	LANL	Bikini Island	- - - -
71	Zuni	05/27/1956	LLNL	Bikini Island	- - - -
72	Yuma	05/27/1956	LLNL	Enewetak	- - - -
73	Erie	05/30/1956	LANL	Enewetak	- - - -
74	Seminole	06/06/1956	LANL	Enewetak	- - - -
75	Flathead	06/11/1956	LANL	Bikini Island	- - - -
76	Blackfoot	06/11/1956	LANL	Enewetak	- - - -
77	Kickapoo	06/13/1956	LLNL	Enewetak	- - - -
78	Osa!le	06/16/1956	LANL	Enewetak	- - - -
79	Inca	06/21/1956	LLNL	Enewetak	- - - -
80	Dakota	06/25/1956	LANL	Bikini Island	- - - -
81	Mohawk	07/02/1956	LLNL	Enewetak	- - - -
82	Apache	07/08/1956	LLNL	Enewetak	- - - -
83	Navajo	07/10/1956	LANL	Bikini Island	- - - -
84	Tewa	07/20/1956	LLNL	Bikini island	- - - -
85	Huron	07/21/1956	LANL	Enewetak	- - .. -

CHAPTER ELEVEN
ATOMIC HEYDAY - PART THREE - ARMAGEDDON

Image of the largest thermonuclear detonation in history, the Russian Tsar Bomba test on October 30, 1961, https://deaduxx.blogspot.com/2009/11/tzar-bomb-soviet-hydrogen-bomb-king-of.html

The Castle Operation's Bravo test of the first practical hydrogen or thermonuclear device proved to be the most massive detonation the United States ever conducted. It, in fact, became a disaster, exploding with far more force than intended at 15 megatons. Only the Russian Tsar Bomba test in 1961 proved bigger at 57 megatons. EG&G found themselves a part of the infamous Castle Bravo event, which signaled the beginning of a prolonged arms race with the deadliest weapons ever made. In relation to that time, Al O'Donnell recounts:

In 1954, Herb Grier came out and stayed permanently in Las Vegas. He began running the operation here. He left the Boston facilities under the direction of Barney O'Keefe. One of the main reasons for Dr. Grier's coming to Las Vegas was in support of the NTS, leaving the Boston facility and staff caring for the overseas operations. That was great. We knew we were home and did not have to leave for long periods of time. That went along fine until 1954, when Dr. Teller began experimenting with the hydrogen bomb. It was such an event that it required the involvement of 10,000 people. This required our group in Nevada to leave our post at NTS and join the crew from Boston at the

Eniwetok/Bikini operation. I had to go home and tell my wife, "I'm leaving you again." [91]

O'Donnell once said that to understand the significance of what Operation Castle accomplished, you have to go back to the Operation Ivy tests and the Greenhouse operations before that. O'Donnell took part in all these far-off tests. He explained that Operation Greenhouse in March and April of 1951 at Enewetak used two out of four tests to explore the concept or preliminary physics of a thermonuclear device. The following year, Operation Ivy consisted of two shots, "Mike" and "King," which took place in October and November of 1952. Mike was a key test involving the first trial of a two-stage fusion device. This was all part of the creation of a true hydrogen bomb in what was perceived to be a race with the Soviets, and it closely involved the EG&G team. Concerning the Mike shot, O'Donnell recalled:

They selected a small island away from the [Enewetak] atoll itself, on the northern tip. They wanted to get as far away from the southern control point island as they could. They selected the island of Elugalab. . . The shot went off that morning. All the cameras and equipment operated properly. . . You could stand on the beach on the control island and look up north and see this huge, horrendous fireball. The next thing we saw was about 90 percent of all the water in the lagoon being sucked up into this funnel, going straight up. God it was fantastic. . . The recovery team went back up to get the cameras from the towers for Dr. Edgerton. They were in helicopters and when they flew over the spot where the bomb detonated, they radioed back to the control point saying, "Elugalab existed no more."[92]

O'Donnell always said the best way to understand the operations is to realize that Operation Ivy tested the first primitive hydrogen bomb. That bomb, however, was literally the size of a building. The concept worked and worked well, yet the apparent challenge then arose to make a hydrogen bomb that could be delivered by an aircraft, and that is what Operation Castle hoped to develop. There was also a lot of politics and internal dissension leading up to the development of the hydrogen bomb, which, as a result, impacted Operation Greenhouse, Ivy, and then Castle.

Al O'Donnell always stressed that EG&G had a unique insight, or as he called it, a "front-row seat" to all those historic times. Barney O'Keeffe, unlike O'Donnell, widely published his memoirs in what has become by far the best book ever written on the early days of nuclear testing. Titled *Nuclear Hostages*, O'Keeffe's book provides revealing insights that few others offer. This is by far the best-written description ever produced on the events surrounding the coming of the H-bomb, and it was done by a man like O'Donnell, who worked center stage for the events.

This passage also explains how Lawrence Livermore, the second National Laboratory, emerged and related to the witch hunt against Robert Oppenheimer:

Although there had been a successful thermonuclear experiment in 1951 [Operation Greenhouse] and a full-scale thermonuclear explosion on November 1, 1952, [Operation Ivy] Edward Teller and Ernest Lawrence were still not happy with the progress of H-bomb development. They felt it should be given the highest priority. However, Los Alamos had other responsibilities.

---

[91] Al O'Donnell DRI Oral History Project, University and Community College System of Nevada, 2004, p. 17.

[92] Ibid., 18.

It had to worry about production problems with the [fission/A-bomb] stockpile, the development of the half-megaton fission weapon, and the development of the lower-yield tactical devices for the Army and Navy. Tensions built up between Teller and Norris Bradbury to the point where Teller did not attend the 1952 test. There had been considerable agitation in Washington, particularly from the Joint Committee on Atomic Energy and the Air Force, to open a second weapons laboratory. In October 1952 a second lab was established under the aegis of Ernest Lawrence of the University of California's Berkeley Laboratory. The new weapons lab was located in the small town of Livermore, thirty miles east of Berkeley.

It was a good compromise. Los Alamos was clearly overworked; the University of California would run both laboratories; Berkeley had been the center of cyclotron research and electromagnetic uranium separation. A cadre of scientists from Berkeley, headed by Herbert F. York, the new Livermore director, had worked at Eniwetok on the Greenhouse tests. Although Teller became chief scientist, the laboratory was not organized soon enough to play a significant part in the first thermonuclear weapon explosion, scheduled for test at Bikini in 1954. While the opening of a second laboratory calmed down the scientific bickering, it had no effect on a larger battle going on in Washington. As the emphasis on thermonuclear research increased, the importance and power of the Air Force grew with it. If the big bombs were to be used, they would be used by the Air Force, particularly the Strategic Air Command, with its new bomber, the B-36.

Although tactical weapons were being developed, they were not yet available, nor was there any means of delivering them. The Army and the Navy were being squeezed out. The first public indication of this interservice rivalry came in 1949 with the "revolt of the admirals" when the Navy publicly protested overreliance on the B-36 as opposed to the super aircraft carriers they wanted to build. The admirals' revolt was quickly stifled, but the controversy continued inside the Pentagon. For the most part, hearings and reports were classified and unavailable to the public. It flared up again in the hydrogen bomb controversy, where the pivotal character became J. Robert Oppenheimer. He was not directly involved in the battle between the services, nor was he in a position to influence the allocation of budgetary resources. But he was able to exert influence over the size and purpose of the atomic weapons being designed. He was convinced that atomic weapons with smaller yields could be designed for armed services other than the Strategic Air Command, and for conflicts short of total war.

He believed that atomic weapons had a potential tactical as well as strategic role. This conviction by Oppenheimer and others led to the fateful recommendation of the General Advisory Committee, chaired by Oppenheimer, that a crash program for the development of the Super weapon is not pursued. The big bomb men saw Oppenheimer as an enemy.

Within the White House, the new President, Dwight D. Eisenhower, was having budgetary difficulties. He had campaigned on the pledge of a balanced budget, but his Treasury Secretary told him that to balance the budget he must not only end the Korean War, but make drastic cuts in the peacetime services. The quickest and cheapest way to his objective was to emphasize the cheapest and most powerful weapon, the atom bomb... Secretary of State John Foster Dulles conceived the concept of the "massive retaliation," where any provocation would lead to a full-scale atomic blitz. It was effective because the Soviets had not yet built up their atomic arsenal, nor did they have delivery capability comparable to our B-36. It was cheap; nuclear weapons cost only a fraction of the cost of a standing army or a fully equipped navy. But it was a dangerous doctrine. A simple border clash could bring about total war. Only a man of Eisenhower's reputation could convince the people that massive retaliation made sense. The Air Force had won the battle.

Even though the administration policy seemed solid on the outside, it was not so tranquil in the secret world of military planners and their science advisers. Tempers grew short and tolerance narrowed. Nine days after Truman announced the [first] Russian [nuclear test] explosion, [in 1949] Senator Joseph R. McCarthy, in a speech at Wheeling, West Virginia, [started a witch hunt.] . . . As time went on, the internal battle continued; externally, McCarthy's witch hunt widened. A jittery President saw the investigations spread into his beloved Army. Scurrilous articles against Oppenheimer appeared in the press. Finally, William Borden, the former staff director of the Joint Committee on Atomic Energy, composed a letter to J. Edgar Hoover, accusing Oppenheimer of being an espionage agent for the Communists. A copy was sent to the Joint committee.

The system moved quickly. Hoover brought the letter to Eisenhower, who ordered a "blank wall" between Oppenheimer and classified information. He instructed the Atomic Energy Commission to investigate. A three-man board of inquiry was set up to investigate the charges. . . The inquiry, the closest thing to a trial for heresy ever seen in American politics, was a travesty. . . Dozens of witnesses spoke for Oppenheimer, to no avail . . . To a lesser extent, Teller's testimony constituted a second element of tragedy at the trial. Teller never questioned Oppenheimer's loyalty; he did question his judgment. . . Oppenheimer was excommunicated as a heretic against the dogma of massive retaliation.

With massive retaliation firmly established as policy, the test series [Operation Castle] at Bikini Atoll scheduled to begin on March 1, 1954, took on added importance. The Greenhouse George test in 1951 had been a "thermonuclear experiment," showing only that a deuterium-tritium reaction could occur. Ivy Mike in 1952 had been a full-scale, ten-megaton thermonuclear explosion, but it was a sixty-five ton monster with liquid tritium and deuterium cooled to a temperature close to absolute zero, certainly not a weapon. Bravo shot, the first in Operation Castle, had two major revisions that made it the prototype of an aircraft-deliverable device. It used lithium deuteride, a solid with the tritium derived by neutron bombardment of the lithium. It also used the "Teller-Ulam" configuration, an assembly scheme that held the weapon together an extra hundred-millionth of a second until the hydrogen isotopes could fuse. Operations had shifted back to Bikini after the Mike detonation had erased the island of Elugelab in Eniwetok Atoll from the face of the earth. Bikini was large enough to allow twenty miles' separation between the shot island of Namu and the control island of Enyu. The firing party could stay ashore in a heavy, sand-covered concrete bunker, eliminating the need for radio-controlled ship-to-shore firing mechanisms. Once again our company, now employing 150 persons and operating simultaneously at the Continental Test Site in Nevada, was awarded a contract to do the timing and firing and the technical photography. [93]

With our post-Cold War mindset, it is hard for us today to appreciate the mood of that time. The year 1954 represented a high state of tension in the Cold War. A CIA top secret document titled "Estimate of the

[93] Bernard J. O'Keefe, *Nuclear Hostages* (Boston: Houghton Mifflin Company, 1983), pp. 158-163.

Effects of the Soviet Possession of the Atomic Bomb on the Security of the United States and Upon the Probabilities of Direct Soviet Military Action" provides a reflective window into that period. It's a long name for a serious document that clarifies the perception of the time. The study concluded that a surprise nuclear attack by the USSR was a real possibility. With historical hindsight, we now know that the Soviets had no more desire to launch an unprovoked nuclear war than we did.[94]

Then, however, the memory of Pearl Harbor still resonated fresh in everyone's minds, along with a deep paranoia of Communism and the Soviet system. As a nation, we were determined never to go to war again as unprepared as we were when the Japanese attacked us in 1941. The CIA estimate reinforced this nationally-entrenched paranoia of being taken by surprise. The recent 1949 testing of the Soviet's first atomic bomb years ahead of what had been anticipated, followed by the arrest of key Manhattan Project member Klaus Fuchs as a Soviet spy, startled the world. The West feared further Russian expansion beyond the lands the USSR had acquired at the close of the war, and appeasement would not be tolerated this time.

The CIA's document speculated that the Soviets may have two hundred nuclear devices that could explode over 200 American cities and thus effectively neutralize the United States. Of course, no one fully realized then that they did not have that capacity, and we, in fact, barely approached such numbers of nuclear weapons ourselves. Besides that, the Soviets then only had a poor copy of our B-29 called the TU-4, which was reverse- engineered from three B-29s that had made emergency landings in Vladivostok and interned during our bombing effort of Japan in World War Two. **TU-4 Illustration by Basil Zolotov, http://basilzolotov.com/side-views/tupolev-tu-4d/.**

Stalin kept the prized planes and had them copied nut for nut and bolt for bolt, eventually building 850 TU-4s or "Bulls," as we called them. The TU-4 became operational in 1949, the same year the Russians test exploded their first atomic bomb, also thanks to stolen American technology. By 1954, the Soviets were on the road to building bigger bombers and eventually intercontinental ballistic missiles or ICBMs, just as we were working to develop. Of course, both sides wanted the hydrogen bomb, so it became an intense arms race. Nuclear testing played a vital role in that buildup.

Operation Castle signified a watershed in the escalation and development of nuclear weapons. Castle consisted of six tests in all. Castle "Bravo" served as the first test in the series. Bravo took place on February 28, 1954, from a short test tower erected on an artificial island reef 3,000 feet off of Nam Island in the Bikini Atoll. It proved by far the most significant and historic of all hydrogen bomb tests. The shot utilized a two-stage thermonuclear configuration called a "dry" or solid lithium deuteride fueled H-bomb nicknamed "Shrimp." A boosted fission design that acted as the trigger on this device had been tested in the last series in the Operation Upshot- Knothole Climax in Nevada the previous year.

---

[94] Public record information states that the CIA produced a long analysis in 1950, focusing not only on Soviet nuclear capabilities but also on Moscow's intentions and the extent to which a nuclear weapons capability increased the risk of US-Soviet conflict. The first report was entitled "Estimate of the Effects of the Soviet Possession of the Atomic Bomb upon the Security of the United States and upon the Probabilities of Direct Soviet Military Action" and was dated 6 April 1950.

The yield or explosive force of Bravo greatly exceeded its estimated maximum yield of 5 to 6 megatons and ended up being 15 megatons in force. This tremendous blast made Bravo the largest nuclear explosion the United States ever produced, albeit an accidental one. The explosive force was approximately 1,000 times that of Hiroshima, with a fireball exceeding a diameter of four and a half miles. The unexpected surprise came from what was called a "tritium bonus," which involved an isotope in the lithium called lithium-7. Lithium-7 was supposed to be inert, but instead, it created a substantial reaction in the tritium-deuterium fusion process, which produced high-energy neutrons.

This unexpected yield took the test personnel by surprise. Navy ships as far as 30 miles south of Bikini were hit with significant fallout. The support ship USS *Curtis* (a veteran of many Pacific tests) stood 23 miles from the blast and received a violent shock wave that tossed the ship side to side. Its crew were ordered below for several days while emergency decontamination took place by a wash-down system that the ship had in place. Worst of all, a civilian Japanese fishing vessel 80 miles from the test called the Daigo Fukuryu Maru, or "*Lucky Dragon no. 5,*" became heavily contaminated. Its 23 crewmen received high exposures, and one died from complications associated with the incident. The unfortunate involvement of the fishing vessel created an embarrassing diplomatic situation with Japan.

Barney O'Keefe and Herb Grier were again center stage, serving on the EG&G firing party for the Bravo shot in a concrete bunker on Eneu Island 20 miles from ground zero. Their heavy bunker literally moved with the shock wave. O'Keefe recalls:

"Bill says it looked like a good one to him," he reported jubilantly. "The shock wave may hit pretty hard. Let's get ready for it.". . . His look of flushed pleasure had slowly changed to one of puzzlement, then his expression changed as his face turned white with concern. I felt it too as I talked, the words coming more and more slowly and finally stopping as all thoughts of home quickly vanished. Something was wrong! Grier spoke the words first, as he reached out to steady himself at the workbench. "Is this building moving or am I getting dizzy?" "My God, it is. It's moving." Grier reached for the bench to steady himself as I stood bewildered in the center of the room. The whole building was moving, definitely now, not shaking or shuddering as it would from the shock wave that had not arrived yet, but with a slow, perceptible rolling motion like a ship's roll, I began to feel a nausea akin to seasickness.[95]

The radiation levels rose so dramatically, so fast, that they had to seek protection in the most secure room in the bunker but were still subjected to elevated rates. They were evacuated as the levels dropped enough to fly in a helicopter eleven hours later. The EG&G firing party wrapped their hands and feet in rags and pillowcases as they left the bunker and calmly boarded vehicles to make a short drive to the helicopter landing area. As they rode in the vehicles, they elevated their feet from the floorboards as much as possible to make as little contact with the radioactive surfaces of the vehicles.

Twenty-eight US weather personnel on Rongerik Island, 133 miles from the detonation, received significant exposures. They also had to be evacuated. At least 154 Marshallese Islanders 100 miles from the detonation were evacuated as well, but not until three days later, and records indicate the natives on the islands of Rongelap, Ailinginae, and Utirik received significant doses of radiation. The natives remained evacuated

---

[95] Ibid., pp. 179-181.

for three years; however, their return was brief as the islands proved too contaminated to reestablish agriculture.

At least fifteen islands were finally deemed permanently contaminated. The islanders were compensated in 1956, and in 1995, the United States awarded 43.2 million dollars to the Marshall Islands. A US medical study called Project 4.1 surveyed the radiological effects on the Marshall Islanders and some US service personnel. The study's conclusions were published in 1955 in the "Journal of the American Medical Association." The report stated that 239 Marshallese on the islands of Utirik, Rongelap, and Ailinginae were exposed to significant levels of radiation, along with 28 Americans stationed on Rongerik Atoll. All the Castle shots were considered very "dirty" tests because they were conducted low to the ground, and the explosions turned the surrounding sand and water into gas, which created eradiated elements that carried the fallout high and far.[96]

Herb Grier and some EG&G engineers knew after the Bravo shot that those procedures had to be changed. Because of the unpredictability of this new thermonuclear technology, all following firing operations took place from a Navy ship called the USS *Estes*. The second test, Castle "Romeo," fired on March 26 from a barge off of Bikini, proved just as unpredictable and went a little better. Just as in Bravo, the Romeo test had a much higher yield than expected. Although a smaller device, it exceeded its original explosive force estimate and came off at 11 megatons. This made it the third-largest US detonation in history.

Castel "Koon," a surface shot off of Bikini, came next on April 6. Again, not all went as planned. This time, however, the explosive yield was much smaller than the expected one-megaton-designed device. The actual yield came out at just 110 kilotons, which is still a very big bomb, but it made the test classified as a "fizzle." The fourth test, Castle "Union" on April 25, fired from a barge near Bikini, came off with better predictability. Union had a yield of 6.9 megatons. The fifth test on May 4, Castle "Yankee," was also fired from a barge close to Bikini and served as a proof test using natural lithium instead of enriched lithium. Yankee involved a big thermonuclear bomb with a yield of 13.5 megatons, which made it the second largest hydrogen bomb the US tested. Yankee and Romeo led to the deployment of the first deployable thermonuclear bombs called the Mark 17 and Mark 24.

The 17 and 24 were not just city killers but large city killers, designed for strategic targets spanning many square miles. People tend to think of Hiroshima and Nagasaki when they envision nuclear war. Those were, however, relatively small cities. Urban centers like Tokyo and Berlin were huge in comparison and would have needed much more destructive fire power, exceeding the individual force of the first atomic bombs. Of course, the idea of bombing cities or area bombing has always been controversial, yet the Second World War demonstrated that a military force is not what makes a nation able to wage an extended war. The factories and infrastructure sustain a nation's ability to field armies and their hardware continually. The factories and the people who make up that infrastructure are civilians. Civilians live in cities or often large urban areas. To win a global-sized war, it has been believed you have to kill large numbers of civilians and the machines and the infrastructure they service.

---

[96] Barton C. Hacker, *Elements of controversy: the Atomic Energy Commission and radiation safety in nuclear weapons testing, 1947- 1974* (Berkeley, CA: University of California Press, 1994).

H-bomb image from GolbalSecurity.org.

The 17 and 24 could be carried by the vast, new yet ungainly B-36 intercontinentally ranged strategic bomber. Five Mark 17 bombs were built, and ten Mark 24s. They differed only in their internal firing mechanisms. Both entered service in 1954 in a program called "Emergency Capacity" to rush H-bombs into production and were replaced in 1957 by more advanced types. In 1957, a Mark 17 accidentally fell through the closed bomb-bay doors of a B-36 on approach to Kirtland AFB. Protective safeguards disarmed the bomb on impact, but the device spread moderate radioactive contamination over a wide area, resulting in an expensive clean-up operation.

Castle "Nectar," a 1.69 megaton detonation test on May 13, took place off Enewetak. Exploded from a barge, Nector, like all the Castle tests, served as a weapons test but focused on a new design. The goal was for a lighter-weight bomb, and this design created a fusion-boosted fission bomb. It led to the Mark 15 bomb, of which 1,200 were built between 1955 and 1957. Three versions of the Mark 15 were produced, with the third model yielding about 3.8 megatons. It was a Mark 15 bomb that was lost during a training mission off the coast of Savannah, Georgia, in 1958, which, as far as is still officially reported, has never been recovered. A seventh test called "Echo" failed to take place because its design of a liquid fuel configuration became obsolete as the other Castle tests proved the success of the more practical dry fuel concept.

A little-known fact is that five liquid-fueled or cryogenic liquid deuterium fusion devices were briefly employed in the stockpile. That massive and highly impractical bomb design was called Jughead or TX-16 or EC-16, the "EC" standing for emergency capacity. Some sources say only two Jugheads were ever built, but this desperate measure came in response to what was at the time thought to be a need to compete with Soviet thermonuclear bomb development, which was progressing rapidly. (If any readers are fans of the science fiction series *Lost,* they will recall a memorable and purely fictional episode that made Jughead one of the more famous nuclear weapons in atomic culture.)[97]

When Al O'Donnell returned from the Pacific in 1954 following Operation Castle, he found the company at a turning point. By then, Grier was living full-time in Las Vegas and would, from that point on, devote most of his attention to the work being contracted with the laboratories and the Nevada Test Site. O'Donnell was also living fulltime in Las Vegas with his family and felt happy to be a part of this work and a new life in Nevada. However, he continued to interact with the Boston home-based office.

O'Donnell knew that in Boston, O'Keefe and Germeshausen wanted to focus more time on non-defense-related activities. They seemed to sense that nuclear testing may be entering a period of unpredictability in

---

[97] For more information on that memorable *Lost* episode, see: https://www.huffpost.com/entry/jughead-is-real-the-truth_b_204061.

terms of world events. It was a new world these men were facing as the arms race began to intensify while simultaneously public controversy world-wide developed over nuclear testing. Even Winston Churchill warned of humanity approaching a dangerous precipice as he said the bombs are simply becoming too big. Certainly, Europe, with its closely-spaced population compared to much larger countries like the United States and the Soviet Union, saw little hope of surviving a nuclear war, which would surely be fought over their territory as well. Repeated talk came out of Washington concerning a testing moratorium. Indeed, a moratorium on nuclear weapons testing began in late October of 1958 and would last for almost three years.

So O'Donnell became aware of this uncertainty as early as 1954, yet nuclear testing in Nevada not only continued but intensified for a period. By 1955, with O'Keefe's commercial diversification in Boston, the continental test work rested primarily with Grier in Las Vegas. The testing program still remained EG&G's chief responsibility. At the Test Site, Grier supervised a contract for fifteen detonations where the timing, firing, and high-speed photography work expanded to include the diagnostic measurement of the rate of the multiplication of the neutrons in a nuclear detonation. Meanwhile, O'Keefe took over the responsibility for further weapons tests in the Pacific.

CHAPTER TWELVE
ATOMIC HEYDAY - PART FOUR - OVERVIEW

**Manikins at Nevada Test Site *Operation CUE*, Apple-
2 Shot, Operation Teapot.**

The story of nuclear testing is a fascinating reflection of the atomic age, and the arms race itself. At the Nevada Test Site, exactly 100 above-ground or "atmospheric" tests took place from 1951 to 1962. In that same time frame, approximately 100 US nuclear tests took place among the Marshall Islands, Christmas Island, Johnston Atoll, and the South Atlantic Ocean. Those were all atmospheric, underwater, or space-based tests.[98] In the Fall of 1963, the Limited Test Ban Treaty was signed by the US, United Kingdom, and USSR, which moved all testing underground. (See Appendix I.) In most of the US, approximately 828 remaining underground tests occurred between 1962 and 1992 at the Nevada Test Site, and there were a handful of underground tests in other US states. In October of 1992, all US nuclear testing ended to date when President George H.W. Bush declared a unilateral test moratorium.[99] In total, since 1945, the US detonated 1,054 nuclear devices, and other countries combined have tested over 1,000. Nuclear testing

---

[98] For an interim period, nuclear testing had ceased between October 1958 and September 1961 when the United States entered a unilateral testing moratorium initiated by President Eisenhower. The Soviet Union honored that pause but resumed testing in September 1961, which was followed by renewed atmospheric tests by the United States.

[99] Although nuclear tests involving the critical release or chain reactions of fissile material have largely ceased, subcritical, or in other words, tests with fissile material not producing a chain reaction have continued. Such subcritical tests have continued to take place at the Nevada Test Site, now called the Nevada National Security Site.

around the world, as a whole, has ceased, barring a series of tests by North Korea over the past decades.[100]

Most of the history of atomic testing involves very scientific data and tedious details. Doing a complete account of that story is not for this book. However, the year 1955 is of note, symbolizing the last of the really dramatic atomic tests or heyday period to which EG&G and Al O'Donnell had a front stage seat. The year began with Operation Teapot, which encompassed 14 nuclear tests and one non-nuclear test taking place from February to May. Like the Tumbler-Snapper tests in 1953, troop maneuvers took place in and around the detonations.

Public record AEC image of US troops at the Nevada Test site circa 1955.

The purpose of those troop movements centered around acclimating soldiers to a tactical nuclear ground war. Approximately 11,000 Defense Department personnel took part, with about 8,000 of those being ground troops acting as observers and employed in tactical exercises during the various tests. The tests, as in previous operations, multitasked, and all involved proof tests of fission weapons in the ever-progressing development of a multi-use nuclear stockpile. As a whole, while large strategic thermonuclear bombs were being developed with the benefit of tests in the Pacific, other tests, like those in Operation Teapot, were trying to develop small, lightweight devices for tactical use. This would also include the development of highly specialized nuclear weapons that could be used in air defense and anti-submarine roles.

As in Tumbler-Snapper, the Teapot tests included weapons effects tests. This time, they would also concentrate on how nuclear explosive waves affected aircraft both in flight and on the ground. In addition, the tests were designed to gain more information on cratering effects. Every type of nuclear weapon imaginable was, in fact, being imagined by 1955. Al O'Donnell once commented that 1955 signified a pivotal year, as the Operation Teapot tests demonstrated. The Atomic Energy Commission's nuclear bomb production logistics would soon consume almost seven percent of the entire power output of the United States. The nuclear weapon fuel production facilities at Oak Ridge and Hanford had doubled in size. A new complex at Savannah River would produce tritium through heavy water reactors. Historian Richard Rhodes documented that in 1955, nuclear production facilities utilized 11 percent of the annual US nickel production and 34 percent of all stainless-steel resources, and combined with other purchases, totaled $9 billion. Rhodes points out that this AEC investment exceeded the capital investments of General Motors, Bethlehem Steel, US Steel, Alcoa, Dupont, and Goodyear all combined.[101]

The vast expansion in the AEC's infrastructure had one goal in mind, and that was to increase the number and variety of nuclear weapons in the stockpile. By 1955, the US nuclear stockpile went to 2,422 nuclear

---

[100] There is debate over how nuclear tests were counted because a handful did not fully detonate, and some may count those that "fizzled."

[101] Richard Rhodes, Dark Sun: *The Making of the Hydrogen Bomb* (New York: Simon & Schuster, 1995), p. 561.

weapons from several hundred bombs a few years previous. General Curtis LeMay, the effective architect, and head of the Strategic Air Command or SAC, had taken over the targeting of Soviet territory, and his staff's target list was considered so secret that it was not even shared with the Joint Chiefs of Staff. By 1955, SAC controlled nuclear target planning that even General Lauris Norstad, US commander in Europe, did not have access to.

LeMay's list included more than 5,000 key logistical areas in the USSR. By 1961, LeMay would have almost 19,000 strategic nuclear weapons at his disposal for those targets. The Teapot operations, however, concentrated on tactical nuclear weapons and defensive-tipped nuclear devices. This would serve a need should the Soviets ever attempt a land war in Europe. On a trip to America in January 1953 to visit the new incumbent US President, the reinstated British Prime Minister Winston Churchill and President Eisenhower reasoned that tactically-applied nuclear weapons could serve as a useful deterrent against the Soviets. Small fission or A- bombs and nuclear artillery shells negated the cost of matching the USSR man for man and tank for tank if they charged through the Fulda Gap.[102] Teapot would provide proof-of-concept tests for everything from new kinds of nuclear artillery shells to nuclear depth charges and even nuclear anti-aircraft weapons. Teapot also involved a lot of civil defense studies that examined all sorts of scenarios and effects of nuclear war on civilian and military structures, and vehicles.

The first test in Operation Teapot comprised a Lawrence Livermore Scientific Laboratory (LASL) experiment. (Later, their name would change to Lawrence Livermore National Laboratory.) Named "Wasp," as LASL named their tests during Teapot after flying insects, the detonation took place on February 18, 1955, over Area 7 of the Nevada Test Site. Wasp was classified as a weapons effects test of a 1.2 kiloton device dropped in a Mark 12 bomb casing from a B-36 bomber. The device utilized a light-weight implosion-system fission design. This test proved unique because it used an airdrop delivery that was set to go off at a relatively low altitude of 762 feet. The test provided a good indication as to what would happen

Public record AEC image of "News Knob" press area

to ground targets subjected to such a close-ranged arial blast wave. Today, the concept of an aircraft dropping such a device over the United States during a nuclear test would be unthinkable. Then, although air drop tests were not that common, they had occurred and did not raise much concern during the intense days of the early Cold War.[103]

Troops took part in numerous special projects set up for the Wasp shot, which involved 697 Army, 146 Navy, 28 Marine, 105 Air Force, and 71 other special troops, partly from the Sixth Army. The number of actual troops who were sent from the tent camp called Desert Rock totaled 1,000. They observed from trenches some 4,500 meters south of ground zero. Due to changing winds, their observation point

---

[102] The Fulda Gap is a lowland corridor running southwest from the German state of Thuringia to Frankfurt and was feared to be a possible invasion route that could have been used by the Soviets to spearhead through the American occupation zone. It roughly corresponds to Napoleon's course of retreat after the Battle of Leipzig.

[103] Conversation with Dr. Robert Brownlee, August 15, 2015.

moved to the "News Nob" area, which usually facilitated the media and was located near the control point operations building. The troops had been scheduled to march into an "equipment display" area containing military hardware after the shot. That maneuver was canceled because of the changes in the wind and the strict safety procedures that had been put in place by that time to safeguard the troops. Film badges to record the effects of ionizing radiation were also being issued to some units, usually one badge to each squad of troops.[104] Al O'Donnell spent most of his time in the control point during this test, and EG&G participated in and facilitated a lot of the accompanying blast wave photography required in the weapons effects tests in the Wasp shot.

Increasingly sophisticated radiological measuring tests accompanied all of these experiments and focused on neutron and gamma radiation, which had caused such contamination at the Operation Crossroads tests. The Air Force had a team on the Wasp shot that was doing early experiments in the measurement of electromagnetic effects. After the detonation, AEC records indicate a B-50 was assigned to track the remnants of the mushroom cloud at 20,000 feet as it moved southeast of Las Vegas. Cloud sampling followed up to three hours after the detonation. Within four hours, the AEC had air sample analysis delivered to its Washington, DC, laboratories by courier plane.[105]

The next test, "Moth," took place on February 22 in Area 3 from a 300-foot tower and involved the detonation of a 2-kiloton device, although the predicted yield had been twice that. This was another LASL experiment. The nuclear device involved a proof test of an air defense weapon. In theory, the blast and radiation from such a weapon could knock down or disable approaching enemy aircraft. The physics of this design involved an upboosted fission device using an external neutron source utilizing a tube to pulse neutrons. Effects tests on military equipment also served as a component of Moth. This time, however, it extended to structures and military ordnance. Troops from Camp Desert Rock took part. During Moth, the troops remained in their trenches to observe the shot dug 2,290 meters southwest from ground zero. Official AEC records detail these troop deployments:

One hour before the detonation, orientation and safety instructions were given over a public-address system, and communications systems and attendance rosters were checked. Two minutes before the detonation, personnel turned away from the shot-tower, crouched in the trenches, and shielded their eyes. Observers maintained this position during the detonation and until the blast wave had passed, when they stood to view the rising fireball. Within ten minutes after the detonation, the observers left the trenches to prepare for the return of the transport vehicles, which had been parked eight kilometers to the south. The trucks arrived about 15 minutes after the detonation and departed from the shot area for Camp Desert Rock within 30 minutes after the detonation.[106]

---

[104] *Operation Teapot,* 1955, Defense Nuclear Agency, publication DNA 6009F, November 23, 1981, https://www.osti.gov/opennet/servlets/purl/16016884.pdf.

[105] As an interesting footnote, the Wasp test became the first nuclear detonation witnessed by a young astrophysics graduate, Dr. Robert Brownlee. He had applied to Los Alamos and soon went to work for them in the J-Division, or the nuclear test division under Dr. Alvin C. Graves and Dr. William E. Ogle. Dr. Brownlee shared vivid recollections of his first sight of a nuclear detonation at that test. I was fortunate to have had a number of fascinating conversations with Dr. Brownlee in 2015 as he recalled that test.

[106] *Operation Teapot,* 1955, Defense Nuclear Agency, publication DNA 6009F, November 23, 1981,

A special military project during Moth involved 37 soldiers from Battery C of the 532nd Field Artillery Observation Battalion. The troops used this test to train in locating the direction and yield of a nuclear weapon under battlefield conditions. Other Army Signal Corps personnel stationed at the News Nob observer area trained in using long-range cameras that employed infrared film. They conducted intensive effects tests on military equipment. Hardware positioned in the display area included three M48 tanks, one M59 armored infantry vehicle, one T97 self-propelled gun, six jeeps, six two-and-a-half-ton cargo trucks, and four five-ton trucks. Other military equipment, including more trucks and a 155mm self-propelled gun, were placed almost right at ground zero and varying degrees back from it.

The third test, "Tesla," took place on March 1 but was scheduled for February 25. Weather conditions were carefully monitored, and delays were called twice until ideal conditions prevailed. The detonation occurred from a 300-foot steel tower in Area 9 and yielded a 7-kiloton explosion. The predicted yield had actually been only two kilotons, although no accidents resulted, as did with the Bravo test in the Castle series.

That was a University of California Radiation Laboratory test, which named their experiments after famous inventors. The physics of the Tesla shot involved a linear implosion device nicknamed "Cleo I." Troops from Camp Desert Rock were brought out again and participated in four projects—one involved troop observations as before. The troops in Tesla were placed in trenches 2,220 meters from the tower. In the second operation, Battery C of the 532nd Field Artillery Observation Battalion rejoined an observer position.

Another observation assessed additional military equipment that the detonation had subjected to its effects. Tesla also incorporated fallout and shielding tests. It included a "missile detonation locator experiment" as well. For this experiment, personnel set up broad-band receivers on special platforms 110 miles away in California. The equipment measured electromagnetic pulses to determine if it could adequately measure the location and range of a nuclear detonation from a significant distance.

The fourth test was on March 7 in Area 2, called "Turk." It was another University of California Radiation Laboratory test. The 43-kiloton device, code-named "LINDA," served as a mockup of the XW-27 radiation bomb design. On this test, the yield was expected to be 45 kilotons, yet safely detonated, it was mildly weaker at 43 kilotons from a 500-foot tower.

LASL conducted the next test, called "Hornet," on March 12. This took place in Area 3a from a 300-foot tower. The device served as a boosted version of an XW-30 air defense warhead, and estimates allowed for a yield as high as 10 kilotons, although the yield was only four kilotons. "Bee" came next on March 22. This was another Lawrence Livermore test of a boosted XW-25 air defense warhead utilizing an unheard of light weight design of only 130 pounds and measuring only 17 inches in diameter. The yield was anticipated to be as high as 20 kilotons but came out at 8 kilotons as it was fired from a 500-foot tower in Area 7.

As we examined, Operation Teapot involved 14 nuclear tests spanning four months. O'Donnell spent most of his time during the tests in the Control Point located near the News Nob area. This is where the devices were remotely detonated. During this series, he had a cousin participating in the many military exercises that were planned around the various detonations:

---

https://www.osti.gov/opennet/servlets/purl/16016884.pdf.

You can get an idea of the size of Camp Desert Rock as the perimeter is marked by power poles. The poles brought power into the camp from Camp Mercury, which is just a little to the north. Camp Desert Rock housed, at its maximum, approximately 6,000 army personnel at one time. Their function was to participate in tests on Yucca Flat and Frenchman Flat. Unfortunately, the personnel were restricted to the Camp Desert Rock area. They were transported into the forward areas in carrier trucks. Only on given days could they enter into Camp Mercury, because of this restriction, I wasn't able to meet with my cousin, who was one of the army personnel stationed at Camp Desert Rock during the testing. I can understand why the commanding officers wanted to keep their personnel under control. It was a security issue; not all of the army personnel carried Q security clearances for the forward areas. They didn't want them roaming around. When Camp Desert Rock was first established in the early 50s, it was primarily tents. Later, the tents were replaced by permanent type structures of which only cement slabs remain today.

They are scattered throughout the area. I would imagine that they segregated this camp into sections (possibly A, B, C and D). Each man was assigned a section designation. The accommodations at Camp Desert Rock did not have air conditioning as we did at Camp Mercury. The desert environment was hot during the summer days and you just felt so sorry for those fellows down there. I would look down on Camp Desert Rock from Camp Mercury and wonder how the army boys were surviving.[107]

Lawrence Livermore conducted the seventh test, named "Ess," on March 23 in Area 10. That detonation took place 67 feet below the surface in what was called an Atomic Demolition Munition (ADM) cratering device. It employed a 1.2 kiloton Mk-6 HE warhead, making a crater 300 feet wide and 128 feet deep.

"Apple-1," the eighth test, occurred six days later on March 29. It involved a Los Alamos device on a 500-foot tower in Area 4 and served as a testbed for a lightweight bomb casing. The device was designed as a thermonuclear weapon with a radiation implosion primary system and used small quantities of fusion fuel. The resulting 14-kiloton explosion proved much smaller than the 40-kiloton anticipated yield. No reaction had occurred in the secondary stages, so something apparently went wrong. Five hours later, on that same day at 10 a.m., another test took place in Area 7 called "Wasp Prime." This involved a repeat of a Los Alamos design used in the first Wasp test on February 18. This time, a higher yield 3-kiloton detonation resulted from an air drop from a B-36.

---

[107] Al O'Donnell DRI Oral History Project, University and Community College System of Nevada, 2004, p. 21.

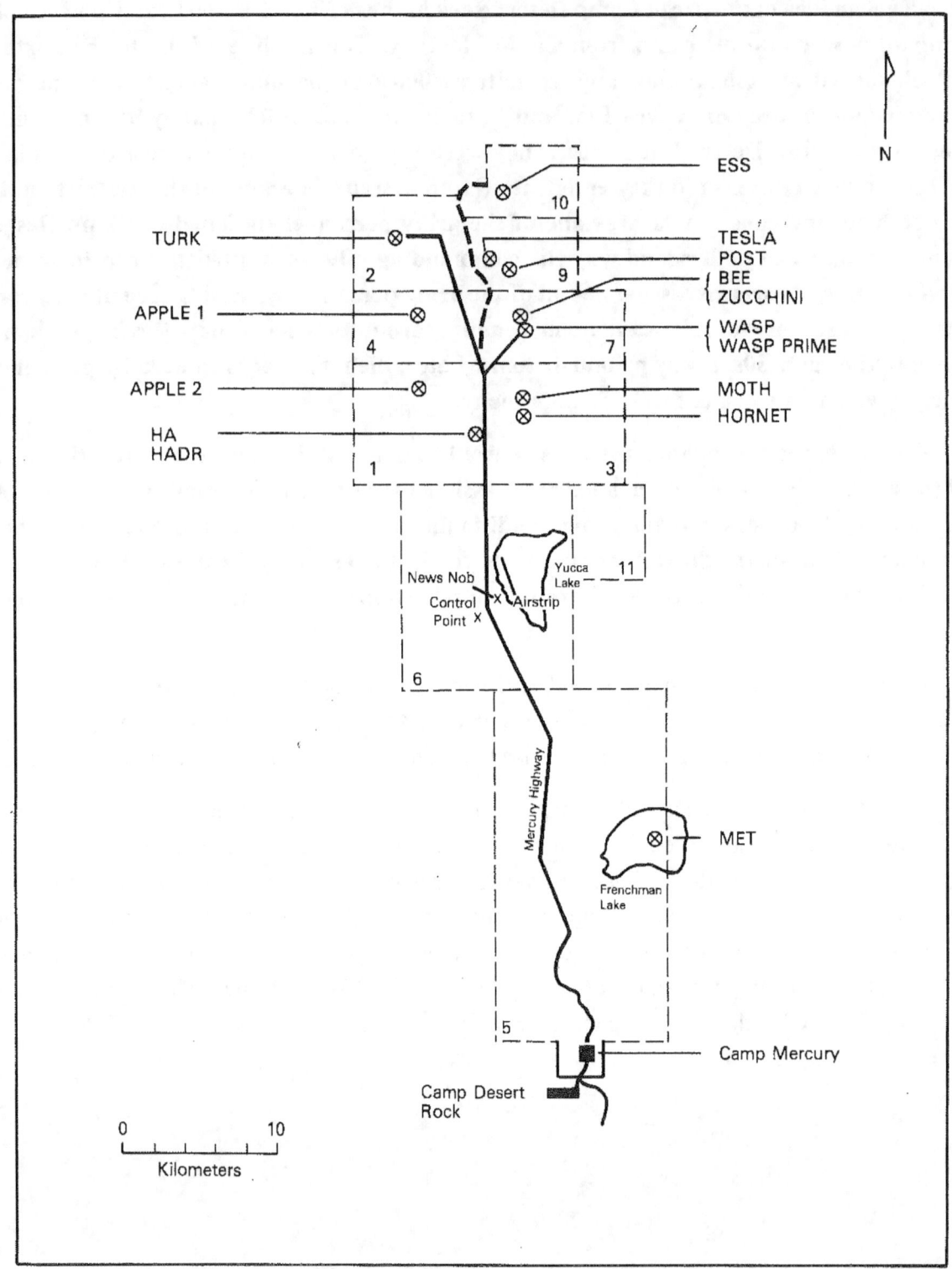

Figure 1-2: NEVADA TEST SITE SHOWING LOCATIONS OF SHOT GROUND ZEROS IN TEAPOT SERIES

Operation Teapot, 1955, Defense Nuclear Agency, publication DNA 6009F, November 23, 1981, p. 33. https://www.osti.gov/opennet/servlets/purl/16016884.pdf.

On April 6, another airdrop test took place over Area 1, the 10th in the series, and has since only been referred to as the "HA" or High Altitude Test. Los Alamos explored in this test the use of an atomic device as an air-to-air weapon that could knock down enemy aircraft. The 3.2-kiloton detonation was set for 36,620 feet and utilized a parachute so as to allow the B-36 aircraft sufficient time to escape the blast wave. This proved the only parachute weapon drop ever conducted at the Test Site. Proof of concept of this device led to the development of nuclear-tipped antiaircraft missiles.

The 11th test in Operation Teapot, "Post," took place on April 9 from a 300-foot tower in Area 9c. It involved a University of California lightweight linear implosion device of only 322 pounds called "Cleo II," which yielded two kilotons. The 12th test in Teapot is known only as a "MET" or Military Effects Test. This April 15 test involved a Los Alamos device that exploded from a 400-foot tower in Area 5 on Frenchman Flats, producing a 22-kiloton blast that fell far short of its predicted 33-kiloton yield. It is unique in that the device used a previously untried plutonium 233 core in a lightweight 800-pound casing.

The most notable events in Operation Teapot centered around the 13th test on May 5, called "Apple-2." This involved a Los Alamos shot fired from a 500-foot tower in Area 1. The predicted yield of the test was calculated to be 40 kilotons but fell short of that at 29 kilotons. What is often overlooked or forgotten about Apple-2 is that unfavorable winds and weather conditions repeatedly delayed the shot for ten days.

Radiative fallouts traveled in widely varying degrees depending on the shot, with the prevailing downward winds; however, veterans like Al O'Donnell confirmed that extreme caution was always taken when evaluating the weather. That fact, however, did not stop a local newspaper from printing the sensational headline "Las Vegas narrowly misses nuclear cloud." The story proved ridiculous because there was never any intention of running the test until the wind conditions were favorable. Conditions had to be ideal with this test in particular because so many special projects were planned around it. Apple-2 was also, as O'Donnell described it, a "public" or "open" shot. Nick Aquilina, Nevada Test Site Historical Foundation Board Member and former Department of Energy Nevada Operations Office manager, recalls: "Of course, after all the delays, not as many attended after ten days of waiting. Attendees were mostly civil defense, DOD, and reporters. Many ran out of money and left before the shot finally went off." The observers, including press officials and some foreign VIPs, watched the blast from the previously described News Nob area. The press often called this Media Hill."[108]

News Nob, which was near the Atomic Energy Commission Control Point where O'Donnell was stationed that day, was about 13 kilometers south of the Apple-2 shot tower. Another set of observers, which included a small and more select group of news media, some civil defense officials, and a few NATO observers, were placed in a trench on an elevated area known as "Mine Mountain." This trench had been placed only 4,480 meters southwest of ground zero. That must have been quite an experience, but nine select Army volunteers and one civilian occupied a six-foot- deep trench only 2,380 meters from the blast. Another 783 Army observers, including two female Army observers, were stationed in trenches 3,200 meters south.

The Apple-2 test is, in fact, so notable because it had such a large number of military and civilian effects projects connected with it. The military aspect comprised a collection of 12 projects called Exercise Desert Rock VI and involved 11,700 personnel. In one of these, the Army Armored School conducted an exercise of an armored unit. The armored column passed near the Apple-2 ground zero area shortly after the detonation

---

[108] Conversations with Nick Aquilina.

had occurred. This was no small maneuver and involved a thousand troops and 238 vehicles, of which 89 were armored. It was all in conjunction with a helicopter exercise involving Air Force, Navy, and Marine Corps personnel. This very rare photo illustrates that maneuver.[109]

Public record AEC image from Teapot tests

The Department of Defense authored the official history of Apple-2, providing some interesting insights.

During the night of 4-5 May 1955, the tanks and armored personnel carriers of Task Force RAZOR were manned and positioned as close as 2,835 meters south-southwest of the Apple-2 shot tower. Other vehicles were positioned between 3,570 and 6,400 meters south of the tower. In addition, non-armored support units and infantry units waited with their helicopters at Yucca Lake airstrip, near the AEC Control Point, about 13 kilometers southeast of the Apple-2 site. Early in the morning of May 5, the ten Army volunteer observers, nine officers and one civilian, awaited the detonation in their trench 2,380 meters south of the tower. Most of the remaining 783 DOD observers crouched in long trenches 3,200 meters south of ground zero, east of the armored task force, or in trenches on Mine Mountain, 4,480 meters southwest of the Apple- 2 ground zero. Many of the Federal Civil Defense.[110]

Administration Operations CUE observers, including news media personnel, manufacturer's representatives, Civil Defense volunteers, and officials, waited in the FCDC observation area at News Nob, near the AEC Control Point and Yucca Lake airstrip. A special FCDC volunteer group waited in a trench 3,200 meters south of the Apple-2 shot-tower with Desert Rock observers. Instruments and equipment of the many military effects and diagnostic projects had been placed around ground

---

[109] *Operation Teapot,* 1955, Defense Nuclear Agency, publication DNA 6009F, November 23, 1981, https://www.osti.gov/opennet/servlets/purl/16016884.pdf.

[110] Ibid.

zero. **In the air, aircraft participating in operational training projects and support activities positioned themselves for the detonation.**[111]

In addition to numerous other military exercises in Apple-2, at least 40 separate civil defense effects tests took place. Private industry and civil defense agencies sponsored many of these tests, which were collectively called Operation CUE. Iconic to these civil defense effects tests, the "Apple Houses" were ten different 1950s-style home structures built to observe the impact of atomic blast waves on various types of brick, wood, and concrete buildings. This included double-story and single-story structures, power lines, transformers, a complete electrical substation, a completely equipped radio station, two types of radio towers, a propane tank filling station, a weigh station, and numerous construction experiments utilizing

reinforced areas or shelters built into the test buildings in bathrooms or basements.

Trailer homes were also staged for the test. Houses and structures were filled with stylishly dressed mannequins, mainly to test different types of textiles under the extreme radiated heat conditions rather than to simulate blast effects on people. Even canned and frozen foods were placed in the homes for the detonation. Some canned food was buried at different distances from the blast to determine if contamination took place. The food was later sent for laboratory testing. Some news accounts from the time claim

Public record AEC image.

conscientious objectors, presumably from the Fort Detrick, Maryland program, volunteered to eat some of the food to learn more about the health sciences involved.

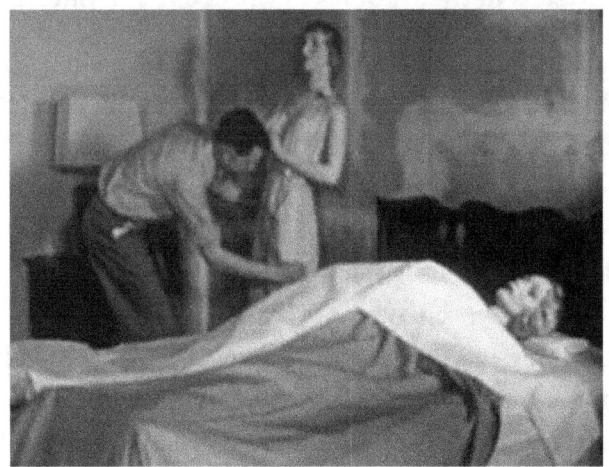

The key to these building effects tests and previous experiments involved an understanding of the best building practices that could maximize a structure's use as a shelter in the event of nuclear war. The test proved that using simple reinforcing techniques, incorporating extra wooden beams, or upgrading material from 2x4s to 2x6s greatly increased a structure's stability in a blast wave. Venetian blinds were placed in the windows of the houses to test the deflection of thermal energy. Because of limited funds for these effects tests, no electrical wiring or plumbing was installed in the houses, and the interiors were plastered and not painted. Wood trim, doors, and floors were left unfinished.

Fully furnished structures were positioned at different distances from the tower where the nuclear device was detonated. Two-story wooden homes were built 6,600 feet from ground zero. Brick homes were positioned 8,000 feet from ground zero. The pressures exerted on the structures during the detonation were 1.7 pounds per square inch or greater. Similar structures had been built for earlier such tests in Operation

---

[111] Ibid.

Upshot-Knothole, but they were demolished either by that operation's blasts or removed after the tests, so all new structures were used in the Apple-2 CUE operations.

Only two houses, known as the Apple Houses, survive to this day from the Apple-2 test in Operation Teapot. The demolished test tower remained in place for a period, and some steel anchors are still visible. Currently, the old Apple-2 site is being repurposed to train hundreds of first responders each year in radiation detection and safety methods, thanks to the background radiation remaining from Operations Upshot-Knothole and Teapot, which is extremely mild and safe to work in.[112]

Following is an excellent article on Operation CUE by Amy Shira Teitel from

Popular Science:

In the 1950s, tourists visiting Las Vegas could sometimes look out into the desert to see mushroom clouds hovering in the air. Sixty-five miles from the city was the Nevada Proving Grounds, a 1,375 square mile site near the Nellis Air Force Gunnery and Bombing Range that hosted a series of atomic weapons development and fallout tests. Among them was Operation Cue, a test within Operation Teapot designed to see how a nuclear blast would affect everyday objects.

BEFORE OPERATION CUE.

(US Federal Civil Defense Administration)

For the purposes of Operation Cue, "everyday objects" meant the things we use but might not think about as being useful: our houses, the clothes we wear, the appliances we cook with, and the food we store in our cupboards. These were the objects being tested; the Federal Civil Defense Administration needed to know what would happen to citizens in their homes in the event of a nuclear attack.

Public record AEC image.

This very specific test demanded a very particular set up. Ten residential houses were built specifically for this test, each one a common type and built with common materials like wood, brick, lightweight reinforced concrete blocks, and lightweight precast concrete slabs. There were two of each type of house: two single story ramblers built on concrete slabs; two two-story brick masonry houses with basements; two houses made of concrete blocks; two single story ramblers made of precast concrete; and two reinforced two-story frame houses. Some houses also featured common types of shelters in bathrooms or basements.

DESTRUCTION OF OPERATION CUE

---

[112] I had the unique experience of participating in many tours of these historical areas of the old Nevada Test Site and never failed to be overwhelmed by the remaining artifacts of that heyday of nuclear testing.

Within the test area, one house of each pair was placed closer to the detonation site in a pressure zone where major structural failure and collapse was expected. The second was placed further away in an area where damage was expected but wouldn't cause a fatal collapse.

But collapse was part of the goal. The point of testing these houses was to find their weak points so they could be reinforced before rather than after a nuclear bombing attack.

To add to the realistic environment of the test, each house was furnished and fitted with appliances. Mannequins dressed in clothes of all different types of materials were placed inside the homes, seated in front of TVs or being tucked into bed. Some were less lucky, placed in a line facing the blast to see what would happen to someone caught outside when a nuclear strike came. Canned and packaged food was stocked in the kitchen pantries and buried underground, the question being whether this food would be safe to consume after exposure to a nuclear blast. Other structures on the test site included radio towers, power lines, transformer stations, and gas tanks. These were all things people would need for communication after an attack, so knowing how they would withstand the blast was vital. Once everything was in place, the test was delayed for weather; strong winds made a detonation unsafe. The holds went on so long that personnel involved in the project nicknamed the blast Operation Mis-Cue, and it was during one of these holds that six Army personnel went into Las Vegas and crowned showgirl Linda Lawson Miss Cue in honor of what felt like endless delays on the program.

Finally, at 5:10 on the morning of May 5, 1955, the 30 [29] kiloton bomb was detonated from the top of a 500-foot steel tower at the Yucca Flat at the Nevada Test Site. About 500 people were on hand that morning, including media, who drank hot coffee and donned thick black goggles as they broadcast the test

live to the nation.Public record AEC image.

Just 24 hours after the blast, media and military personnel alike were allowed to wander through the test area to survey the damage and begin gathering data. The damage was extensive. The masonry house was little more than a heap of rubble. The reinforced concrete house fared pretty well, though the windows were blown out. The two story frame house was damaged but standing. Both the basement shelter and the reinforced bathroom shelter were standing beneath the rubble of their houses so anyone inside these structures could theoretically have survived. Of course, this is assuming they would have had enough time to move into the shelter. On the whole, Operation Cue was deemed a successful test for the survival of homes and families in the nuclear age.[113]

The final and 14th test, "Zucchini," took place on May 15 in Area 7 from a 500-foot tower. This Los Alamos weapons development test had a yield of 28 kilotons. Like the Apple-2 shot, it had been calculated to produce a much higher yield of 40 kilotons but failed to reach that intended mark. Zucchini proved a

---

[113] https://www.popsci.com/blog-network/vintage-space/operation-cue-aka-nuking-houses-emergency-preparedness/. Also see video of Operation Cue blasts" https://publicdomainreview.org/collection/operation-cue-1955/.

unique test in experimenting with a very lightweight aluminum casing for the device.

Looking back on Operation Teapot, Al O'Donnell recalled what a key set of tests it represented for 1955. Others agreed. President Eisenhower's official biographer, Stephen Ambrose, stated that Ike wanted further nuclear testing like this to determine how to make "the bomb" more of a military weapon and less one of mass destruction. In Eisenhower's view, these were not simply new weapons but new weapon systems that could be utilized as integral parts of more modern-styled armed forces. That is why Operation Teapot became so significant in the history of early nuclear testing because the tests explored that very concept of using nuclear weapons in a tactical sense. That gave leaders more options than just fighting a war of mass retaliation. The principle of deterrence in the strategic sense remained valid and would be developed by other types of nuclear tests, but Teapot and Operation Upshot-Knothole before it retains a unique place in Nevada Test Site history.

In O Donnell's final reminisces, he felt that all the nuclear tests collectively represented an important vast body of work that "provided us with the peace and liberty that survived the Cold War." In his final years, O'Donnell proclaimed he would never hesitate to go back and do it all over again if needed. [114]

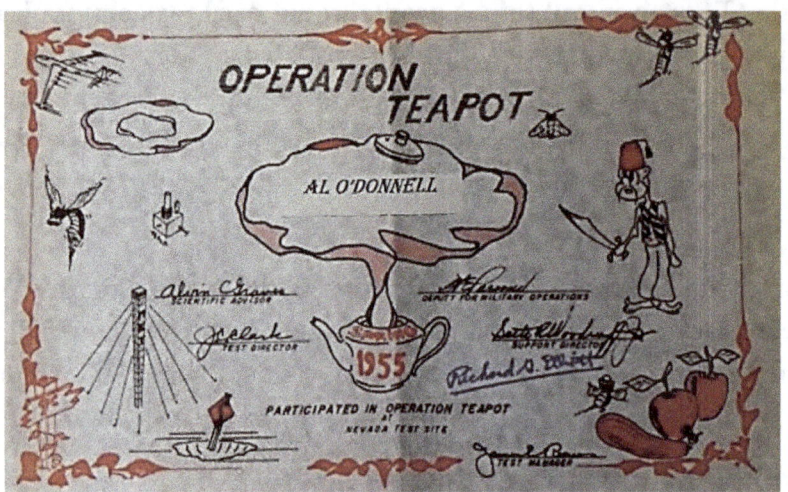

All of the nuclear tests had a special certificate offered to their participant.

Image from Authors' collection.

---

[114] Al O'Donnell DRI Oral History Project, University and Community College System of Nevada, 2004. Recommended further reading of the book *Nuclear Weapons Testing At The Nevada Test Site* by John C. Hopkins and Barbara Germain Killian.

CHAPTER THIRTEEN
IRON CURTAIN

Winston Churchill delivering *The Sinews of Peace* speech in Fulton, Missouri.
Authors' collection from the Gold Coast Railroad Museum, Presidential Rail Car archives.

On March 5, 1946, when responding to an invitation from President Truman to speak at Westminster College in Fulton, Missouri, Winston Churchill brought the phrase "Iron Curtain" into use again. "From Stettin in the Baltic to Trieste in the Adriatic, an iron curtain has descended across the Continent."[115] I used the phrase *again* because, at different times in history, both Kaiser Wilhelm II and Joseph Goebbels had bantered that same metaphor around, warning of a growing threat and even a "cold war" with Russia, but this time, history proved Churchill right.

Winston Churchill had been replaced by Clement Attlee as Prime Minister the previous year, just prior to the defeat of Japan in August 1945. Churchill, nevertheless, had become the most respected British statesman in history by 1946. A day after the delivery of The Sinews of Peace speech (ever since known as the "Iron Curtain Speech"), Churchill gave a similar talk at the Virginia General Assembly in Richmond. He was accompanied to the Virginia capitol by General Eisenhower, who had become US Army Chief of Staff. Both men retained a close relationship since the Second World War. Churchill's friendship with Eisenhower would continue into his presidency and Churchill's second prime minister ship. Churchill understood, like Eisenhower, that freedom provided people with dignity, a dignity that was so important to human wellbeing that it made it imperative to stand up to totalitarian regimes whenever necessary.

President Truman, of course, was the man he had to deal with first, and Churchill knew him well from the Potsdam Conference in July of 1945 while the atomic bomb was being tested. Churchill and Truman then became very well acquainted during the 1946 visit on what became a long train journey from Washington to Fulton and Westminster College. That trip took place on Truman's presidential railcar, *Ferdinand Magellan*.

---

[115] https://winstonchurchill.org/resources/speeches/1946-1963-elder-statesman/the-sinews-of-peace/.

**Authors' collection from the Gold Coast Railroad Museum,
Presidential Railcar Archives.**

As an interesting side note, a number of years ago, I had the great pleasure to oversee with my son the restoration of the United States Presidential railcar Ferdinand Magellan which took Truman and Churchill to Fulton, Missouri for that historic speech. The railcar actually served numerous presidents all the way from Herbert Hoover to Roosevelt to Truman to Eisenhower, and briefly Ronald Reagan. Aside from providing the platform for Truman's famous "Whistle Stop Campaign," the most significant event to take place on that train car concerned the famous journey to Westminster College. While Churchill's *The Sinews of Peace* speech was being finalized on board the Magellan, a very foretelling incident occurred. Just before the trip, in the aft observation lounge of the Magellan, Truman had commissioned a miniaturized version of the presidential seal to be prominently installed on the leather-paneled rear facing wall. The President had proudly redesigned the seal himself by turning the eagle's head away from the talons of war, to instead face toward the olive branch of peace. Truman took great pride in this in hopes of a prosperous postwar period. More the skeptic, and in his seasoned characteristic manner, Churchill made a telling comment while the two men sat and chatted in the *Magellan's* lounge. With the still unfinished speech sitting in draft form on his lap, Churchill looked up at the new presidential seal and frankly told his friend that he should have the eagle's head put on a swivel. This he said, would allow the head to face either way, "depending on the way the winds of world events changed."[116]

A few days after Churchill's speech to the Virginia General Assembly, which followed his Fulton,

---

[116] Michael Hall, *Miami-Dade County's First National Historic Landmark Interior Restoration Funded by National Railway Historical Society Heritage Grant Presidential Railcar Ferdinand Magellan*, Gold Coast Railroad Museum, 2010.

Missouri, premier, he found his way to New York City. There, the newly appointed US Ambassador to the Soviet Union, Walter Bedell Smith, visited Churchill. Churchill was in a plush Manhattan hotel suite soaking in a hot bath when he received Smith. The flamboyant Churchill, as usual, paid little attention to formalities. (Churchill had talked to President Roosevelt late into the night in 1941 under similar circumstances. Following his bath and standing stark naked in front of him, Churchill said to FDR, "clearly, I have nothing to hide.")

During their meeting, Smith and Churchill commented about a group of demonstrators just outside protesting over the rhetoric of the Fulton speech. Smith, General Eisenhower's ulcer-ridden, hard-driving, wartime chief-of-staff, and future CIA director, paid little attention to the picketers. He agreed with Churchill that those same crowds in a year's time would be applauding the former British Prime Minister. That indeed happened.

Churchill's loss of office after the Second World War aggravated him. During that time, he was not given any intelligence on the further evolution of the atomic bomb, to which he, with Roosevelt, had given birth. Certainly, Britain could claim as much credit for contributing to the success of the Manhattan Project as anyone. In terms of atomic ascendancy, Churchill felt content with keeping all the technology they could and letting the United States retain ownership of this profound new weapon. Attlee, however, ordered the construction of an atomic pile by 1946 and, thus, the production of plutonium. This decision was made because of the post-war formation of the Atomic Energy Commission by Harry Truman and a resulting AEC policy stemming from its originating McMahon Act, which stopped the sharing of nuclear technology, even with the UK. In addition, the US Congress at that time had significant anxiety about giving any more aid to Britain after spending billions in assistance to that island nation during the war.

Unfortunately for Britain, the English team on the Manhattan Project did not take part in the production of the fissile material, so they knew little about it. Building the logistics for a nuclear enrichment program from scratch became a gargantuan undertaking. Attlee did not reach the decision lightly. He called a confidential meeting of the key leaders and scientists of the day, and they all felt Britain needed its own atomic bomb to stay competitive in the post-war world. So, across a period of seven years, Britain, much to Churchill's dismay, would spend considerable sums from its debt-ridden treasury in a new arms race.[117]

William Penney, the former Los Alamos British team leader, became England's new head of armaments. He had been an observer in the photographic plane over Nagasaki and also observed at the Bikini Atoll Crossroads tests in 1946. Los Alamos scientists treated him with great appreciation and respect. The United States also stood firmly behind Great Britain as an ally. The fact remained, however, that Congress would not then tolerate any new assistance to Britain to support a nuclear program of their own—at least not yet. By the time of Churchill's Fulton, Missouri, speech and United States tour, Britain made the commitment to an all-out, expensive nuclear program. As British Foreign Secretary Ernest Bevin stated, "We have got

---

[117] Kevin Ruane, *Churchill and the Bomb in War and the Cold War* (London: Bloomsbury Publishing, 2016), pp. 155-171.

to have this thing whatever it costs, and it's got to have the bloody Union Jack on it!"[118]

Authors' collection, UK archives, https://nuclearweaponarchive.org/Uk/G2.jpg.

Penney led that postwar British nuclear project. Future chief British nuclear bomb designer Ken Johnston has since recounted that time, stating that they concentrated on a plutonium device. The first experimental British A-bomb design is depicted below in this extremely rare picture that has only recently been released.

The British effort became known only as "High Explosive Research" or HER. Of all the British scientists who participated in the Manhattan Project, only William Penney worked on HER, although research from British Los Alamos veterans was utilized.

Their experience at Los Alamos had pretty much given Britain the ability to make a bomb. The hard part, as already stated, came in recreating the enormous logistics that go into the production of fissile material. The village of Aldermaston, 60 miles from London, slowly became the center of Britain's nuclear program. This had been the site of a key airfield supporting the D-day landings and was about as remote or rural an area as you could get in the highly populated English countryside. There simply was no comparable location to Los Alamos or the Nevada Test Site in Britain.

By the time Conservative leader Winston Churchill had returned to power as Prime Minister in a general election over Labor Leader Clement Attlee in October of 1951, Britain had almost become a nuclear power. At heart, Churchill still did not see the need for a nuclear stockpile but favored possessing the technology. He stated, "We should have the art rather than the article."[119] In contradiction, Churchill had been very bombastic in the early Cold War days and freely threatened Russia with a nuclear attack, even though Britain did not then have a nuclear weapon, nor had he even been in power at the time of the statements. Such details, however, never concerned one who had such oratory powers.[120]

Yet, to Churchill's consternation, the situation became even more complex in the years between his two terms as prime minister. Once the USSR threatened Berlin by blockade and then a year later test exploded its first atomic bomb, "Joe One," in the fall of 1949, the Cold War went into high gear. England and America should have then reignited their nuclear cooperation. They probably would have if not for a recent revelation that British Manhattan team member Klaus Fuchs had passed nuclear secrets to the Soviets.

Animosity over the wartime association with Britain rekindled all over again, and many in Congress felt that the English simply could not be trusted. Despite this, there is now a feeling among some Aldermaston veterans that by that time, William Penney, still leader of Britain's nuclear program, began receiving useful

---

[118] Winston Churchill's remarks on unveiling a bust of Bevin in the Foreign Office, quoted in "Sir W. Churchill on 'a great Englishman," The Times (5 November 1953), p. 5.

[119] Conversations with *Steve Fisher, United Kingdom US/UK Mutual Defense anniversary exhibit committee, 2015.*

[120] Michael Hall, *1958 US & UK Mutual Defense Agreement, Background History*, National Atomic Testing Museum, 2016.

assistance from the Americans, even though this would have been in clear violation of the McMahon Act.[121] Such an unofficial and very quiet exchange would have likely involved technical information from US nuclear testing. The evidence for this is in recently-released data about the early British nuclear tests, which clearly indicate similarities with pre-1952 US technological breakthroughs in nuclear testing. Churchill may have even been involved in this early back-channel dealing, which facilitated cooperation long before a "US/UK Mutual Defense Agreement" came into effect in 1958.

Regardless of how much help America may have covertly started to lend, one would think that Churchill would have then become an even more determined Cold War warrior. Surprisingly, Churchill began a dramatic, although slow, metamorphosis into a nuclear peacemaker. Churchill's evolving behavior may have stemmed from his desire first to try to mend relations with the United States and work toward another World War II-like partnership. This would also include, in his mind, direct talks with the Soviets via a Big Three-like, WWII-style conference. Churchill repeatedly pleaded with then-President Eisenhower several years later for just such a collective summit with Stalin.

Public record image of Churchill and Eisenhower, https://www.catawiki.com/nl/l/16216403-unknown-pix- camera-press-winston-churchill-and-dwight-eisenhower-1942.

Also, on his mind may have been his vast experience dating back to the beginning of an earlier and equally dangerous arms race when he served as First Lord of the British Admiralty. He always said that the further you look back in history, the further you can see into the future. In the ten years before the First World War, he lived through a time when Britain and Germany spent enormous sums of valuable national treasure on arms. This had originally been intended for important social programs, but diverted to building great steel battleships that were affectionately christened "dreadnoughts." That arms race led to an equally disastrous war in which afterward, even the victors suffered from decades of debilitating debt. This was not just a lesson that Churchill took to heart but one that future President Dwight Eisenhower also became deeply mindful of from the perspective of his own experiences in the Great Depression of the 1930s. Churchill, himself, had been Chancellor of the Exchequer leading up to that world-wide Depression, and both men knew the security risks of economic turmoil.

Churchill also knew by that point in history that his many years of building a great alliance with America had not been wasted. The US would continue to secure Britain's security even if they, at that particular time, were not then willing to share nuclear technology openly. Churchill understood deterrents and knew America would share its nuclear weapons if and when the need came. President Truman was much on the same page and had not only asked General Eisenhower to take over the new NATO command but also based B-29s in England under the Attlee government. The B-29s were never nuclear-armed, but no one knew that.

---

[121] *Britain's Nuclear Bomb—The Inside Story, https://youtu.be/9vAX7EujOYI.*

In Britain, however, Churchill's hard-learned patience did not spread virtue. Despite the famous statesman's great influence and words of wisdom, the momentum toward becoming a nuclear nation was simply too great by that point. England could not resist. Churchill inherited, in 1951, a full-grown and largely homegrown British nuclear project. The life-long warrior only grudgingly accepted ownership of the bomb. Churchill took a full month before giving the required permission needed to proceed with plans for their first nuclear test. To his horror, he had learned upon assuming office that Britain's bomb project had already cost 100 million pounds since he had warned against beginning such a program back in 1946. Churchill sincerely felt that if he could have remained in power in 1945, the cost of gaining a nuclear deterrent could have been paid for with US dollars instead of English pounds. He pointed out that America was there whenever the need came, such as in the Berlin airlift in 1948. America, in return, was generously offered bases by the British. So, in Churchill's eyes, cooperation existed already. He felt confident if he had been given the chance, he could have gotten Britain the bomb for free or barter. On New Year's Eve, shortly after re-assuming the prime minister's seat, Churchill voyaged on the *Queen Mary* as 1952 dawned with his top advisors, just as he did ten years earlier on the HMS *Duke of York*, bound for America. That first trip, right after Pearl Harbor, cemented an alliance with Roosevelt. A decade later, almost to the month, the British leader courted Harry Truman and then soon courted Eisenhower, hoping to reestablish close ties between the two nations again now that he was officially in power once more.

There were important similarities to his wartime dealings a decade earlier. The US and UK were again allies in a war in Asia, this time in Korea. The Middle East held the attention of both countries, and Britain was an important partner, as it had been in 1941. Also, as in 1941, enemy bombers could reach London but not Washington. This restriction was as true in 1952 for Soviet bombers as it had been in 1941 for German bombers, so England, not America, was again on the front line in this new Cold War.

As in the Second World War, America was highly dependent on Britain for land to establish bases, which became urgent by the 1950s. Truman had been allowed under Clement Attlee to base B-29s at East Anglia airfields. So, Britain was a major player and partner in all things (potentially) nuclear, whether the US Congress liked it or not. Now that the Cold War was heating up, Churchill knew if the Americans ever launched a nuclear attack from England, it would mean that they, in return, would be under direct fire. London, not Washington, would be a target.

Thus, Churchill was going to demand a mutual consent agreement as he had with Roosevelt in World War II. Truman, in fact, honored that original mutual consent agreement, signed at the Quebec Conference, before he used the bomb on Japan. Truman and Churchill thus had a history of cooperation. Churchill would not get a second mutual consent agreement, but Truman made it clear he would be consulted. This was becoming an increasingly thorny issue back in England, and Churchill once again proved himself as the one man with the clout to make Britain a respected voice again with the mighty US.

Now that Britain had returned to centerstage, the two leaders had a lot to talk about. First on their agenda was an amendment to the McMahon Act, allowing the AEC greater interchange in atomic physics. Both leaders wanted more official sharing of technology, but unfortunately, this amendment did not include the exchange of weapons technology. So, by the strictness of the law, Churchill and his staff could not even receive a detailed briefing from the Strategic Air Command, who could only talk to them in generalities while they were in Washington. In contrast, Churchill freely discussed Britain's first nuclear test scheduled for later that year.

The British were transparent about their program. The United Kingdom tested its first atomic bomb on October 3, 1952, in a shallow bay on Trimouille Island in the Monte Bello Islands chain off the coast of western Australia. They called the test Operation "Hurricane." Prior to Churchill's second term, the British had requested to use the American Pacific Proving Grounds, but that request was rejected. Then negotiations started in 1951 to use the Nevada Test Site. The idea of sharing space at the Nevada Test Site was for Britain to share its test information with the Americans. President Truman gave tacit approval, although the AEC wanted to charge Britain a fee for the use of the Test Site and demanded payment in US dollars. Still suffering economic hardships from the Second World War, it was finally decided by the time Churchill had regained power to test their new bomb in their own dominion lands. Clement Attlee had already secured permission from Prime Minister Robert Menzies of Australia to do so.

That first British nuclear test was very focused on Britain's unique geography. Operation Hurricane involved the detonation of a 25-kiloton plutonium device on a ship anchored in shallow water. At the time, the British were extremely worried about the possibility of a foreign nation smuggling a bomb, hidden in the hull of a ship, into the Thames estuary. To test such an effect, the old destroyer HMS *Plym* was used to detonate the device. The nuclear package was positioned in *Plym's* hull about three meters below the waterline as the veteran ship anchored 350 meters off of Trimouille Island. The resulting explosion left a saucer-shaped crater on the harbor bottom about 6 meters deep and 300 meters across.[122]

British scientists at Aldermaston had developed their own high-speed camera technology to record the blast and measure the yield. In England, there were no specialized contractors with EG&G's expertise, so this had to be developed from the ground up. Detailed images of that special film taken at the first British nuclear test have only recently been released from Aldermaston.

By 1952, Britain had become the world's third nuclear power. Weeks later, America tested the first concept for a thermonuclear device using the principle of fusion. The Soviets were soon to follow, and even

---

[122] Churchill closely oversaw the Operation Hurricane test with William Penney and in coordination with the Admiralty. Churchill formed the APEX committee under his science advisor of many decades, Professor Frederick Lindemann. Churchill had provided the name APEX, and the X stood for unknown. The first part, APE, came from a satirical cartoon of the time. In the cartoon, apes are reflectively looked on as sole survivors of a nuclear war. Viewing their devastated landscape, they wisely decided not to evolve into man again. This is an interesting anecdote because Lindemann became the hawk and Churchill the dove in nuclear issues. Lindemann, unlike Churchill, in previous years had lobbied the Labor government for an atomic bomb project and called for a literal equivalent to America's Atomic Energy Commission. As we have seen, Churchill acted more conservatively but was devoted to Lindemann. He never forgot that Lindemann had been the first to recognize the evidence of the V-2 rockets when all the other experts said such a weapon could not exist.

China became a nuclear and then thermonuclear power.[123] A nuclear arms race had begun.[124]

Churchill understood that Europe, and especially Britain, with the majority of its population concentrated in just twelve cities, could not survive thermonuclear weapons. The loss of those 12 critical cities would mark the end of Britain, whereas the destruction of 20 key American cities would cause the US to lose its ability to function as a nation. Churchill's dozen cities were then within easy reach of Soviet bombers, whereas none of the American cities were yet able to be threatened in the early 1950s. He warned in his speeches of "humanity approaching a precipice," but he was really talking about Western Europe approaching Armageddon. Emergency plans were put in place to evacuate the Queen in the event of a nuclear strike.[125]

---

[123] Churchill certainly understood that the genie had long since been let loose from the bottle, and a nuclear arms race was not a contest you could simply stop playing before the other side did. The big concern to Churchill by 1952, and the whole declining British Empire, centered on the economy. We must remember that the Second World War actually stimulated the American economy and finally took us out of the Great Depression. It led us into one of the most prosperous periods in our history by the late 1950s. In contrast, two world wars had bled Britain dry. Rationing continued for nine years after 1945 when life in Britain became even more severe than during the war. Britain, like America, was also trying to address the starvation diets in war- torn western Europe and was by then taking part in the Korean War. Even its small participation in Korea forced Britain to increase its defense spending once again despite its enormous war debt. That debt, measured by 1950-era dollars, amounted to billions. As much as $200 billion was then still due from their First World War debts. Economic facts such as these are what so impressed conservative leaders of their day, like Churchill and Eisenhower. We cannot adequately appreciate such figures with today's blind complacency with our own national debt—which far exceeds what England once faced in comparative numbers. We do not realize the historical significance of our economic predicament yet, but our grandchildren will. It does, in part, date back to a spending battle during the Cold War, which eventually destroyed the USSR—proving in Churchill's own words that economics is as mighty a weapon as the sword. If Churchill lived today, he would be pleased that in 2014, Britain finally paid its First World War debt off, thanks to many years of determined work by leaders inspired by his example.

[124] An excellent resource for further study in this area is a recent book by historian Kevin Ruane titled *Churchill and the Bomb in War and the Cold War*.

[125] Churchill had to face reality by 1952. He put in motion a rough, quasi-unofficial equivalent to the Atomic Energy Commission to manage all research and development. To head that effort, he chose the aging professor of physics, Frederick Lindemann. Lindemann had been Churchill's supreme scientific advisor for more than three decades and the leader of Britain's Second World War science endeavors under the cabinet office of Paymaster General.

CHAPTER FOURTEEN
DUCK AND COVER

Public record image, authors' collection, of New York State Civil Defense Commission publication, circa 1952.

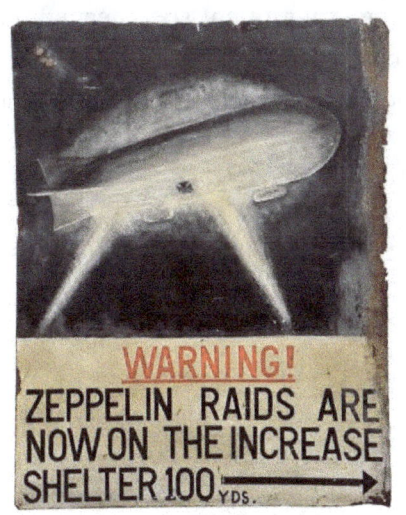

Public record image British air raid poster.

On the night of May 31, 1915, a then state-of-the-art weapons delivery system appeared over London, England. Using the outline of the Thames River to navigate the darkness, the enemy intruder positioned itself for the first bombing raid on one of the world's largest cities. Four times the length of a modern B-52 and bigger than any aircraft now flying, the bomber opened its lower doors to release almost 100 explosives and incendiaries.

This was not a story from H.G. Wells's 1908 novel *The War in the Air*. Wells foresaw such events, and his visions certainly came true, but this bombing was not fiction. The huge German Zeppelin that bombed London that night brought modern war beyond the front lines and onto the civilian population. Civilian casualties in Liege, Antwerp, Paris, London, and numerous towns in Great Britain became part of the reality of the face of modern war. This type of bombing became considered strategic in nature, while the media of the time called the German Zeppelins "baby killers."[126]

---

[126] *World War One: How the German Zeppelin wrought Terror*, BBC News story, Aug. 4, 2014, https://www.bbc.com/news/uk- england-27517166.

The broad concept of civil defense was initially organized out of those First World War air raids. Even prior to the war, the mere threat of German Zeppelins had created just as much apprehension as Kim Jong Un's missiles have in recent years. Certainly, the history of civil defense starts in those early times. Once America became concerned about the European war, the focus centered on "anti-saboteur vigilance." America's eastern seaports by 1915 were handling large quantities of munition shipments bound for England. President Woodrow Wilson formed the Council of National Defense in August of 1916 to address that. The term was then "Civilian Defense," and when we entered the war a year later, the vast Atlantic Ocean still formed an effective defensive shield.

Public record image First World War American civil defense poster.

The Council of National Defense disbanded in 1921 after the finalization of the Treaty of Versailles, which ended the First World War. However, by 1940, President Roosevelt reactivated it and created the Division of State and Local Cooperation to assist.[127] By May of 1941, little doubt existed that America would soon be involved in an already two-year-old Second World War. Accordingly, the president went further with an executive order that created the Office of Civilian Defense. The OCD primarily boosted morale and confidence, originally headed by flamboyant New York City Mayor Fiorello La Guardia.

After the Pearl Harbor attack on December 7, 1941, the primary fear centered around air raids, while fears of gas attacks were also widespread. The OCD had four divisions. The government spearheaded the first blackouts, special fire protection procedures, and air raid drills. It also created the Civil Air Patrol which assisted in spotting offshore submarines. An Industrial Protection Division advised factories about the dangers of enemy sabotage.

---

[127] https://www.gilderlehrman.org/history-resources/spotlight-primary-source/civilian-defense-home-front-1942.

Eleanor Roosevelt pushed to get women involved in war work, and they contributed substantially to OCD initiatives. She became an assistant director and founded the Civilian Participation Branch. This division oversaw childcare, health, housing, and transportation. The OCD is comprised of nine regional offices, each with a paid regional director and numerous volunteers serving as state, county, and city directors with block leaders. Although at its height, the OCD only had 75 paid staff members. By 1943, it had 14,000 local defense councils throughout the United States, utilizing 11 million volunteers.

Public domain image from author's collection.

Most reminiscent of this period were the "Blackout Drills," which millions of Americans participated in. Such drills were staged more like rehearsals where a warning signal would sound. That first signal was called "Yellow." Volunteers would then have fifteen minutes to make sure all lights were extinguished or curtains drawn in their neighborhood. Then, the condition "Blue" sounded, indicating the blackout was underway. If actual enemy aircraft had ever been spotted, a condition "Red" would be sounded, and all area motor traffic stopped.

American shores remained relatively safe during WWII. Despite some minor incidents of artillery bombardments from German and Japanese submarines and random attacks by Japanese Fugo Balloon bombs, the United States mainland went through a Second World War, still primarily protected by its geographic isolation. As a result, by the end of the war, OCD fell under some criticism, which arose into a political issue with accusations of wasted resources. In fairness, the organization had effectively mobilized a civilian population like no other program in American history. Its model would soon be needed again.[128]

What is often forgotten is that the war could have easily gone in the enemy's favor. For a period, it did. The eastern and southern seaboard merchant lanes saw more than 400 cargo ships sunk by German U-boats. The oceans were becoming less of a barrier. Nuclear weapons and long-range bombers also emerged out of World War II, although for a period, the US had an exclusive monopoly on both.

At the end of the war, the OCD civilian defense "structure" remained in place, although OCD technically disbanded in 1945. In other words, its core structure became the foundation of "Civil Defense" during the Cold War. In 1947, the National Security Act created the National Security Resources Board, among many other organizations.

---

[128] Research conducted by Michael Hall using Department of Energy files for Smithsonian-affiliated National Atomic Testing Museum tours, 2015-2022.

Public domain image from author's civil defense collection.

Its eight-member board would oversee mobilization if a war situation arose. By 1953, it evolved into the independent government Office of Defense Mobilization, or ODM. World War II taught us that defense is about far more than air raid procedures. Leading up to our entry into the war, President Roosevelt had great logistical challenges and faced resistance to mobilization. These Cold War-era organizations were formed, in part, to help speed up mobilization if a new war came quickly. They also allocated resources.

People tend to forget that consumer and price controls existed even during the Korean War. For example, television production had to be controlled because the scientists and engineers required for their development were needed for defense projects. During the Korean War, my mother owned a department store, and she had as many challenges ordering suitable varieties of inventory as her mother had when she owned the store during the rationing days of World War II.

Yet civil defense, in the strictest sense, was another matter during the immediate post World War Two years. America, after all, had a nuclear monopoly that consensus concluded could last for decades. Opposition also formed to any civil defense agency because no one wanted the budget obligations. The military, in particular, resisted any suggestions that they should be responsible for civil defense as their regular budget requests were already being reduced in the peacetime atmosphere. Peace also hindered any

Public record image.

federal support for civil defense, and the Truman administration concluded it should be up to state and local authorities to shoulder the responsibility.

The slow birth of the Cold War would gradually change that thinking. The years 1948 and 1949 witnessed the Berlin Air Lift. In August of 1949, the Soviets tested their first atomic bomb. It shocked everyone, especially the nuclear experts who predicted it would take many years before they would be capable of such a feat. Russia also started producing an exact copy of the American B-29 bomber called the Tupolev Tu-4. Then, in June of 1950, North Korea invaded the South. A newly constituted communist China would soon give its support to the North, which already had strong Russian support behind the scenes. The public started to wonder if Russia might even have its sights on the United States. Fear and paranoia spread along with an irrational fear of Communist subversion within the United States.

By the end of 1950, The Nevada Test Site was soon to be established near a then very small city called Las Vegas. America had to stay competitive with Russia or any other potential adversary. A nuclear arms race had begun even though no one had yet realized it. In response to a growing Cold War, President Truman formed the more comprehensive Federal Civil Defense Administration, or FCDA, in December of 1950. It basically mirrored the World War II-era OCD. Congress then formally passed the Federal Civil Defense

Act of 1950, giving the FCDA vague guidelines. The Defense Production Act was passed, which provided guidelines for industrial defense contingencies. The FCDA had the authority to draw up plans to provide guidance in civil defense matters for states and their counties. Fifty percent of federal matching funds were to be offered to states for supplies and materials.

The concept of a nationwide shelter system was also envisioned for "critical target cities." The projected cost was 403 million dollars; however, Congress only approved $31.75 million. This would become a recurring theme of civil defense in the United States. Congress could never commit to sufficient sums to make it an effective program. Many congressional leaders wanted the states to assume full responsibility, while others thought the FCDA should only offer training, planning, and guidance. It did that, of course, and there were many civil defense publications and films throughout the period.

Public domain image from author's civil defense collection.

Then came a significant disaster for civil defense. Harry Truman named Millard Fillmore Caldwell Jr. the first director of the FCDA. He was a former Florida governor who had sincere intentions but displayed a combative attitude at Congressional hearings. The fault was not his because the FCDA was never given a definitive mandate. Caldwell spoke with sincerity, which, unfortunately, is not always a wise tactic at a Congressional hearing. The problem was that the questions always revolved around what it would take to provide effective shelters, and Caldwell told it honestly: "To provide adequate shelters for the population would take at least 300 billion dollars."[129] Today, that dollar figure would not shock many; however, in 1952, such an amount, although a fair estimate, was **the most expensive** project in US history up to that point had been the B-29 program during World War Two, which at three billion dollars even exceeded the cost of the Manhattan Project's two-billion-dollar price tag. The mind then simply could not fathom Caldwell's $300-billion-dollar shelter estimate. Some in Congress were not even necessarily against spending such vast funds for defense by the 1950s. However, if that kind of money was to be on the table, they wanted the funds for Army, Navy, and Air Force appropriations. That became a far more proactive philosophy than a bomb shelter mentality. That entrenched notion advocating a shelter system also became called the "Maginot Line Mentality" after the expensive and outdated system of French fortifications built prior to the Second World War, which proved to be of little value against Hitler's mobile warfare tactics.

By 1953, President Eisenhower agreed with the prevailing winds of the Congressional Committees and conceded that the main responsibility for civil defense rested with the states. However, he felt the federal government should provide technical information and guidance. This was also a period when thermonuclear, or H-bombs, were first being tested. These devices were not measured in kilotons like A-bombs but in megatons. Many thought thermonuclear bombs made the idea of bomb shelters pointless. The focus in the

---

[129] Civil Defense: The Truman Administration: US, 1947-1952.; Wayne B. Blanchard. *American Civil Defense 1945-1984: The Evolution of Programs and Policies*, (Washington, District of Columbia, United States of America, U. S. Government Printing Office, 1986); Lawrence J. Vale, *The Limits of Civil Defense in the USA*, (Switzerland, Britain and the Soviet Union, New York, New York, United States of America, St. Martin's Press, 1987).

MICHAEL AND JAMES HALL

new thermonuclear age then briefly went to evacuating the cities instead of hardened shelters. However, a particular nuclear test 1954 called Castle Bravo changed the thinking once again. Operation Castle signified a watershed in the escalation and development of nuclear weapons.

Castle Bravo's explosive force was approximately 1,000 times that of Hiroshima, with a fireball exceeding a diameter of four and a half miles. The test thus resulted in a great deal of long-range radioactive fallout. After that, the idea of shelters seemed relevant again in light of the anticipated levels of fallout that could likely occur in a war using such weapons. The term however became not "bomb shelter" but "fallout shelter," as the goal became to shield against radioactivity. Most realized nothing could protect one against the blast of such a great weapon if you were directly under it, but if you lived in the outlying suburbs, your fallout shelter could give you a chance. At least, that became the prevailing idea.

Despite this, President Eisenhower brought in a new Civil Defense Director, by the name of Val Peterson, who continued to support the idea of evacuation over shelter systems. Gradually, congressional committees started politicizing the issue as they looked more and more into a system for providing shelters. Under Eisenhower's first term, civil defense appropriations rose to 65 million dollars compared to Truman's 50-million-dollar allocation. Under his second term, civil defense rose again as Congress further politicized it.

Thus, the renewed attention to a shelter system was partly a backlash to the earlier evacuation stance. In other words, it became a political football. The question of where the actual responsibility for civil defense rested returned. Senator Estes Kefauver's Armed Services Subcommittee and Representative Chet Holifield's Military Operations Subcommittee undertook lengthy investigations into civil defense during Eisenhower's presidency. The official history on Civil Defense published by the National Emergency Training Center in 1985 and authored by Dr. B. Wayne Blanchard states the following:

Representative Chet Holifield introduced legislation H.R. 2125 which called for (1.) the reorganization of civil defense into a Cabinet-level Executive Department; (2.) the establishment of civil defense as a primary Federal, rather than primarily State, local, and citizen responsibility; and (3.) the creation of a nationwide shelter system. . . estimated price tag of $32 billion.[130]

That bill did not get very far and did not sit well with the administration. President Eisenhower appointed a committee to study the issue, which became known as the Gaither Committee. Basically, the administration felt that increased spending should come in defense measures, including ballistic missile technology, as opposed to spending vast amounts on shelters. In addition, Eisenhower wanted to ease the growing cold war—not intensify it. A moratorium on nuclear testing followed from 1958 to 1961 as nerves calmed somewhat.

Official wording became important to controlling fears. Building shelters around a comprehensive civil defense plan could be perceived as mobilizing for war. That was not a message Washington wanted to send. Eisenhower is also unique in that he was one of the last presidents to appreciate the importance of a balanced budget. His chief advisor, John Foster Dulles, agreed and opposed an expensive shelter program because of its huge costs. In addition, President Eisenhower saw the threat of "massive retaliation" using strategic and

---

[130] Dr. B. Wayne Blanchard, *The Official History of Civil Defense* (Emmitsburg, MD: National Emergency Training Center, 1985).

tactical nuclear weapons as a more proactive move and viewed it as a deterrent strategy. This was not yet referred to as "mutually assured destruction," as it came to be known in later years under a mutual deterrent concept. Eisenhower's policy of "massive retaliation" was rather unilateral. In his day, the arsenals were not that big yet, and American nuclear weapons did not need to pose a strong deterrent because they outnumbered the Soviet nuclear arsenal. So, all this seemed much more sensible than investing in a "Maginot Line" mentality.

By 1958, FCDA merged with ODM and became the Office of Civil and Defense Mobilization, or just OCD for short.[131] By the time John F. Kennedy got on the political scene he advocated an increase in funding for civil defense, even though he often expressed skepticism over the practicality of a shelter system. When he became president, the congressional pendulum would swing once again. He submitted a $207.6 million request to Congress, which was fully approved! With these funds, the OCD began surveying adequate sites for a system of fallout shelters.

A whopping 59 million dollars went into stocking existing government buildings which were then being used as shelter sites. This included a five-day food ration and a two-week water supply. A further 10 million dollars was earmarked to increase the size of existing government or school basements for shelters. Then, 7.5 million dollars covered the construction of shelters in the new government. Finally, 38 million dollars became available for Geiger counters.

The following year, President Kennedy requested $695 million to continue the survey on shelter sites. The money would also include federal funds for non-profit organizations to promote educational activities and further funding for the construction of fallout shelters or the modifications of existing buildings for shelters. One survey of the day recorded 1,565 private or home fallout shelters had already been built in 35 states. By then, however, the familiar cycle of disinterest in civil defense repeated, and Congress had delayed or canceled much of the follow-up funding.

By 1963, Kennedy had lost his interest in civil defense and, like previous presidents, saw the proactive choice to concentrate on new weapon systems and overall military defense as a better use of funds. He advocated turning civil defense over to the Defense Department. Perhaps his experiences during the Cuban Missile Crisis, by which time the US arsenal reached 27,000 warheads, reinforced the concept that a nuclear war must never be allowed to happen in the first place. Investing huge sums of money in a national shelter program was still perceived as provocative, and fears arose that this could send a signal to the Soviets that

---

[131] Both known by the acronym OCD, the World War II Office of Civilian Defense and the Cold War era Office of Civil Defense were established to prepare and protect civilians in times of conflict, but they operated in distinct contexts and addressed different primary threats. The WWII OCD focused on air raids and bombings, while the Cold War era OCD primarily addressed the threat of nuclear war and fallout. Both the World War II Office of Civilian Defense (OCD) and the Cold War era Office of Civil Defense (OCD) used a triangle logo as part of their organizational symbols. However, the specific designs and variations of the logo differed between the two periods. The logo used during World War II was a blue triangle with a white border. Inside the triangle, there was typically a white stylized letter "C" (representing "civil") along with a white star in the center. This logo was often seen on various civil defense materials and publications from that era. The logo used during the Cold War era also featured a triangle, but the design varied over time. In the United States, the Federal Civil Defense Administration (FCDA), which was the agency responsible for civil defense during the Cold War, used a yellow triangle with a black border. Inside the triangle was a black radiation symbol known as the "trefoil," representing nuclear dangers. This trefoil symbol became synonymous with civil defense efforts during the Cold War and was often seen on fallout shelter signs, educational materials, and other related items.

Americans were willing to risk a nuclear war. President Kennedy clearly had skepticism of shelters from the beginning of his term and summed it up well in a speech on July 25, 1961:

Civil defense …cannot be obtained cheaply. It cannot give an assurance of blast protection that will be proof against surprise attack or guaranteed against obsolescence or destruction. And it cannot deter a delineanuclear attack. We will deter an enemy from making a nuclear attack only if our retaliatory power is so strong and so invulnerable that he knows he would be destroyed by our response. If we have that strength, civil defense is not needed to deter an attack. If we should ever lack it, civil defense would not be an adequate substitute.[132]

President Johnson gave little support to civil defense. This, despite the fact that he had inherited Kennedy's Secretary of Defense, Robert McNamara, who felt the developing antiballistic missile system necessitated a large system of fallout shelters. McNamara felt that way because the developing plans were then to blow up incoming Soviet ICMBs with a nuclear-tipped missile system or ABM. The fallout in such a scenario would have been tremendous.

The coming war in Vietnam diverted almost all attention away from civil defense to that troublesome, prolonged, and ever-expensive conventional war. Richard Nixon did order a study of the nation's then-aging civil defense-sponsored shelters and also advocated that new government buildings should be designed in a way to maximize protection in the event of a nuclear war. Of course, making buildings resistant to nuclear blasts was challenging despite the offer of government grants.

In 1972, President Nixon formalized the shift in civil defense back primarily to local emergency planning by eliminating the OCD and transferring responsibilities to a new Defense Civil Preparedness Agency. Government funds were directed at the states, but more money was spent on natural disasters from that point on, displacing any meaningful interests in civil defense. There were some significant and still very classified expenditures on high-level bunkers designed for the continuity of government. Those, however, were for the government and military elite and not part of any civil defense program. As the years proceeded, federal funds decreased year by year as we entered an era of détente with the ABM Treaty and then the eventual fall of the USSR.

A lasting legacy of the days of civil defense is a plethora of publications and films. We think of civil defense more in terms of bomb or fallout shelters, but as we have examined, the funding for that level of preparedness continued to lack consistent support from Congress as the interest came in irregular waves. There was, however, ample funding for literature and films which promoted public awareness. In 1950, the National Security Resources Board created a document called "Blue Book" that outlined civil defense measures for the next 40 years, even though Congress never met the budget requirements for the programs described.

The Federal Civil Defense Administration sponsored education efforts like "Duck and Cover." That

---

[132] John F. Kennedy, "Speech on Civil Defense," (July, 25, 1961).

concept tried to warn people of the initial blinding flash of a surprise nuclear detonation and the need to seek immediate cover in any way available. It used a cartoon figure, Bert the Turtle, who defused anxiety in films and pamphlets to teach children and adults alike that nuclear war could be dealt with. Many booklets detailing what to do during a nuclear attack circulated, such as "Survival Under Atomic Attack." Radio, still much more common than television in the 1950s, often hosted public service announcements on civil defense.[133]

Drills took place at times, and evacuation plans were drafted. In 1955, the City of Portland, Oregon, evacuated its city center in 19 minutes flat in a drill called "Operation Greenlight." Then, the government considered it may be more important for people to shelter in place and try to rebuild the cities after an attack. That was the point, as previously explained, when shelters were dubbed "fallout shelters" because the point of nuclear disaster preparation became offering protection from the radiobiological effects that would occur after the attack had ended. President Kennedy initiated a plan of "fallout" shelters that even included upper floors on skyscrapers where people could ride out the three to five days thought necessary to evade the radioactive residue left from a nuclear attack.

No direction, however, existed as to where people would go if a nuclear bomb destroyed the fallout shelters. New York City invested millions of dollars in identification tags for school children in the horrible event that bodies needed to be identified or lost children located after an attack. Other cities followed.

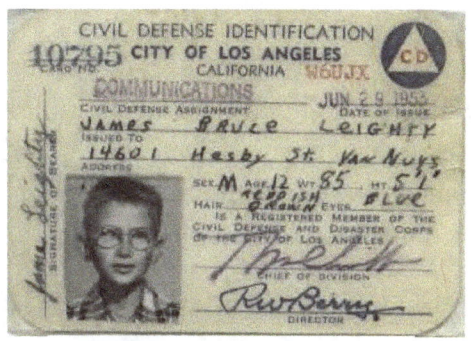

Public domain image.

Years later, President Reagan had a unique nuclear plan. He considered a $10 billion program to create "rural host areas" where up to 80 percent of the population could go in the wake of a nuclear attack. The plan never received funding.

Public record image of George E. Valley, Jr. https://www.aps.org/programs/honors/prizes/valley-bio.cfm.

At this point, a word should be said about the state of air defense in the early days of the Cold War. Once the Soviets tested their first atomic bomb in the late summer of 1949, the United States Air Force realized a nuclear attack across the North Pole from Russia was a possibility, albeit an unlikely scenario in those early days. The World War II era radars that were then in use around the country could not give a wide coverage of North America and were not designed to detect intercontinental ranged bombers.

The Air Force asked the Massachusetts Institute of Technology (MIT) for advice. That led to the formation of a special MIT project or division called Lincoln Laboratory. An associate professor in the MIT Physics Department, George E. Valley, Jr., took a leading role, consulting with Theodore von Karman of the Air Force Scientific Advisory Board or SAB. They decided a report should be made. That led to the formation of a committee with Valley as chair, which became known as the Air Defense Systems Engineering

---

[133] Kenneth D. Rose, *One Nation Underground: The Fallout Shelter in American Culture* (New York: New York University Press, 2001), p. 128.

Committee, or ADSEC. Their work confirmed that for all practical purposes, the United States and Canada had no adequate radar coverage to detect a Soviet bomber force coming out of the USSR.

They initiated the planning of a modern digital-based radar system across Canada. To say that was no small task is an understatement. A construction project followed, exceeding that of the entire Manhattan Project in terms of costs and logistics.[134] This became the Semi-Automatic Ground Environment air defense system or SAGE. SAGE was basically a system of large computers processing data from a wide network of new radar stations and funneling the information into one processing center, much as the British had done during the Battle of Britain with their Chain Home system.

Critical to SAGE was the radar network built across Canada, known as the Distant Early Warning Network or DEW. The DEW line of radars became operational by 1957. Built across the Arctic horizon, it became an immense engineering feat to build the sites and maintain them in an environment with such harsh weather conditions. Over the years, the system was upgraded and supplemented with additional radars forming three lines of networks stretching from the DEW Line near the 70th parallel to the Mid-Canada Line near the 55th parallel to the Pinetree Line on the 50th parallel. Today, our early warning defense detection systems largely rely on satellite technology. That system is called the Space-Based Infrared System or SBIRS. SBIRS is a constellation of integrated satellites in geosynchronous orbit and high elliptical orbit that sends its information to the Air Force Space Command in Colorado and to the North American Aerospace Defense Command (NORAD). However, a line of radars stretching across northern Canada, known as the North Warning System, or NWS, still exists and would be used by NORAD to track incoming missiles after detected by SBIRS.

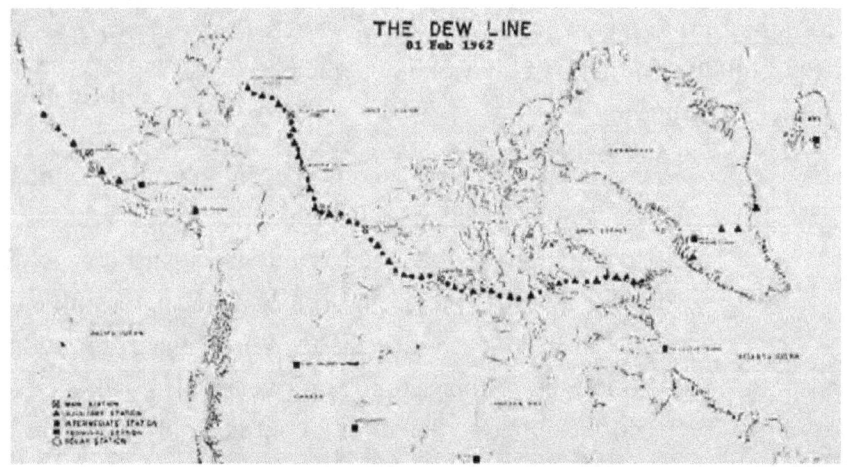

Larry Willson and Paul Kelley have an excellent online overview of the early air defense radar system. They write:

"The Distant Early Warning (DEW) Line began on 15 February 1954 when President Eisenhower signed the bill approving the construction. The DEW Line was designed and built during the 'Cold War' as the primary line of air defense warning of an 'Over the Pole' invasion of the North American Continent. The actual construction of the 58 sites took place between 1955 and 1957. Many tons of

[134] MIT, https://www.ll.mit.edu/about/history/sage-semi-automatic-ground-environment-air-defense-system, p. 18.

supplies and equipment were moved to the Arctic by air, sea and river barge. One such carrier, the USAF 62nd Airlift Wing, moved over 13 million pounds of material in this monumental effort. The DEW Line was declared fully operational on 31 July 1957. The DEW Line was centered along the 70th parallel. It was the first line of air defense for the North American Continent. There were three different lines of air defense constructed in Canada during the Cold War. The Pinetree Line was the first to become operational, during the 1952-1953 time period and it was, more or less, centered on the 50th parallel - with a section covering southern Ontario, and another extension covering the Labrador coast - going as far north as Frobisher Bay. The Pinetree Line was composed of 44 long range radar stations, and six USAF manned Gap Filler radar stations. Some of these locations had a very short operational life - while others remained operational for more than 35 years. Sandwiched between the DEW Line and the Pinetree Line, the Mid Canada Line (MCL), also known as the McGill Fence was a series of military sites designed to function as the second line of detection. Its purpose was to detect enemy aircraft that had penetrated south of the DEW Line into the heart of North America. This system operated on the 'Doppler' principle. Conceived during the 'Cold War' in 1951 it wasn't until January of 1958 that the line became fully operational. The MCL sites were strategically located along the 55th parallel from the Alaska border to the Atlantic Ocean. At its peak, it consisted of 8 Sector Control Stations and approximately 90 unmanned sites about 30 miles apart. With the 1960's came improvements in technology and jet aircraft design in particular which rendered the MCL no longer economically feasible or strategically required, so it was shut down. The western sites were decommissioned in January 1964 and the eastern sites in April 1965."[135]

An early warning information system for civil defense was also critical. In 1951, President Truman established the Control Electromagnetic Radiation Plan (also called Key Station System), which allowed only key radio stations (called Basic Key Stations) to broadcast an alert. Radio and television stations throughout the country would have access to and monitor these key stations.

If an alert came, people could be directed to those few designated key frequencies at 640 and 1240 kHz. The idea behind this convoluted system was that enemy bombers would have difficulty trying to home-in on radio stations if only a few stations were broadcasting. The Key stations not only had to lower power signals but also would shut off and on for periods of five seconds to further aggravate attempts by enemy aircraft to acquire a directional signal. One has to wonder how effective such a low-profile warning alert system would have been if a nuclear attack had come.

Once the age of intercontinental ballistic missiles, or ICBMs, arrived by the 1960s, bombers were almost obsolete. A Russian ICBM can reach orbit from its launcher in five minutes, fly to North America in eighteen minutes, and reenter to a target in one minute. So, since that point, early warning for the public became very problematic. The earlier public alert systems ceased with the establishment in 1963 of the Emergency Broadcasting System or EBS. Under that system, all radio and television stations would broadcast the same civil defense alert. In some cities, air raid sirens were installed, although sirens were already in place in most areas of the country where the need existed for severe weather warnings. The sirens served both purposes but had different sound durations; however, the Emergency Broadcasting System was not used for

[135] http://www.c-and-e-museum.org/Pinetreeline/misc/other/misc5d.html; and https://lswilson.dewlineadventures.com/.

cases of natural disasters until 1976. The Emergency Alert System replaced the EBS in 1997.

Initially, in theory, with advanced warning, people could benefit from hardened shelters to protect themselves from the actual blast and radiation of a nuclear attack. At the Nevada Test Site, a great deal of testing occurred of all types of wooden, brick, and concrete structures. The engineers learned from blowing up structures in actual nuclear tests that some relatively simple modifications to building codes could significantly strengthen even wooden frame houses. Archival films of these tests are now famous and are on YouTube in the public domain.

Of course, that was the day when nuclear bombs were measured in kilotons, such as those used on Hiroshima and Nagasaki. By the 1960s, nuclear stockpiles fielded weapons measured in multiple megatons. Russia had mass-produced devices up to 20 megatons, and we had bombs in the arsenal as big as nine megatons. In the wake of the growing nuclear arsenals with ever-increasing yields, interest in bombs and even fallout shelters waned. Protests organized against civil defense and a nuclear arms race began. Antinuclear demonstrations, in general, began as early as 1958. The earliest organized anti-nuclear groups formed in Britain and used the famous crow's foot or peace sign symbol which is today only identified with a much later group of demonstrators protesting the Vietnam War. It was, however, originally an anti-nuclear symbol when, in 1958, British artist Gerald Herbert Holtom designed the symbol for the Campaign for Nuclear Disarmament.[136]

In the post-Cold War and post-9/11 world, the focus of civil defense moved to terrorism and the increasing severity of natural disasters. The Federal Emergency Management Association, or FEMA, is what we look to today. Nuclear war has not been a focus of FEMA in recent years. Today, fears of North Korea have reignited the issue of civil defense. Recently, possibly as a result of nuclear saber-rattling over the war in Ukraine, the New York City Emergency Management system has issued a new public service announcement advising citizens on what to do in a nuclear emergency: https://youtu.be/zznmdUJbeU8. Los Angeles has now followed as of the spring of 2024.

This renewed concern over nuclear war is familiar to this author. I grew up in the Cold War, and the fear of nuclear attack is nostalgic. My generation simply lived with it. When I was a youth, my father served as a volunteer civil defense warden in our small hometown in central Illinois. He had a pistol locked away that had been issued through some sort of civil defense fund. It served two purposes: he was also a volunteer guard for the local bank, where he had the full-time paid job of the bank president. As crazy as all of this sounds, it did tie together because if a nuclear war came, the basement of the bank would become the "fallout shelter" for the local population. The basement held stores of government-issued food and water that, by the time I was a young boy, had not been upgraded in some years. As a youth, I would love to explore that basement with all the amazing artifacts of the earlier "atomic age." I recall seeing chemical toilets, canned food, medical supplies, and water cans.

Despite the fun, it proved a sobering experience even for a naïve youngster. The basement would have held about 50 of our town's population of 5,000 people. Even as a very small boy (circa 1965), I realized that to be quite a peculiar discrepancy. I recall then that my father would express concern about it when

---

[136]   https://www.news.com.au/national/nsw-act/gerald-holtom-created-a-nuclear-disarmament-symbol-that-became-the-global-sign-for- peace/news-story/11b20b81566fb3086dc3b89fede8c5bc.

discussing the subject because, in 1965, there was still some thought that somehow a nuclear war could be survived. Although, he clearly had no idea what his role would be if a nuclear air raid actually came, and apparently, no one had ever thought about it enough to tell him. It was simply a title that fulfilled an underfunded civil defense program, and it never really was rescinded as much as it was forgotten about with the passing of time. As far as I know, that basement still exists with its civil defense artifacts in place, or at least it did, to my knowledge, as late as 1987.

That year, my father's bank constructed a new building and vacated the old combination bank/civil defense shelter, which became a city hall. I recall asking him at the time if anyone would move the old emergency supplies and foodstuffs. His only comment by then was, "What is the point?"

Well, the point is that by that time, no one really thought you could survive a nuclear war. My father had by no means lost his patriotism. He was as patriotic as they came. Even the most loyal citizens by then appreciated the commonly accepted policy of MAD, or Mutually Assured Destruction. In short, nuclear war ceased to be practical or even thinkable because it came to equate the end of civilization. The arsenals of both America and the USSR had grown to numbers exceeding 70,000 combined devices, which, if used in an all-out war, would have clearly been a civilization-ending proposition. That, in fact, was what everyone hoped because the whole idea was, and still is, that it would have made war too costly to start.

Another defining moment in my memory of the Cold War/civil defense era concerned the November 20, 1983, ABC network airing of the film *The Day After*. The program reached 100 million people in 39 million households. I saw that film with my father and mother the night it debuted as we all ritually sat in front of our single television set. To this day, I recall the sobering reality that the program left on my local community and me. It did not matter if you came from a very conservative environment as I did. The film almost universally impressed people with the horrors of a nuclear war, and I recall that after that, I never heard anyone in my town even suggest there would be any way to prepare for such a scenario.

Public record image of ABC advertisement for *The Day After*.

I see the Cold War era as a strange paradox and recall the phrase former EG&G president Barney O'Keefe used. He came up with a very insightful title to his highly praised 1983 book on the Cold War era. He paraphrased it as a weapon that holds all sides as "Nuclear Hostages."[137]

Of course, times have changed, and there are no longer only two great sides with nuclear weapons. China is now building a modern nuclear arsenal. Russia and the US, with its NATO allies, still have more than fifteen hundred warheads available for immediate use. And what do we do now—now that we have

---

[137] Bernard J. O'Keefe, *Nuclear Hostages* (Boston: Houghton Mifflin Company, 1983), pp. 219-243.

increasing nuclear proliferation in other countries that do not necessarily play by the Cold War/old-school rules? Countries like India and Pakistan each have about 100 nuclear weapons aimed at each other, and the consensus is that they will not hesitate to use them if those highly ideologically and philosophically motivated foes feel significantly threatened.

Nor, if their rhetoric can be believed, does a country like North Korea appear to be deterred from the continued development of nuclear weapons. Kim Jong Un sees them as legitimizing his position as a world player, and what is more, the weapons are now part of his country's national identity.

The Russians, who have always had a keen insight into North Korea, say without exaggeration that the Korean people will eat grass before they give up their nuclear weapons. They are not talking about the elite, but the mass lower class populace who will not stand for their country giving up this advancement after so much of their own economic sacrifice.

And could North Korea or a new nuclear player sell its technology to a terrorist organization? That group would have little inhibition, having nothing tangible to lose if they ever detonated such a weapon. So, the rules are changing. In response to these scenarios, we must ask what the lesson of civil defense is. How do we answer the concerns of those who fear we may still need such a program?

There are simply so many variables and uncertainties as to what countries like North Korea or Russia could or could not do, and there is almost no way to plan. However, the broader aspects of defense, such as missile defense systems, are still a much more productive focus than a bomb shelter mentality. Certainly, there is an attractive nostalgia for that age, but there is no easy way to solve such a huge logistical problem, despite modern-day Russia and China continuing to fund an active civil defense program of their own.

That is actually a good way to understand the civil defense of past years. Little funding went into it because the focus shifted to bigger issues, such as regular defense spending. Indeed, the day may come when people realize that the best nuclear civil defense is to prevent the proliferation of such weapons in the first place. That is exactly where the modern-day global security specialists now focus. Another issue is the subject of nuclear deterrence in general. For better or worse, it has worked for many years, but is it still effective in preventing large peer-level nuclear powers from going to war? I hope that some of our following chapters can address that question.

CHAPTER FIFTEEN
UFOS/UAVS AND NUKES

From the article: *We flew Above the Flying Saucers* by 1st Officer William B. Nash and 2nd Officer William H. Fortenberry. That *True Magazine* feature from October 1952 documented the story of a Pan American airliner flying at 8,000 feet when its pilot and co-pilot saw a group of objects shaped like "flying saucers" in formation about 2,000 feet below them. After the airplane landed, the Air Force carefully interviewed the pilots. Their widely publicized account remains as one of the most credible sightings in history.

Since 1945, there has been concerning evidence of a connection between nuclear weapons and the centuries-old phenomenon of unidentified anomalous phenomena. During those years, the sightings of unidentified flying objects have not been taken very seriously by the mainstream media, even though behind the scenes, the military has periodically recognized the phenomenon as real. Today, the subject is losing its so-called ridicule factor as many leading congressional and senate members in a bipartisan fashion are looking at this topic with a serious eye. Former Senator Harry Reid, who was a supporter of the museum I worked at, was one of those individuals who focused attention on the subject. I met him at work many times. He sincerely felt the subject deserved serious attention. Primarily, he was concerned with the national security implications of the phenomenon as opposed to speculation about extraterrestrial life. Like military and government officials for over seventy years, he simply wanted to know if the phenomenon posed a threat.

That interest has existed since the late 1940s when the unknown objects were commonly called "flying saucers" and then "unidentified flying objects," or UFOs. Now, the term is unidentified anomalous/aerial phenomena or UAPs. A great deal of renewed and less critical media coverage comes from just the last two decades, which have witnessed rather dramatic sightings and encounters with unknown aerial phenomena by United States Naval aviators. In 2019, some of those reports were released in the *New York Times* and picked up by the international press.

A subsequent report requested by Congress and generated by the Office of the Director of National Intelligence (ODNI) on June 25, 2021, studied 144 of those recent unexplained incidents. Later, in July of 2023, a House Oversight Subcommittee held an unprecedented public hearing on UAPs, which renewed government attention on the subject and brought media attention from around the world. At this writing, Senate Majority Leader Charles E. Schumer is supporting a bill that, if passed, will call for more government transparency and a further review of records pertaining to the UAP subject in 2024.[138]

Amazingly, this series of events has happened before. At the dawn of the nuclear age, hundreds of reports were filed by United States military and civilian pilots which generated similar reports. As you will see, history, or at least government bureaucracy, repeats itself.

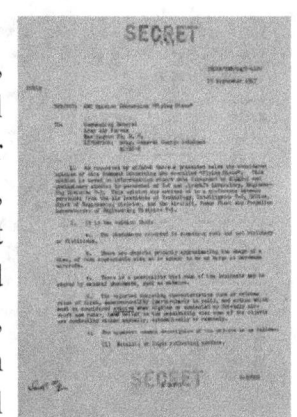

On September 23, 1947, General Nathan Twining, Commanding General of the Air Force Materiel Command and Air Technical Intelligence Center (ATIC) at Wright-Patterson Air Force Base, initiated a classified report or "memo" on what was then called "Flying Saucers."[139](Imaged right.) Seventy-four years later, on June 25, 2021, the ODNI released a very similar public report on what is now called "Unidentified Aerial Phenomena" (Imaged left.)

Both reports have very similar language and similar conclusions. The 2021 ODNI report states that: "Most of the UAP reported probably do represent physical objects given that a majority of UAP were registered across multiple sensors, to include radar, infrared, electro-optical, weapon seekers, and visual observation."[140] In 1947, the then strictly internal General Twining report stated that: "The phenomenon is something real and not visionary or fictitious."[141] It led to a 22-year series of Air Force sponsored

---

[138] Senate news service, July 14, 2023, reports: Majority Leader Chuck Schumer (D-NY) and Senator Mike Rounds (R-SD) are submitting an amendment to the National Defense Authorization Act which would mandate government records related to Unidentified Anomalous Phenomena (UAP) carry the presumption of disclosure. The Unidentified Anomalous Phenomena (UAP) Disclosure Act of 2023 is modeled on the President John F. Kennedy Assassination Records Collection Act of 1992. The report that came out in 2024 was a great disappointment because it provided only a superficial review.

[139] https://archive.org/details/twinning-memo.

[140] https://www.dni.gov/files/ODNI/documents/assessments/Prelimary-Assessment-UAP-20210625.pdf.

[141] [142] http://luforu.org/twining-schulgen-memo/; and The Twining(-Schulgen) Memo was a classified letter written by Lieutenant General Nathan Twining, head of the US Air Materiel Command (AMC) at Wright-Patterson Air Field, in response to a request from an Air Force general (A-2) to provide information on the recent spate of "flying saucer sightings" In the memo, Twining presented the considered opinion of his command concerning the so-called "Flying Discs" based on interrogation report

investigations under three separate phases and three different codenames which began with the Projects "Sign" and then "Grudge." They examined what were called for many years "Unidentified Flying Objects" or "UFOs" as coined by the last and third Air Force study known as "Blue Book." The 1947 Twining report initiated this when it called for further study: "Headquarters, Army Air Forces issue a directive assigning a priority, security classification and Code name for a detailed study of this matter to include the preparation of complete sets of all available and pertinent data which will then be made available to the Army, Navy, Atomic Energy Commission, JRDB, the Air Force Scientific Advisory Group, NACA, and the RAND and NEPA projects for comments and recommendations, with a preliminary report to be forwarded within 15 days of receipt of the data and a detailed report thereafter every 30 days as the investigation develops. A complete interchange of data should be affected."

The memo stated that:

The phenomenon reported is something real and not visionary or fictitious.

There are objects probably approximately the shape of a disc, of such appreciable size as to appear to be as large as man-made aircraft. There is a possibility that some of the incidents may be caused by natural phenomena, such as meteors.

The reported operating characteristics, such as extreme rates of climb, maneuverability (particularly in roll), and action that must be considered evasive when sighted or contacted by friendly aircraft and radar, lend belief to the possibility that some objects are controlled either manually, automatically, or remotely.

The apparent common description of the objects is as follows:

Metallic or light-reflecting surface.

Absence of trail, except in a few instances when the object apparently was operating under high-performance conditions. Circular or elliptical in shape, flat on bottom and domed on top.

Several reports of well-kept formation flights varying from three to nine objects.

Normally no associated sound, except in three instances, a substantial rumbling roar was noted. Level flight speeds normally above 300 knots are estimated.

It is possible within the present US knowledge – provided extensive detailed development is undertaken – to construct a piloted aircraft that has the general description of the object in sub-paragraph {illegible} above, which would be capable of an approximate range of 7000 miles at subsonic speeds.

Any developments in this country along the lines indicated would be extremely expensive, time-consuming, and at the considerable expense of current projects. Therefore, if directed, they should be set up independently of existing projects.

Due consideration must be given to the possibility that these objects are of domestic origin – the product

---

data furnished by AC/A5-2 and preliminary studies by personnel of T-2 and Aircraft Laboratory, Engineering Division T-3. The opinion was arrived at in a conference between personnel from the Air Institute of Technology, Intelligence T-2, Office, Chief of Engineering Division, and the Aircraft, Power Plant and Propeller Laboratories of Engineering Division T-3.

of some high-security project not known to AC/A5-2 or this Command.

The 2021 ODNI report also called for further study. It inspired former Senator Bill Nelson, NASA Administrator, to task the space agency to look into the science of UAP reports. NASA has completed its own report as of 2023. The Department of Defense is also formalizing a system to investigate military sightings. The 2021 ODNI report stated: *"In line with the provisions of Senate Report 116-233, accompanying the IAA for FY 2021, the UAPTF's long-term goal is to widen the scope of its work to include additional UAP events documented by a broader swath of USG personnel and technical systems in its analysis. As the dataset increases, the UAPTF's ability to employ data analytics to detect trends will also improve. The initial focus will be to employ artificial intelligence/machine learning algorithms to cluster and recognize similarities and patterns in features of the data points. As the database accumulates information from known aerial objects such as weather balloons, high-altitude or super-pressure balloons, and wildlife, machine learning can add efficiency by pre-assessing UAP reports to see if those records match similar events already in the database. The UAPTF has begun to develop interagency analytical and processing workflows to ensure both collection and analysis will be well informed and coordinated. The majority of UAP data is from US Navy reporting, but efforts are underway to standardize incident reporting across US military services and other government agencies to ensure all relevant data is captured with respect to particular incidents and any US activities that might be relevant. The UAPTF is currently working to acquire additional reporting, including from the US Air Force (USAF), and has begun receiving data from the Federal Aviation Administration (FAA). Although USAF data collection has been limited historically the USAF began a six-month pilot program in November 2020 to collect in the most likely areas to encounter UAP and is evaluating how to normalize future collection, reporting, and analysis across the entire Air Force. The FAA captures data related to UAP during the normal course of managing air traffic operations. The FAA generally ingests this data when pilots and other airspace users report unusual or unexpected events to the FAA's Air Traffic Organization. In addition, the FAA continuously monitors its systems for anomalies, generating additional information that may be of use to the UAPTF. The FAA is able to isolate data of interest to the UAPTF and make it available. The FAA has a robust and effective outreach program that can help the UAPTF reach members of the aviation community to highlight the importance of reporting UAP. Expand Collection: The UAPTF is looking for novel ways to increase collection of UAP cluster areas when US forces are not present as a way to baseline "standard" UAP activity and mitigate the collection bias in the dataset. One proposal is to use advanced algorithms to search historical data captured and stored by radars. The UAPTF also plans to update its current interagency UAP collection strategy in order bring to bear relevant collection platforms and methods from the DoD and the IC. Increase Investment in Research and Development."*[142]

---

[142] A UAP reporting system is being implemented by the DOD with an online form becoming available as of this writing in 2023.

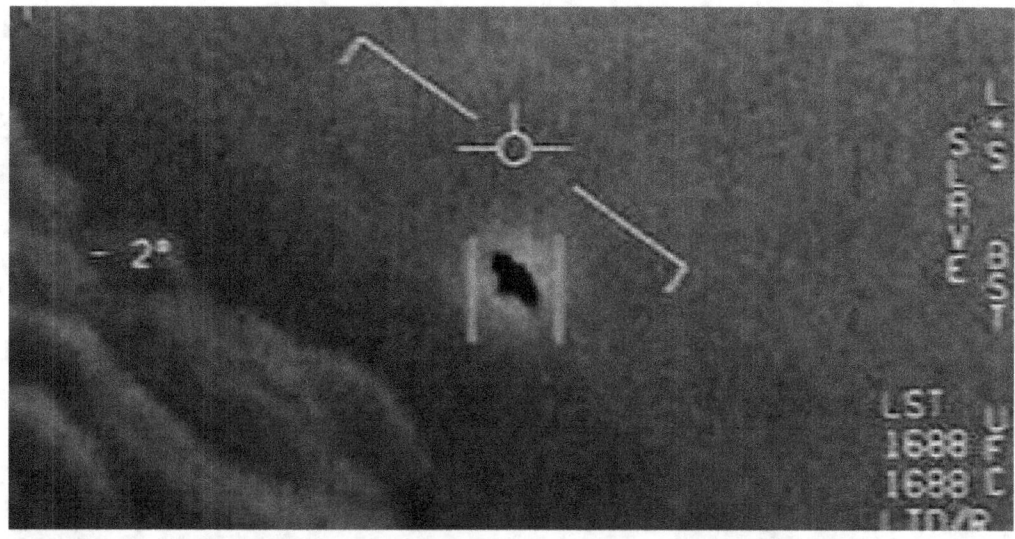

**United States Navy video image frame of a UAP taken
in 2015 near the Florida coast by a Navy fighter jet
from the *USS Theodore Roosevelt*.**

The similarities between the two assessments, now spaced seven decades apart, continue. Both reports attempt to define significant and disturbing characteristics of the phenomenon. The 2021 ODNI report stated that: "*And a Handful of UAP Appear to Demonstrate Advanced Technology. In 18 incidents, described in 21 reports, observers reported unusual UAP movement patterns or flight characteristics. Some UAP appeared to remain stationary in winds aloft, move against the wind, maneuver abruptly, or move at considerable speed, without discernable means of propulsion. In a small number of cases, military aircraft systems processed radio frequency (RF) energy associated with UAP sightings. The UAPTF holds a small amount of data that appear to show UAP demonstrating acceleration or a degree of signature management. Additional rigorous analysis are necessary by multiple teams or groups of technical experts to determine the nature and validity of these data. We are conducting further analysis to determine if breakthrough technologies were demonstrated.*"

In contrast, the ATIC Twining memo of 1947 stated: "*The reported operating characteristics such as extreme rates of climb, maneuverability (particularly in roll), and motion which must be considered evasive when sighted or contacted by friendly aircraft and radar, lend belief to the possibility that some of the objects are controlled either manually, automatically or remotely.*"

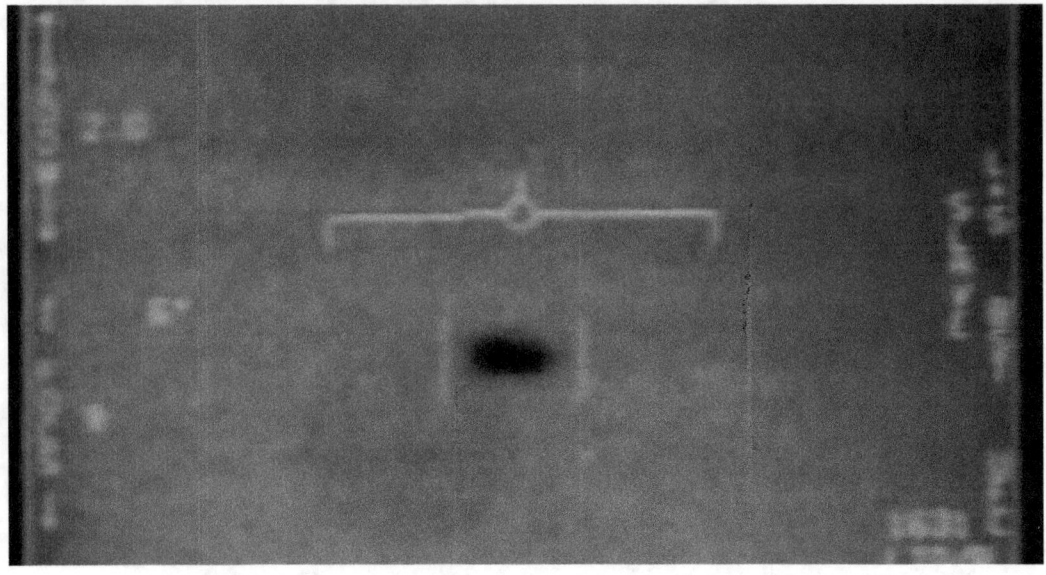

**United States Navy video frame of a UAP sighted in 2004
by a Navy fighter jet from the *USS Nimitz*.**

The 2021 study added: "After carefully considering this information, the UAPTF focused on reports that involved UAP largely witnessed firsthand by military aviators and that were collected from systems we considered to be reliable. These reports describe incidents that occurred between 2004 and 2021, with the majority coming in the last two years as the new reporting mechanism became better known to the military aviation community. We were able to identify one reported UAP with high confidence. In that case, we identified the object as a large, deflating balloon. The others remain unexplained. 144 reports originated from USG sources. Of these, 80 reports involved observation with multiple sensors. Most reports described UAP as objects that interrupted pre-planned training or other military activity. Although there was wide variability in the reports and the dataset is currently too limited to allow for detailed trend or pattern analysis, there was some clustering of UAP observations regarding shape, size, and, particularly, propulsion. UAP sightings also tended to cluster around US training and testing grounds, but we assess that this may result from a collection bias as a result of focused attention, greater numbers of latest-generation sensors operating in those areas, unit expectations, and guidance to report anomalies."

The Twining memo from 1947 similarly elaborated: "The apparent common description is as follows:(1) Metallic or light reflecting surface.(2) Absence of trail, except in a few instances where the object apparently was operating under high performance conditions.(3) Circular or elliptical in shape, flat on bottom and domed on top.(4) Several reports of well kept formation flights varying from three to nine objects.(5) Normally no associated sound, except in three instances a substantial rumbling roar was noted."

Everyone, then and now, want to know what these objects are or who are they. The Twining report stated that: "*Due consideration must be given, the following: - (1) The possibility that these objects are of domestic origin — the product of some high security project not known to AC/AS-2 or this Command. (2) The lack of physical evidence in the shape of crash recovered exhibits which would undeniably prove the existence of these subjects. (3) The possibility that some foreign nation has a form of propulsion possibly nuclear, which is outside of our domestic knowledge.*"

In contrast, the modern-day ODNI study stated in this regard*: "Airborne Clutter: These objects include* *birds, balloons, recreational unmanned aerial vehicles (UAV), or airborne debris like plastic bags that muddle a scene and affect an operator's ability to identify true targets, such as enemy aircraft. Natural Atmospheric Phenomena: Natural atmospheric phenomena includes ice crystals, moisture, and thermal fluctuations that may register on some infrared and radar systems. USG or Industry Developmental Programs: Some UAP observations could be attributable to developments and classified programs by US entities. We were unable to confirm, however, that these systems accounted for any of the UAP reports we collected. Foreign Adversary Systems: Some UAP may be technologies deployed by China, Russia, another nation, or a non-governmental entity. Other: Although most of the UAP described in our dataset probably remain unidentified due to limited data or challenges to collection processing or analysis, we may require additional scientific knowledge to successfully collect on, analyze and characterize some of them. We would group such objects in this category pending scientific advances that allowed us to better understand them. The UAPTF intends to focus additional analysis on the small number of cases where a UAP appeared to display unusual flight characteristics or signature management."*

What is most significant about the recent ODNI report is the conclusion that UAPs pose a potential threat to navigation: "Ongoing Airspace Concerns: When aviators encounter safety hazards, they are required to report these concerns. Depending on the location, volume, and behavior of hazards during incursions on ranges, pilots may cease their tests and/or training and land their aircraft, which has a deterrent effect on reporting. The UAPTF has 11 reports of documented instances in which pilots reported near misses with a UAP."[143]

Both reports do consider the possibilities the phenomenon could represent foreign peer-level adversaries. The June 2021 report stated: "Potential National Security Challenges We currently lack data to indicate any UAP are part of a foreign collection program or indicative of a major technological advancement by a potential adversary. We continue to monitor for evidence of such programs given the counter intelligence challenge they would pose, particularly as some UAP have been detected near military facilities or by aircraft carrying the USG's most advanced sensor systems."

And in contrast, in 1947 the (Army) Air Force stated: "Headquarters, Army Air Forces issue a directive assigning a priority, security classification and Code name for a detailed study of this matter to include the preparation of complete sets of all available and pertinent data which will then be made available to the Army, Navy, Atomic Energy Commission, JRDB, the Air Force Scientific Advisory Group, NACA, and the RAND and NEPA projects for comments and recommendations."

What neither of these landmark reports acknowledged then or now is that the phenomenon has been around not just since the beginning of the atomic age but for many years previously. News media around the world has periodically reported on sightings of unknown aerial craft for well over one hundred and fifty years. Artists and writers have recorded the phenomenon going all the way back to the Renaissance. History definitely repeats itself.

---

[143] A provision in the $2.3 trillion coronavirus relief and appropriations bill that former President Donald Trump signed in 2020 stated that "unidentified aerial phenomena" represented safety of flight issues and potential operational security issues. Parts of the report remained classified, NBC News, June 25, 2021, https://www.nbcnews.com/politics/politics-news/ufo-report-government-can-t-explain- 143-144-mysterious-flying-n1272390.

Since the early days of atomic bomb development and testing, there have been a number of notable sightings near Atomic Energy Commission installations. According to procedures, AEC security guards filed reports when unidentified aircraft or phenomena were sighted. These reports put the AEC in the middle of some of the early investigations.

Not all early reports described conventional UFOs. Numerous accounts from the 1940s and 1950s involved a phenomenon known as "Green Fireballs" which received considerable attention because they were spotted so often near AEC sites. These early UFO reports led to meetings with AEC officials.

One meeting in particular is of note because the National Archives has the transcriptions of the session's minutes. It occurred at Los Alamos on February 16, 1949, and involved Edward Teller and Robert J. Oppenheimer's successor, Norris Bradbury, who attended with other AEC scientists. Also attending was a noted astronomer of the region, Dr. Lincoln La Paz. The meeting hoped to utilize the expertise of the nuclear physicists. They had a productive discussion and reviewed the latest reports on the phenomenon, primarily the green fireball reports of the time. This meeting's records are in the National Archives under file number: NARA- PBB88-398 and NARA-PBB90-1027 through NARA-PBB90-1050 in document code group T1206-88.

Another interesting National Archive file relates sightings involving a key Lockheed aeronautical engineer, Kelly Johnson. He engineered the U-2 reconnaissance aircraft in a very remote area adjacent to the Nevada Test Site called Area 51. The story is notable to recount because Kelly Johnson, pictured right, served as a first-hand witness along with some of his key technical staff. Their noted sighting occurred during daylight on December 16, 1953, near Agoura, California.

Air Force Historical Association Kelly Johnson is known for his development of the U-2 and SR-71 spy planes.

At that time, Johnson observed a very large saucer-shaped high-flying aircraft from his sea-side Pacific home. By sheer coincidence, his top team of engineers was then aloft in a Lockheed Constellation just offshore while testing avionics equipment when they observed the same unknown object.[144]

Johnson reported this sighting to Air Force Intelligence, where it was reviewed by a good friend of his, Lieutenant General Donald Putt. Putt had a long-respected history in aeronautics and intelligence work. Putt, with others, had helped collect and disseminate much of the exotic, advanced aircraft and rocket technology brought back from Germany at the end of World War Two. His team also worked with American industry after the war to commercially utilize a lot of the advances made in Germany.

---

[144] Project Blue Book Files, National Archives, Record Group 341, Microfilm Publication No. T-1206, Roll No.20, Case 2837.

Declassified Pentagon memos from Putt indicated he knew every classified Air Force project then going on but did not know what Kelly Johnson and his technicians witnessed. Johnson's reflections of these sightings are carefully documented not only in Record Group 341 of the National Archive holdings in College Park, Maryland, but the National Archives Record Administration files at Maxwell Air Force base collection ID: MAXW-PBB19-1710.

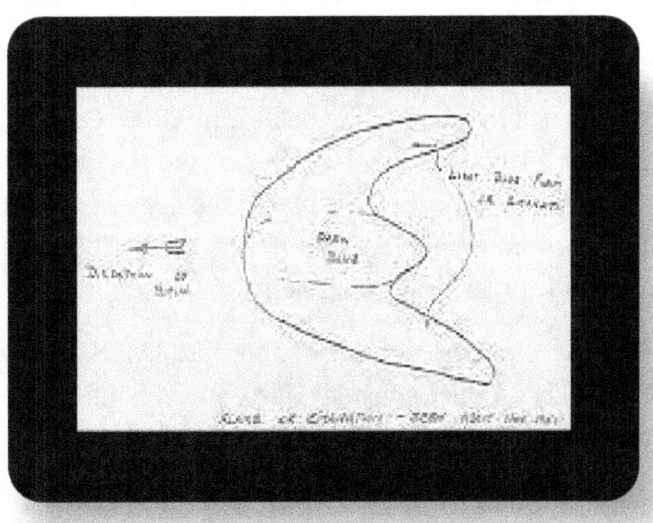

A sketch depicted above, which Johnson and his Lockheed engineers produced to document what they saw in 1953 for United States Intelligence agencies, looks amazingly like a modern-day B-2 Bomber. From that time on, Johnson considered himself a believer in the term "flying saucer," although there is no evidence to suggest he ever publicly postulated on the origin of the many credible sightings like his own.[145]

It should be noted from a proper historical perspective that various sources document that there was a period from approximately 1947 to 1953 when the subject we now refer to as "UFOs" was given considerable attention and reflection by various US Intelligence services. Many of the Air Force sighting reports went to a large variety of agencies, including the Atomic Energy Commission. This was not because the AEC had a specific interest in UFOs but because of a mandate from Vannevar Bush, who directed the post-World War Two Research and Development Board established by Harry Truman. Bush and his board simply thought the AEC was a logical resource for scientists to keep informed, and Air Force Intelligence occasionally consulted their opinions until about 1953. What almost everyone fails to realize is that the urgency surrounding those UFO reports in the early Cold War days centered not on visitors from space but on a true fear that they might represent advancements in Soviet aircraft technology. That theory was taken extremely seriously.[146] By January 1953, a special panel convened and was named after the panel's chairman, Dr. Howard P. Robertson. The Robertson Panel emerged from a recommendation to the

---

[145] Ibid

[146] Project Sign files, which are in part detailed in the National Archives Project Blue Book Files, National Archives Records Group 341, Microfilm Pub. No. T-1206, Roll No. 1-9; and History of Air Materiel Command Intelligence, T-2, Historical Study No. 228, Vol. 1, Prepared by Doris A. Canham, Historical Office Executive Secretariat, Air Materiel Command, Wright Patterson Air Force Base, August 1948, declassified 23 January 1990. (This document is courtesy of Rob Young, historian for the National Air Intelligence Center, and Ms. Jean August of the Materiel Air Command History Office,1999.)

Intelligence Advisory Committee in December of 1952 after a Central Intelligence Agency review of the US Air Force UFO files. The CIA changed the narrative to downplay the subject, supposedly out of fears of creating confusion between delineating possible UFO reports as opposed to possible Soviet incursions should a nuclear war ever come.[148]

Years later, from the 1960s through the 1980s, a different series of incidents transpired during the very peak of the nuclear arms race. These involved unidentified objects hovering directly over ICBM sites. Such incidents occurred in both the United States and the USSR. The incidents involved compelling evidence suggesting that electrical interfaces occurred within the nuclear missiles. One case in particular comes from Captain Robert Salas, a former US nuclear launch officer. Salas went public with his account and stated that in 1967 a UFO appeared at Malmstrom Air Force Base in Montana. At that moment, ten of the US's nuclear Minuteman missiles that Salas was overseeing moved into what is described as a "no-go" setting, meaning they could not be launched even if the order had been given.

Salas stressed that for a period, they lost control of the missiles. Salas confirmed similar incidents happened at nearby silo control bunkers. I once spoke to Harry Reid about such incidents, and he confirmed to me that such occurrences have happened. A June 2021 *BBC World News* article elaborates:

The Pentagon has been quietly gathering data since 2007 as part of the military's little-known Advanced Aerospace Threat Identification Program. Money for the programme came at the request of Nevada Senator Harry Reid,

. . . Other US military and intelligence officials have detailed the odd sightings, with some of the more credible reports coming from pilots who have personally observed UFOs near military weapons and training facilities from within their cockpits.

In March, Mr Trump's former director of national intelligence John Ratcliffe

- who previously oversaw all 18 US intelligence agencies - summarised the phenomena, telling Fox News: "Frankly, there are a lot more sightings than have been made public."

"We are talking about objects that have been seen by Navy or Air Force pilots, or have been picked up by satellite imagery, that frankly engage in actions that are difficult to explain."

"Movements that are hard to replicate, that we don't have the technology for or are travelling at speeds that exceed the sound barrier without a sonic boom." . . .

Former Senator Reid has defended the $22m (£16m) he had obtained for the original UFO programmed at the Pentagon, saying that other countries were also studying the issue.

"We know that China is doing it," he told Nevada Newsmakers in 2019. "We know that Russia, which is led by someone within the KGB, is doing it, too, so we better take a look at it, too."

Research done by the US Defence Department "showed that not two people, four people or six people or 20 people but hundreds and hundreds of people have seen these things, sometimes all at the same time," he argued.[147]

---

[147] *Pentagon won't rule out aliens in long-awaited UFO Report,* BBC World News, June 25, 2021,

The Advanced Aerospace Threat Identification Program, or AATIP, which Harry Reid initiated, was an unclassified but unpublicized investigatory contracted effort funded by $22 million over five years until the available appropriations were ended in 2012.

According to the Department of Defense, a similar military program looking into aerial phenomena called the Airborne Object Identification and Management Group, or AOIMSG, had already been formed under Pentagon control. However, in July of 2022, the Defense Secretary's office announced that AOIMSG would have its role expanded and was renamed the All-domain Anomaly Resolution Office or AARO. That is now the officially recognized office within the United States Office of the Secretary of Defense that investigates unidentified aerial phenomena.[148] Unfortunately, that office had not initially been granted the security level classifications to conduct records research on UAPs. AARO has announced a new website for information and as a way for the public to report sightings of "unidentified anomalous phenomena."[149] The new website can be reached at https://www.aaro.mil/. As the AARO website premiered, it featured a message of introduction by the agency's director:

Welcome to the website for the All-domain Anomaly Resolution Office (AARO). Our team of experts is leading the US government's efforts to address Unidentified Anomalous Phenomena (UAP) using a rigorous scientific framework and a data-driven approach. Since its establishment in July 2022, AARO has taken important steps to improve data collection, standardize reporting requirements, and mitigate the potential threats to safety and security posed by UAP. We look forward to using this site to regularly update the public about AARO's work and findings, and to provide a mechanism for UAP reporting. Thank you for visiting. - Dr. Sean Kirkpatrick, Director

The next passage is illuminating in that it shows the agency has expanded its interpretation of the phenomenon, which heretofore had been confined only to just aerial objects:

Unidentified Anomalous Phenomena (UAP) means (A) airborne objects that are not immediately identifiable; (B) transmedium objects or devices; (C) and submerged objects or devices that are not immediately identifiable and that display behavior or performance characteristics suggesting that the objects or devices may be related to the objects or devices described in subparagraph (A) or (B). (Per the NDAA FY23 Section 1673(d)(8)). The DoD considers Unidentified Anomalous Phenomena (UAP) as sources of anomalous detections in one or more domain (i.e., airborne, seaborne, spaceborne, and/or transmedium) that are not yet attributable to known actors and that demonstrate behaviors that are not readily understood by sensors or observers. "Anomalous detections" include but are not limited to phenomena that demonstrate apparent capabilities or material that exceed known performance envelopes. A UAP may consist of one or more unidentified anomalous objects and may persist over an extended period of time.

On April 19, 2023, the Senate Armed Services Committee, Subcommittee on Emerging Threats

---

https://www.bbc.com/news/world- us-canada-57559179.

[148] Canada has also launched a UAP program called the Sky Canada Project, under the Office of the Chief Science Advisor of Canada.

[149] https://www.space.com/pentagon-ufo-office-new-website-report-sighting.

and Capabilities, conducted an open hearing.[150] Little new information came out at the televised event, although a number of new anomalous videos of UAPs were discussed. A far more interesting development occurred on July 26, 2023, when the House Oversight Subcommittee, chaired by Tennessee Republican Congressman Tim Burchett, met to address the UAP subject. As stated at the beginning of this chapter, that hearing has illuminated the perception that the phenomenon is now being taken much more seriously and in a largely bi-partisan manner.

The subject title for the hearing read: Unidentified Anomalous Phenomena: Implications on National Security, Public Safety, and Government Transparency.[151] The subcommittee interviewed several noted witnesses to sightings and a highly placed intelligence officer by the name of David Grusch, who had served as a former National Reconnaissance officer. Grusch, along with a number of other so-called government insider whistleblowers, have made some amazing claims about defense establishment contractors and federal agencies who are said to be aware of important data and possibly even technology related to UAPs.

Committee members expressed concerns in that hearing over an inefficient process for military and commercial airline pilots to report UAP sightings. In its own fact-finding activities, House members encountered an intentional lack of cooperation from active military officials at high levels and Department of Defense bureaucracy. A clear lack of transparency on UAPs at a time when the number of sightings is increasing is a central focus of House members. The US Senate has similarly become concerned about a long history of failure in UAP transparency. Senate Majority Leader Charles E. Schumer, a Democrat from New York, and Senator Mike Rounds, a Republican from South Dakota, strongly supported legislation within the overall defense budget, calling for more government oversight and a further review of UAP records. They specifically wanted to see the disclosure of any records on "technologies of unknown origin and non-human intelligence."[152] Although there is generally bipartisan support, the actual degree of declassification will remain to be seen. A significant resistance from Defense Department interests and lobbyists are apparent. It seemed that in 2024, the bill would allow a limited declassification of additional UAP-related documents. As of early 2024, a released report that was supposed to detail a careful review of the history of the investigations into the phenomenon has proved to be widely disappointing and superficial.

I have had a very intense personal interest in the subject of aerial phenomena for over forty years. I wrote and published a number of serious articles and books on the subject. My favorite piece is a rather whimsical essay titled: *What Would Have Happened If the Martians Really Invaded Grover's Mill, New Jersey?* I am reprinting that 2006 article here in its entirety because I think it documents some surprising facts about the UFO/UAP issue as a whole:

---

[150] https://www.aaro.mil/.

[151] Committee on Oversight: https://oversight.house.gov/hearing/unidentified-anomalous-phenomena-implications-on-national- security-public-safety-and-government-transparency/.

[152]*News Nation* reporter Joe Khalil, December 7, 2023, https://www.newsnationnow.com/space/ufo/uap-ufo-panel-recovery-rules- cut/.

Public record image of *The War of the Worlds* radio broadcast which was written by Howard Koch and inspired by H. G. Wells' 1898 novel of the same name. Performed by *Mercury Theater On the Air* actors and directed and narrated by Orson Welles, Sunday, October 30, 1938, 8 pm to 9pm EST.

On October 30, 1938, Orson Welles shocked the world with a dramatic radio play based on the popular novel by H.G. Wells, *The War of the Worlds.* The show convinced many with its news-like broadcasting style that a Martian invasion had taken place at Grover's Mill, New Jersey. Of course, it was just a fictional story, but what would have happened if the Martians had really landed in Grover's Mill, New Jersey? Believe it or not, a young lieutenant in United States Air Force Intelligence was actually tasked with answering that question in the spring of 1952.

We do not know the conclusions he reached, but the thought process behind his assignment is very interesting to speculate about. UFO buffs are very familiar with the man involved in that assignment. He then headed the UFO investigative group called Project Blue Book for the first two and a half years of its eighteen-year tenure. That earlier period in UFO investigations differed somewhat from officialdom's lackluster attitude of later years. At the time, the Director of Air Force Intelligence in the Pentagon had been asking some critical and sincere questions about the phenomenon.

The New York Times.

NEW YORK, MONDAY, OCTOBER 31, 1938.

Radio Listeners in Panic, Taking War Drama as Fact

Many Flee Homes to Escape 'Gas Raid From Mars'—Phone Calls Swamp Police at Broadcast of Wells Fantasy

Maybe it was due to that increased "official" interest in UFOs in 1952 that Blue Book chief Edward J. Ruppelt was tasked with such a curious assignment. The odd job entailed none other than writing a report on the fallout caused by the famous radio drama. A few details are available to give us insight into his task. In fact, the only evidence at all of this assignment comes from a brief passage in Edward Ruppelt's unedited manuscript to what became a best-selling 1956 book titled *The Report on Unidentified Flying*

135

Objects.[153] I thank Professor Michael Swords of Western Michigan University for the rare access to that unedited version. In 2001, I had the pleasure of writing the first biography to be published on Edward Ruppelt. He had a notable Second World War service record. Thanks to the generous assistance provided by his surviving family members and his personal papers, I feel I gave a complete picture of his very interesting life.[154]

Edward Ruppelt traced the story back to Halloween eve of 1938 when the dramatic stage and radio artist Orson Welles narrated a CBS radio drama based on H.G. Wells' 1898 book, *The War of the Worlds*. Like the classic account of a Martian invasion, the radio play was a frightening success. Unfortunately for many East Coast listeners, it seemed so real that thousands flew into a "panicked frenzy." Well, not really.

The actual "panic" was exaggerated. Orson Welles and his producer, John Houseman, were being told even before the radio play had ended that bodies were littering local roadsides from the many accidents caused by the radio-play-induced panic. Welles may have thought his career was over at that moment, although in actuality, it was being made. There were some notable and almost humorous accounts of how people reacted to that realistic newscast-like radio play. The details have been examined in numerous scholarly articles over the years and even Ph.D. dissertations.[155] Yet, despite many sensational stories, no serious harm seems to have ever come to anyone from that event. Nevertheless, it made world headlines and is widely remembered to this day. As we will see, there were other radio plays based on *The War of the Worlds* in other countries in the years that followed, and those, in a few instances, did cause what is described as real panic.

The point of how many people reacted in an irrational manner or even believed the 1938 radio drama to be a real newscast is irrelevant. What is relevant is that it became a popular impression that the drama did cause a vast panic. That is the *perceived* legacy of this incident. That perception, after all, is why Ruppelt relates that the Air Force had tasked him with making a study of the repercussions of the radio play. In keeping with popular lore, military officials believed there had been a dangerous panic, and they wanted a study to find out how extensive it was.

Ruppelt does not relate the contents of the report he eventually filed with the Air Force Directorate of Intelligence. It probably was not all that of an extensive study, as his time was soon being overwhelmed by a wave of UFO sightings during the summer of 1952. His "report" probably filled less than two pages at most and might have resembled more of a hurried high school term paper. It does not even seem Ruppelt did his homework very well because he cites in his private notes that two suicides were caused by the radio play, for which there is no documentation.

As interesting (or amateurish) as his report may have been to read, the report itself is not important. The whole point of this article is that the Air Force would not have given Ruppelt such a task if they did not believe that there was something of substance that could actually recreate events akin to *The War of the*

---

[153] Edward Ruppelt's unedited manuscript to *The Report on Unidentified Flying Objects*, courtesy of Professor Swords.

[154] Michael Hall, Wendy Connors, *Captain Edward J. Ruppelt, Summer of the Saucers-1952* (Rose Press: Albuquerque, New Mexico, 2000).

[155] Herbert Strentz for his Ph.D. dissertation titled: *A Survey of Press Coverage of UFOs, 1947-1966*, —Northwestern University, 1970.

Worlds broadcast. Now, I am not talking about an invasion or anything that dramatic. However, it appears they were serious enough about the subject of UFOs to fear that panic could occur if the reported phenomena proved to be of an extraterrestrial nature. That sounds crazy to say today, yet it characterized how they were thinking.

Documentation has already been provided in many scholarly articles of recent years that there were high-ranking officials in the military in 1952 who believed in an extraterrestrial/off-world connection to some of the reported phenomena. I, in fact, interviewed a mid-level Air Force officer while researching my biography on Edward Ruppelt in 1999. He confirmed this perception.

I have never met anyone I had as much respect for as Colonel Nathan Rosengarten. He had a long and distinguished career, serving many years at Wright Patterson Army and then Air Force airfield bases. Colonel Rosengarten served in Air Technical Intelligence and attended a spirited Pentagon meeting in 1951 focusing on a troubling report from senior Air Force pilots who had encountered an extremely exotic-performing flying object or "flying saucer" over Sandy Hook, New Jersey. [156]

This case involved a T-33 jet trainer piloted by Lieutenant Wilbert S. Rogers, with Major Edward Ballard, Jr. in the rear seat. At 11:35 AM EST on September 10, 1951, those two pilots were flying northward at 20,000 feet over Point Pleasant, New Jersey—headed for Sandy Hook. At that moment, Rogers spotted off to his left a dull, silvery object passing far below on an opposing parallel course. It was southward bound in from the coastline peninsula of Sandy Hook and appeared to be about 12,000 feet below them. Rogers had been on a direct approach heading toward a landing into Mitchel Air Force Base, New York, but wanted Major Ballard to have a look at it.

Ballard, however, was on the radio, so Rogers turned slightly to the left to linger and waited for him to complete his radio communications. Forty-five seconds later, Ballard caught sight of it, and the UFO entered a descending arc-like turn that was about to cut under their flight path. At that moment, ground control heard the pilots' conversations via an open mike. As the object banked, it revealed a "discus-like" silhouette while continuing its turn. So Rogers kept turning left with it to keep it from going under his wing and thus out of view. While the object descended further, Rogers nosed his jet down, completing a hair-raising 360-degree, 3,000 feet descending maneuver to keep it in sight. Both Rogers and Ballard estimated the craft to be around 30 to 50 feet in diameter and perhaps moving as fast as 700 miles per hour, which was extremely fast for anything flying in 1951. By that point, the pilots knew that they were not chasing a balloon because it was not only banking left but was by then outpacing their jet, which Rogers had throttled up from 450 to 550 miles per hour. By that point, the object had completed a 90-degree turn and was heading away from the coastline, traveling out over the ocean in level flight near the speed of sound at around 5,000 feet. Rogers vainly attempted to parallel its course from his current altitude of 17,000 feet as the UFO continued to increase its speed out to sea, covering 35 miles during the short two-minute span of the sighting. [157]

Just prior to that incident, a young Fort Monmouth student Army Signal Corps radar operator, PFC

---

[156] Personal interview with Col. Rosengarten by Michael Hall at the USAF Museum, 13 March 1999.

[157] Project Blue Book Files, National Archives, Record Group 341, Microfilm Publication No. T-1206, Roll No. 8, Case 977. (Air Force records document that a meteorological balloon had been present in the vicinity at the time of the sighting.)

Eugent A. Clark, had picked up an unknown low-flying target moving faster than the automatic setting mode on his AN/MPG-1 radar set could plot. As a matter of coincidence, a number of visiting Army officers happened to be standing behind Clark at the time, witnessing the strange event unfold. They watched in amazement as, in just moments, the curious radar blip traversed the coastline at an estimated speed of at least 700 miles per hour. It was lost off the scope near the Sandy Hook coastal peninsula, not far south of New York City.[158]

Air Force Intelligence privately investigated the sighting, but word of the phenomenal encounter made its way to the press. Unfortunately for the lower ranks, the head of Air Force Intelligence, Major General Charles Cabell, heard about Lieutenant Rogers' and Major Ballard's encounter from the newspapers before hearing about it from his own intelligence channels. Needless to say, the General, a combat-forged veteran from decades of service, demanded answers and did so loudly.

A high-priority Air Force Intelligence meeting followed at the Pentagon, presided over by General Cabell. Colonel Rosengarten was flown to Washington from Dayton, Ohio's Wright Patterson Air Force Base, to that meeting to provide answers for the General. In my interview with Colonel Rosengarten, he confided in me the details forty-eight years previous. He recalled it being a very confidential discussion among Air Force Intelligence staff members. During that meeting, Rosengarten said he had a clear impression that General Cabell had become convinced his pilots were indeed seeing unexplainable objects which, as General Twining had concluded a few years earlier, were "real and not visionary."

He also had the clear impression that General Cabell believed that "flying saucers came from outer space."[159] The General, a highly distinguished veteran of the Second World War, did not consider UFO reports a joke, and he sent word down to his intelligence operations that they were to be taken seriously. He knew his pilots, and he believed them. He let his staff know that he was to be awakened even during the middle of the night if this matter needed his attention. He tasked Colonel Rosengarten to revamp a then stagnate UFO investigation, which was quartered at Wright Patterson, and get it back on track. Colonel Rosengarten played a part in assigning Edward Ruppelt to a new and reorganized effort, which became named Project Blue Book.[160]

Numerous sources document that other senior Air Force officials during that period believed in the extraterrestrial nature of UFOs. Ed Ruppelt clearly stated as much in his own book recalling those years. His private notes named names.[161] So, there is no disputing the point. Now, we may never know what led certain officers to that conclusion. Maybe they knew far more about the phenomenon than has been reported. Maybe they even had a warehouse full of crashed saucers. Or maybe they simply made a serious assessment of the many good sighting reports that were then compiled by the Blue Book office, which we see today in

---

[158] Ibid.

[159] Personal interview with Col. Rosengarten by Michael Hall at the USAF Museum, 13 March 1999. That statement was also confirmed in the personal papers of Edward Ruppelt.

[160] Edward Ruppelt's personal papers.

[161] For a detailed account of the people behind the scenes during those early UFO investigations, my biography of Edward J. Ruppelt details those times largely based on his personal papers: Michael Hall, Wendy Connors, *Captain Edward J. Ruppelt, Summer of the Saucers-1952* (Rose Press: Albuquerque, New Mexico, 2000).

the declassified files.

As someone who has spent a lot of time looking at the files created by the early Air Force UFO research projects Sign, Grudge, and then Blue Book, I can state that there are some amazing reports filed by highly credible observers detailing some highly incredible phenomena. The reports, in many cases, come from very experienced pilots, both military and civilian. It should also be noted that although a significant group of senior officials in the Air Force in 1952 took the subject seriously, there were also those in the Pentagon and Wright Patterson AFB who worked hard to derail or diminish the reports. And it wasn't that these people always disbelieved in the reports; rather, they just seemingly wanted to make them go away.

Thus, it is interesting to speculate why officials worried about a panic scenario. I am convinced from doing a number of interviews with Air Force veterans from the time that "panic" simply became a catchword. I do not know why, but the one common denominator of every source I interviewed always came back to the catchphrase "panic." That became the excuse sources would fall back on when asked why there was no more direct action publicly exhibited by the Air Force in most UFO investigations. I believe their perception was sincere, and I believe it originated from a 1950s-era unspoken policy that military officials quietly observed. In other words, I do not think the military or government ever had any answers to what UFOs were, but I do believe they concluded that they were a real phenomenon and that if they gave too much attention to that line of talk, it could unintentionally lead to a panic which had been thought to have happened in 1938.

Beyond that, as far as why the Air Force did not do anything about UFOs, there simply was nothing they could do because it was such an unpredictable and evidently non-threatening phenomenon that it became easier to ignore. Also, there were serious concerns that the subject could obstruct or confuse focus on a real enemy at hand, the Soviets. We know, in fact, that the CIA and Air Force convened selected scientists together in January of 1953 in the form of the already detailed "Robertson Panel."

Their report led to an unofficial but effective debunking and de-emphasizing of UFO reports, which prevailed through the Cold War period. There certainly were studies already available in the 1950s that suggested contact with extraterrestrial intelligence could challenge many social conventions. Several Ph.D. dissertations have used the *The War of the Worlds* broadcast as a basis for studying human society. Some of these were published by 1952. Today, the concept of extraterrestrial life is not even novel to a skeptic.

However, it is difficult with our 21st-century view of popular culture to consider a time when no extensive set of preconceptions existed about extraterrestrial life. Without a Steven Spielberg to help us dream, or a *Star Wars* trilogy and a thousand other such productions dating back to the late 1940s, we would not have the present-day mindset that we do. Yet, that is not to say there was not already some basis for the consideration of alien visitation prior to the 1938 broadcast or, for that matter, the first "flying saucer" sightings of 1947.

Some researchers have used the "panic" caused by Orson Welles' radio drama as a foretelling explanation for later UFO sightings. In other words, a belief has arisen that the radio drama planted a seed in the public's mind— a self-fulfilling prophecy for extraterrestrial visitation. Some have even speculated that the *Buck Rogers* and *Flash Gordon* radio serials of the 1930s could have impacted the American psyche in subsequent years. Perhaps popular science fiction was an issue Ruppelt addressed in his study. Since his youth, Ruppelt, an avid reader, should have known from his own background that the concept of extraterrestrial life had already been firmly ingrained in the public's mind long before that 1938 radio drama ever aired.

Flash Gordon's Trip to Mars, **1938 movie poster. Public record image of Flash Gordon pop culture, which combined with Buck Rodgers serials, were popular radio, television, and comic book productions into the 1950s.**

H.G. Wells's original 1898 book may have even been inspired by earlier influences. Percival Lowell is one such example. Lowell founded the Lowell Observatory near Flagstaff, Arizona. By 1894, Lowell believed he had seen signs through his huge telescope of canals on the Martian landscape, proving to him the existence of intelligent life there. Of course, Lowell himself had been inspired by Italian astronomer Giovanni Schiaparelli, who first observed what he thought were straight lines on Mars, which he called *"canali"* and were later misinterpreted for the word canals. Some scientists agreed, while others were skeptical. Yet, until the first Mars probes of the 1960s and 1970s showed just how lifeless the surface was, many people kept an open mind. To this day, there is a segment of people who still contemplate scenarios in which Mars may have at one time hosted life of some sort.

It would be nice to end this article with some sort of great revelation. I cannot do that because firm evidence for UFOs is not there. I can, however, give the reader a few interesting insights into *The War of the Worlds* drama that are often overlooked. For example, the radio broadcast was not a true retelling of H.G. Wells' book, *The War of the Worlds*. The radio play turned out to be something much more unique— only loosely inspired by the original tale. The radio play is the sole creation of Orson Welles' associate Howard Koch. Koch is now remembered as the famous co-screenwriter of Casablanca and later as a supposed communist sympathizer.[162] And he really did not use much of the late 19th century story at all. In fact, being inspired by radio news flashes that summer of the brewing Czechoslovakian crisis, Koch turned the tale into a simulated real-time equivalent of a modern-day CNN "breaking news" story. The real war in Europe was still a year away, but Americans were just as nervous about those radio news stories as we are today when we see breaking headlines about growing tensions caused by North Korea or Russia or others. Americans then feared they would soon be impacted by Hitler's madness.

---

[162] Screenplay of Casablanca was written by Julius J. & Philip G. Epstein and Howard Koch and based on the play *Everybody Comes to Rick's* by Murray Burnett and Joan Alison.

Koch moved the location of the Martian invasion from England, as it was set in the classic story, to a real area in New Jersey. Most scholarly articles on *The War of the Worlds* Broadcast incorrectly state that Grover's Mill is a fictional name. It is very real and still exists as an unincorporated community in New Jersey. Koch carefully studied road maps he purchased from a gas station during his daily commute to New York in order to interject the story with real localities. The implications were clear when CBS attorneys reviewed his final script, as all radio scripts were routinely subjected to before rehearsal.

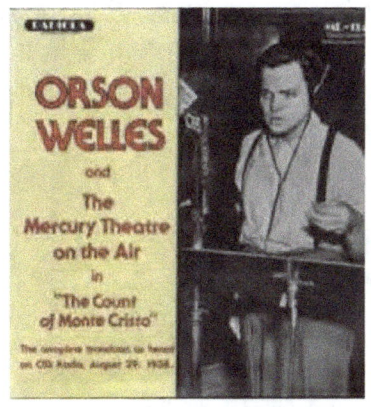

Radio plays like those presented on *The Mercury Theater* were a typical part of the golden age of radio.

Both Orson Welles and his producer, John Houseman, were firmly warned by CBS executives to make numerous changes. Koch's creation simply read too realistic. Welles and Houseman indeed significantly toned down the original script. The problem, however, remained because Koch was a brilliant writer and Welles and Housman were gifted dramatists, and the product they turned out could not help but be a work of dramatic art.

The play continues to this day to stand the test of time. All of the Mercury Theater broadcasts, which retold many classic tales, were, in fact, works of great performing art. The problem was that no one at the time appreciated that fact, and the show was in the process of being canceled. That all changed on the night of the famous broadcast. *The War of the Worlds* episode skyrocketed the weekly radio drama show into stratospheric ratings. Within a few years, Welles, Housman, and Koch, along with most of the Mercury Theater players, were off to Hollywood. Their careers had been made. They did indeed convince many Americans, for a short while, that aliens were real. Unfortunately, Welles' broadcast told us aliens were dangerous and destructive. Hopefully, this will not be the message if the aliens ever do land or have already landed.

So, what good is *The War of the Worlds* analogy? Maybe none. Yet, there is one final point that leaves a chilling message. Little known—there were two other similar radio dramas akin to what Welles did. Both radio dramas were based on the H.G. Wells story but told in the Orson Welles tradition. Those dramas did cause a real panic, and in both instances, people were hurt.

Just a few years later, a second *War of The Worlds* story reached the airways and descended on an unsuspecting listening public. At 9:30 pm, on November 12, 1944, a number of Chilean towns and cities were convulsed with panic when a radio station in Santiago staged their own localized version of *The War of the Worlds*. That script was written by an American named William Steele, who had worked in radio broadcasting and had scripted episodes of The Shadow, a famous radio show that Orson Welles had starred in for a time.

Using actual time and place, Steele and his assistant Paul Zenteno did exactly as Welles' writer Howard Koch had done. He plotted the conquest of Chile using familiar people and place names. The fictional landing site was some 15 miles south of Santiago in the town of Puente Alto. The same device of relaying the action as a series of news flashes was also employed. It met with identical devastating effects to the 1938 broadcast, according to a *Newsweek* report of the time. According to that November 27, 1944 issue, an electrician named Jose Villarroel, a resident of Valparaiso (70 miles northwest of Santiago) was so frightened that he died of a heart attack. Villarroel seems then to have earned the dubious honor of becoming the first person on Earth to be killed in an alien invasion, something that even Welles' Martians failed to do.

Public record image of early artwork created for the H. G. Wells' novel by illustrator Henrique Alvim Corrêa.[1]

The broadcast also hit home because it contained realistic references to organizations such as the Red Cross and actors impersonating well-known voices such as the Interior Minister. Welles did this himself in his 1938 broadcast. For example, he was warned not to use the personality of the President, so he instead inserted an address by a government secretary. However, the mischievous Welles allowed a Franklin Roosevelt impersonation for the voice of the government official. And this convinced many, to their utter astonishment, that FDR was speaking live during the radio broadcast.

In the Chilean broadcast, the Santiago Civic Center was realistically reported destroyed, as were air bases and army barracks. The play was broadcast countrywide on the Cooperative Vitalicia Network, and as the play fictitiously reported roads jammed with refugees—so too in reality, thousands of listeners apparently fled into the streets or barricaded themselves into their homes. It is even said that the governor of one province telegrammed the Minister of War to tell him that he had placed his troops and artillery on alert to repel the invaders.

The broadcasters had given a week's on-air notice of their intentions and mentioned the fictional nature of the broadcast twice during its proceedings, but of course the same blind misconceptions that had engulfed America in 1938 took hold. In an odd coincidence, a law had been passed only a year previously in Chile banning the use of incendiary radio broadcasts. Yet for many of those affected by the broadcast, the fines imposed on the station in no way alleviated their later embarrassment at being fooled by a fictional play.

The other panic came from the year 1949. The capital city of Ecuador, Quito, which was home to some 250,000 people at the time, hosted the radio event on February 12. By the end of that evening, the local newspaper offices would be burned to the ground as people became outraged by the mayhem caused by the realistic play. Many, once again, had been fooled into believing an alien invasion had begun. One group of people became so panic-stricken during the event that a local priest started taking open-air confessions with his parishioners' transgressions easily being heard by all. The sinners were certain they would not survive the night, so they poured out their darkest secrets. When daybreak came, and the world remained in still reasonable condition, their spouses and neighbors were not as forgiving as they had been the previous night.

The best story traces back to Orson Welles himself. Over his long and famous career, which included *Citizen Cane*, he occasionally spoke about the play that had so dramatically catapulted him to national attention. He, however, always downplayed the effect of the play as you would think any great actor would. Welles, after all, became one of the greatest talents of the 20th century. He knew he was good, and his career could have easily made itself even without *The War of The Worlds* drama. So, he justifiably could play it down. Yet, on one occasion, he was asked late in his life how he really could have been so naive to think his dramatic portrayal at the microphone that Halloween eve of 1938 could have been so innocent. How could he have thought that his performance would not shock the world and in the process, save his failing radio show and ignite his reputation? Welles looked thoughtful and then just smiled.

Public record image of Orson Welles who loved H.G. Wells' works, https://www.orsonwelles.org.

CHAPTER SIXTEEN
ALAS BABYLON BY JAMES HALL

**"Alas, alas, that great city Babylon, that
mighty city! for in one hour is thy judgment
come." Revelation 18:10**

As the last few years have dramatically impressed upon us, the old fears of nuclear war are once again with us. Since the start of the war in Ukraine, we have returned to a time of significant global tensions. The fighting has spread to the Middle East and may soon do so to Asia. These days of escalating nuclear rhetoric reminded some of us of one of the best nuclear war novels of the 20th century. This novel remains recognized at the top of its genre. It was written during the early days of the Cold War by Harry Hart Frank under the pen name Pat Frank and was published in 1959. It is certainly on par with the iconic nuclear war novel *On the Beach* by Nevil Shute.

Alas, Babylon deserves note because this sixty-five-year-old story is relevant to anyone living today. The book teaches a tale of the horrors inherent in nuclear war and serves as a literary work of art. The story takes place in a small, remote, racially divided south-central Florida town in the 1950s. The writing style is not unlike that of Tennessee Williams, and every sentence richly conveys the flavor of American life seven decades ago.

The reader enjoys a diversionary trip back in time as the book's first chapters unfold. This is where Frank uses his literary talent to paint an engrossing picture of the old south of central Florida. Then, just as the reader is engrossed in how different life felt in the 1950s, the main plot kicks in as a series of world events are detailed, which chillingly mirror our present time of world tensions.

The book's plot continues to introduce military and political characters who engage in rhetoric that is startlingly familiar to phrases we are now hearing daily on network news. As the novel descends into the holocaust of a nuclear exchange between the US and USSR, the event is simply described as "The Day." All fictional characters' perspectives become based on life before "The Day" and life after "The Day." The novel compares it to the Southerner's perspective that once existed in the 19th century, which divided life between antebellum and postbellum times.

When I consult a recommendation on a book, I do not want to know specific details about the characters or the plot, lest the joy of later exploration be ruined. So, I will not elaborate more nor detail the intriguing meaning of the biblical reference "Alas, Babylon." I will say that, as someone who grew up in Florida, the book provides an engaging trip to another place and another time that can be very well identified by anyone who knows the region. I can also say that this story is unique, unlike many apocalyptic novels in which characters lose their humanity in tandem with societal collapse.

Instead of the all too typical and mindless *Mad Max*-like scenario of so many current post-apocalyptic works of fiction, the *Alas, Babylon* narrative explores how the American character, with all its flaws, can rise to great occasion of "community" at the darkest times. Even in a traditionally segregated southland, humanity finds common ground after the holocaust. The lines of black and white fade and are replaced by those of a far different nature. A distinction with a much more practical nature arises. After "The Day," people are judged simply by those who can work with their hands and those who cannot.

The remarkable ingenuity of common people rebuilds the world in this novel as food, water, and neighborly cooperation become the only commodities of value. Thus, this is not a typical nonstop action thriller catering to the present generation's addiction to constant stimulation and violence. *Alas, Babylon* is filled with carefully orchestrated scenes of human interactions and, more often than not, quiet, reflective, sober dialogue. The novel's very message is sobering. In short, the story concludes that nuclear weapons make global conflict obsolete by making the repercussions of such a conflict unthinkable. The only problem is that the unthinkable did not deter rational leaders from making irrational decisions. In the end the proposition of nuclear deterrence failed. That is a lesson better learned in fiction than in fact, yet it is a message we all need to heed today!

CHAPTER SEVENTEEN
WHEN THE ATOM WAS COOL

**Public record atomic culture art circa
1950s.
PBS, https:// dialogue-and-atomic-culture-
the-bomb.**

Today, nuclear weapons are not exactly a feel-good topic, especially in light of the recently escalating tensions with Russia, China, and North Korea. Once upon a time in America, however, popular culture identified with everything "atomic." By the 1950s, after two decades of turmoil involving the Great Depression and World War, atomic energy allured us with the promise of control over our destiny. Science provided us with the hope of overcoming our past with the magic of the atom. Atomic culture, for a period, became an addictive facet of our national identity. Certainly, the atom and even the A-bomb used to be cool. Of course, our current 21st-century perspective of that time is a little clouded.

For example, in Las Vegas, Nevada, a great deal of interest arose around the above-ground or "atmospheric" nuclear tests going on just north of town. From 1953 to 1957, the remote desert city had vicariously translated those early atomic test detonations into showgirls adorned in atomic costumes, A-bomb-themed beauty contests, local hairstylists offering atomic mushroom-shaped hairdos, casinos with a variety of "atomic cocktails," the local high school yearbook featuring an atomic blast on its cover, and the list goes on.[163]

---

[163] National Atomic Testing Museum archives, Las Vegas, Nevada. I created much of the atomic culture content when I served at the NATM Executive Director from 2015-2022. Today the NATM is called the Atomic Museum

That same atomic testing also found appeal in science fiction novels and Hollywood movies. Yet, there was more to that fascinating atomic culture than just mushroom clouds. Many scholars feel the unique pop culture of the time represented not only an emerging interest in nuclear science but a belief that a "space age" had dawned. A brewing Cold War was also very much a component. It all translated into a nationwide culture craze, which was also reflected in the arts, which complemented a resurgent interest in "deco" styles that carried over from the 1920s and 1930s themes.

By the 1950s and '60s, the term became "Art Deco" as America wanted a sleek look that seemed to reflect the modern discoveries in atomic science. Art specialist Luke Honey has written about how atomic themes inspired mid-century decor and visual design:

Atomic themes inspired mid-century decor and visual design. Geometric atomic patterns were reproduced on textiles, countertops, wallpaper, amoebic coffeetables,futuristicpiecesoffurnitureand toys;   https://clickamericana.com/wp-content/uploads/Vintage-glassy-open-plan-   living-room-midcentury-1950s.jpg Atomic particles became a mainstay of visual design. Geometric atomic patterns were reproduced on textiles, countertops, wallpaper, amoebic coffee tables, futuristic pieces of furniture and toys.[164]

In the enthusiasm and seeming limitlessness of the post-war years, youth particularly marveled at the science of the atom. A favorite 1950s Christmas present included the Gilbert U-238 Atomic Energy Laboratory with a working cloud chamber and Geiger- Mueller counter.

Authors' collection, commercial atomic culture art circa 1950s, Gilbert chemistry set.

Another best seller included the "Atomic Energy Kit," complete with a tiny amount of uranium as well as a piece of paper embedded with a low concentration of radium. In 1957, when Russia launched the first successful orbiting satellite, Sputnik 1, the atomic age became the space age. That, however, meant that rockets and missiles were not just headed to the moon but became instruments that could carry atomic and then hydrogen weapons. Wisdom hoped a civil defense system could address such fears. It may have made people feel better, but civil defense in this country, as we have detailed, never amounted to much more than public

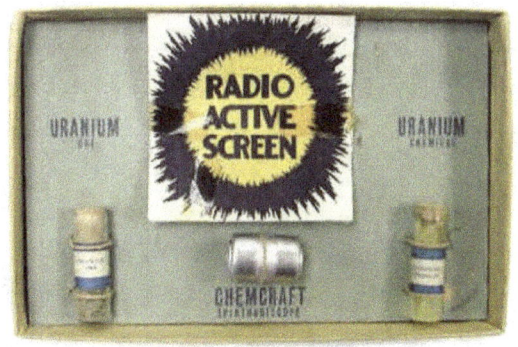

Authors' collection, "Atomic Energy Kit."

---

[164] Ibid

information as the public learned to duck and cover.

The nuclear tensions of the Cold War also led to notable works in literature. The best of these works, *Alas, Babylon*, has just been reviewed. It is a sobering and artful novel that is proving to be timeless. Fine art, in particular, found unprecedented inspiration within the first nuclear blasts. As seen in chapter seven, that was evident via noted California impressionist Arthur Beaumont, who depicted the rich colors inherent in atomic clouds. Few areas of Americana were left untouched by early atomic culture when the atom was still cool.

Excluding the influence of the John Hersey 1946 New Yorker article, we forget that there was a time from 1945 to 1949 when few worried about the scary side of atomic energy. There were those who had regrets about the human toll of Hiroshima and Nagasaki; however, few then probably imagined the United States could ever be threatened by such a catastrophe. That would change after August 29, 1949, when the USSR performed their first atomic test codenamed RDS-1 and First Lightning, which we called Joe-1. Conducted at the Semipalatinsk Test Site in Kazakhstan, it closely resembled our first atomic plutonium bomb test, known as Trinity, in the New Mexico desert. Prior to that, we had a monopoly on atomic energy, and most authorities thought it would be decades before the Soviets could control nuclear fission. They were wrong. What they did not know was that the Soviets had recruited willing spies in our own Manhattan Project. On top of that, they had some very competent physicists of their own who could unlock nature's secrets just as we had done.[165]

Americans began envisioning the new "atomic age" as a Pax Americana using fission for peaceful purposes, assuring the nation's safety and defense. In 1946, the Atomic Energy Act proposed such a mission. It was not altogether unreasoned. President Truman is complimented to this day for taking control of the atomic bomb away from military jurisdiction following WWII and putting it under a civilian organization controlled by elected officials, not generals. That was the purpose of the original Atomic Energy Commission. Despite a long controversial tenure, the modern-day successor, the Department of Energy, still keeps nuclear-related technology under the oversight of civilian authority and is not solely answerable to the military.

That early rather naïve part of the atomic age of the late 1940s created an optimistic feel that carried over into decorative arts and even architecture. The American home reflected the hope of prosperity, inspired by the splitting of the atom and nuclear blasts. Many homes boasted wall clocks designed by "Modernist" founder George Nelson. He often themed his works in starburst designs, which many thought symbolized atomic energy.

---

[165] Reality sank in by the end of the 1940s when Soviet Russia became a nuclear power as the Cold War evolved. It had taken a while for the post-World War Two leaders to absorb that fact and America's loss of the nuclear monopoly. President Truman, in fact, initially boasted after the war that the Russians could never be capable of making an atomic bomb, claiming they could "not even make a jeep." Such statements were not just arrogance. The great costs of the Manhattan Project, totaling two billion dollars, and the tremendous logistical challenges of making a nuclear industry did not seem to be matchable by any nation in the mid-20th century. Or, at least, it seemed so to those who lived through such immense efforts to make America's atomic bomb.

**George Nelson Starburst wall clock,
public record image.**

Authors' collection.

Wallpaper became extremely popular for post-war housewives of the new atomic era and its spawning housing developments in the ever-extending suburbs. Starburst designs reflect the sensation of bursting energy proliferating within the medium. If anyone still has these papers in their home, as seen above, they constitute a pop culture collector's dream.

In architecture, the emerging age of the atom helped influence the "Googie style." The origin of that name dates from 1949, when architect John Lautner designed a Hollywood coffee shop named *Googie's*. Based in Southern California, Googie architecture mirrored the "atomic age," which was equally part of the associated "space age" as atomic rockets promised to take us to the stars. America rode a post-war high, and the future looked promising. Googie used steel, glass, and neon to envision a whole new world. That world would be both affordable and efficient and, of course, stylish.

The Las Vegas welcome sign is a perfect example of the evolving Googie style as well as atomic artistic themes. Journalists felt it a metaphor for the mushroom cloud itself, many of which were often seen from the city's casino rooftops.[166] Las Vegas in those days epitomized an "atomic city." Early Las Vegas casinos had that "over the moon" feeling that only post-war art could provide in the magical desert atmosphere of that atomic age.

All forms of art exemplified the new aerodynamic sensations of rockets, jets, and the splitting of the atom. That craze had actually started in the 1930s with the *Buck Rogers* and *Flash Gordon* serials, which popularized new visionary words like "rockets" and "atomic bombs" long before they became 1950s icons. So, to properly understand the popular cultural influences of the 1950s and 1960s, you first have to understand the 1920s and 1930s. This atomic-inspired pop culture story truly became a progression of themes that spanned the rapidly changing 20th century. Googie, for example, evolved in a sense with the "Streamline Moderne" styles emerging out of the 1920s, whose popularity lasted until the 1970s.

Streamline Moderne's lineage can be traced back to the *Exposition Internationale des Arts Decoratifs et Industriels Modernes* held in Paris in 1925, where a unique style was first exhibited. That initial stylish influence was more commonly referred to as just "Deco" and then "Art Deco" much later and after the fact when named in 1968 by art historian Bevis Hillier. Art Deco is often characterized by sharp-edged geometric styles. It also resembled the ancient straight lines of Egyptian art. Art Deco had a cleaner, more "modern" feel than the elaborate pomp of the older Victorian and French styles. It, however, initially emerged with complex designs of its own in architecture, jewelry, fashions, and decorative arts. It only then gradually became more and more restrained and austere. The style even carried over into fine arts and women's dress with clean, sensual lines, as exemplified in this public record image of a Tamara de Lempicka painting.

Authors' collection.

In pre-war days, Art Deco became "Art Moderne," or modern art of the time. It was the first cool symbol of the advanced industrial era that promised a brighter and happier future, which only science could provide. As the style moved from Europe to America with a sleeker, less complex look, the term became "Streamline Moderne" and later termed "Art Deco in Motion." Where Art Deco was moving and emotional, Streamlining was practical and uniform. To visualize this style, you can recall the dramatic industrial arts involving the

---

[166] Many of the early above-ground or atmospheric tests were detonated in predawn hours because engineers and EG&G photographers wanted to utilize a black sky to get an optimum picture for diagnostic purposes. Casino rooftop parties were held to view the explosions, which for a brief moment turned night into the light of day.

streamlining of locomotives and steamships in the 1930s that were inspired by the sleek aerodynamics of advancing aviation. Even though very objective and austere, this style could still convey great feelings and awe.

Public record image of early New York Central streamlined locomotive, circa 1936 below, and 1939 above.

The Chrysler Building, completed in 1930, symbolizes the pre-war movement. The simple, clean lines of these styles were conducive to the austere economy of the Depression era of the 1930s as well as the post-war efficiency required in the 1950s. That general efficiency is also traced back to a German association of craftsmen called the *Deutscher Werkbund,* which led to *The Bauhaus* design school founded by German architect Walter Gropiusand and the resulting Bauhaus movement, 1907 to 1935. Most forget it was more practical than artistic, emphasizing functionality yet still with a distinct elegance in engaging simplicity.

It is important to understand the many movements in order to appreciate the significance of what came in the 1950s and 1960s. Trending movements in the 1920s and even the Depression era in the 1930s led to a consumer spree over advancing technology and a fascination with sheer speed. That was interrupted by the Second World War.

Yet, by the 1950s, cultural arts resurged and became particularly reflective in the car industry. The by then "jet-age" feel became as cool as the atom with aerodynamic streamlining, and names as outrageous as *Firebird*, *Sting Ray*, *Golden Rocket*, *Romeo*, *Wildcat*, and *Corvette*.

By the 1950s and 1960s, the latest advances in atomic energy and satellites inspired a marvelous resurgent Art Deco-like artistic culture embodying many flavors. Among those flavors included Googie architecture with Streamline Moderne roots and other Art Deco derivatives. Googie and Art Deco represented modernism turned into fashion. It touched all decorative crafts and fine arts.

The great optimism of the post-WWII atomic age is also remembered and epitomized in "Populuxe Design." Populuxe was best described as an American consumer and aesthetic movement of the 1950s and 1960s. It utilized plastics and metals to produce affordable and fashionable consumer goods. Populuxe represents an offshoot of Fordism (mass production and mass consumer consumption) of the early 20th century. In other words, just as Googie overlapped with Streamline.

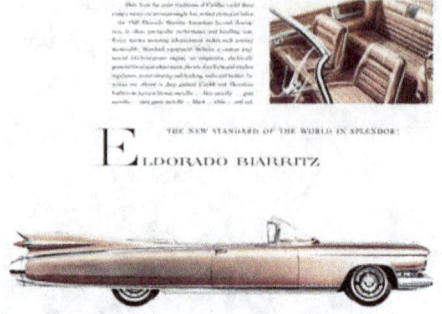

Authors' collection.

Moderne, which overlapped with Art Deco, Populuxe overlapped with all the futurist and modernist cultural trends. The house and garden magazines of the day exemplified the fun feel of Populuxe. Many of that simplistically shaped, colorful, and not always-so-comfortable furniture, which baby boomers remember from their childhoods, are significant collector's items today. Populuxe influence can also be seen in many industrial artifac as well, like the Cadillac *Eldorado* models of that time.

Almost everything in the 1950s and 60s encompassed the sense that the atom and space exploration were bringing a new era. The popular and stylish cartoon *The Jetsons* epitomized the fact that all ages were looking toward space and the atom as the road to the future. We can certainly see now how our technological achievements imprinted those artistic styles.

Authors' collection.

The most important thing to remember is that those icons of days past, including Googie architecture, were by no means just mirroring the rich or new "jet-set" that Deco had catered to in the 1920s and 1930s. Art embodied everyone in the popular post-war culture. That generation, which had grown up in the Great Depression, fought and won the Second World War and now wanted the fruits of its labor and sacrifice. They utilized technology like atomic energy to overcome challenges. Their struggle and their success were reflected by the arts. The performing arts, in particular, would reflect the atomic age when the atom was cool.

CHAPTER EIGHTEEN
ATOMIC TELEVISION

**1951 Motorola television advertisement, authors'
collection.**

The "atomic age" certainly proliferated in the cinema of the 1940s to 1960s. The young medium of television also captured the feel of that age. Television dates back to the 1930s in Germany, Britain, and America. The Second World War interrupted its commercial development. By the late 1940s, however, more and more American and United Kingdom homes were eager to explore the new invention that had captured everyone's interest before the war changed the world.

Television took almost a decade to gain popularity, but the 1950s has ever since been known as "the golden age" of television."[167] By 1960, almost 90 percent of Americans had access to a television. My favorite examples of some very cool atomic-inspired television shows date back to that time. Many are familiar with Rod Serling's *Twilight Zone* six-season series in the 1960s. That show often dealt with nuclear themes in very insightful and unique ways. Although, few recall that Rod Serling was a prolific screenwriter for many earlier 1950s television dramas.

---

[167] The first and classic "golden age of television" is an era in America reminiscent of live television dramas. It is generally accepted that the golden age of television began in 1947 with the first episode of the *Kraft Television Theater* drama series and ended in 1960 with the final episode of *Playhouse 90*.

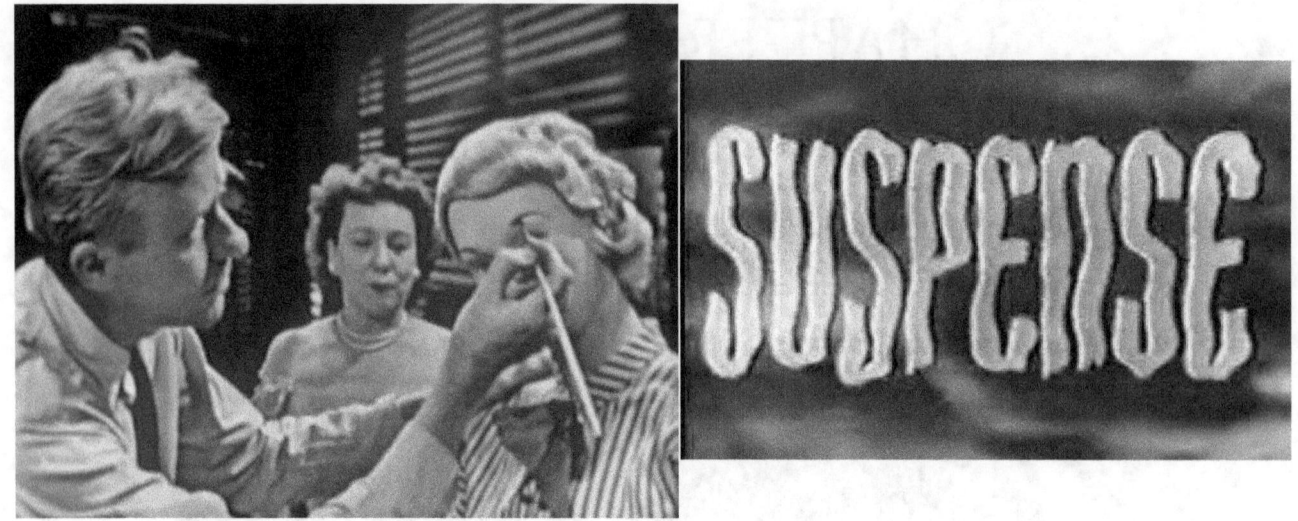

Public record images of "Nightmare at Ground Zero."

One such classic comes from the late 1940s and early 1950s television series called *Suspense*. Season 5, episode 43 of the TV show *Suspense* aired on August 18, 1953, titled "Nightmare at Ground Zero." Staring O.Z. Whitehead, this Rod Serling story unfolded at the then-new and novel Nevada Proving Ground, later called the Nevada Test Site. The United States government tested one hundred nuclear devices in above-ground atomic explosions just 65 miles north of Las Vegas, Nevada during that early period from 1951 to 1963.

The plot for that episode of *Suspense* revolved around an artist whom the government had hired to make mannequins for atomic tests. True to form, the character of this Rod Serling thriller develops a novel idea of how to murder his wife. He decides to drug her and then take her out to a nuclear test with his collection of mannequins destined for an atomic blast in the Nevada desert.

Mannequins were used in actual practice at the early nuclear tests; however, they were not commissioned by artists or used to simulate blast effects on the human body. Rather, they were borrowed from local department stores, and if they survived the blasts undamaged, they were often returned.[168] Their main purpose served to test the new textiles of the time, like rayon and nylon, to determine how they reacted to heat at various distances from the detonations.

This tale from the 1953 *Suspense* episode is like most Serling stories and has a characteristically unique twist to its ending. I will not spoil it for you because this is a classic adventure into 1950s atomic culture that should be pursued and enjoyed.[169]

Another Rod Serling hit from the early days of atomic-themed television drama comes from one of my favorite Golden Age series, *Playhouse 90*. *Playhouse 90* premiered on October 4, 1956, with a story directed by John Frankenheimer and based on Rod Serling's adaptation of Pat Frank's Cold War thriller *Forbidden*

---

[168] In those days the government was much more cost conscious and economical than today.

[169] Public record viewing of this episode can be seen at the following link: https://www.youtube.com/watch?v=-Yi-ODEDnjk.

Area. This first of many *Playhouse 90* episodes truly represents that golden age of television, with superb script writing and all-star casts.

Forbidden Area starred Charlton Heston, Tab Hunter, Diana Lynn, Vincent Price, Charles Bickford, and Jackie Coogan. Jack Palance hosted. The story revolves around an Air Force Colonel, played by Charlton Heston, who sits on a high-level Pentagon advisory board chaired by a defense specialist portrayed by Vincent Price. Heston uncovers circumstantial evidence, based on sabotage attempts, that the Soviet Union will soon launch a surprise nuclear strike against the United States. To his frustration, he becomes unable to convince anyone else of his theory. Again, this story is given that special Rod Serling gift for plot twists, so no spoiler will be given here at its ending.

Public record images from
***Forbidden Area* starred Charlton Heston and Vincent Price.**

One of the most well-known atomic-themed television plays comes from yet another Rod Serling-inspired production. *The Twilight Zone* season three, episode three, aired on September 29, 1961, titled "The Shelter." It had a notable cast, including Larry Gates, Joseph Bernard, and Jack Albertson. The plot involves a typical scene from the 1950s and 1960s in which a respected suburban family is hosting an evening birthday dinner party. The main character and birthday man of honor is named Dr. Bill Stockton and is played by Larry Gates. At the party, he comments to his close friends and neighborhood guests about his recent proactive ingenuity in the construction of a small fallout shelter in his basement.

CHAPTER NINETEEN
MORE SOBERING STORIES BY JAMES HALL

One more television production and a related book-based storyline are of note. Most of you will be familiar with the TNT television series *The Last Ship*. However, very few realize the inspiration for that popular 21st-century television show came from a 1988 book by William Brinkley, dealing not with a pandemic but a Cold War-era scenario of global nuclear war. The only similarity in the two plots concerns the central character, a United States Navy destroyer captain known only as Thomas.

The Last Ship is recognized as noted author William Brinkley's best work, perhaps because he served as a United States Naval officer himself during the Second World War. His knowledge of the sea and the brotherhood of sailors makes this book one of the best sea-going tales ever read. Brinkley writes in the first person, which puts the reader in the very shoes of this fictitious destroyer captain. That character, which comes alive in the pages of this book, not only plays a part in starting a nuclear war but finds himself and his crew among the last survivors of that holocaust. The plot has a unique twist, and any description here would spoil the story. Although seek this book out and you will find it an outstanding work of fiction with an almost too realistic feel for accuracy. Truly a story in the spirit of the Homeric Greek hero Odysseus, this tale is of classical proportions. It is a story that tells the reader what it must have been like for the ancients to look into the mythical Gates of Acheron and how that would feel to us in a real apocalyptical sea-going drama involving nuclear war. The quality of the writing is excellent and reminiscent of a former era that reads more like Herman Melville than a book about the nuclear age.[170]

**Public record image
Admiral Stavridis.**

Another must-read book for anyone trying to understand a world whose culture involves nuclear weapons is the insightful novel *2034*. This futuristic story visualizes what a limited nuclear exchange between China and the United States could look like. Co-written by Elliot Ackerman and Admiral James Stavridis, it is a profound story inspired by Admiral Stavridis' real-world experience.

Stavridis is a United States Naval Academy graduate, pictured left, who rose to the rank of NATO Supreme Commander Europe from 2009 to 2013. Most recently, he served as a vice chair of the Carlyle Group. His recent *New York Times* bestseller is far more than just fiction. This book poses lessons for the future that leaders should consider now.

Admiral Stavridis' book warns of the dangers of cyber warfare and the United States' over-dependency on advanced technology. Without giving the plot away, this book foresees a future Pearl Harbor-like event that is due not to a lack of vigilance but overreaction—specifically, an overconfidence in computerized systems.

---

[170] William Brinkley, *The Last Ship*, (New York: Penguin Books, 1988).

THE SWORD OF DAMOCLES OUR NUCLEAR AGE

When Naval officers can no longer navigate a ship without the ability of computerized navigation aids, it is time to go back to fundamentals. When pilots cannot keep their aircraft aloft without on-board artificial intelligence systems, something very basic is lacking. When a soldier in the field cannot aim standard artillery without the aid of drones or satellite GPS links, basic training in math needs to be reviewed.

This work of fiction theorizes that the United States and its allies are assuming that its advanced weaponry will overwhelm any potential enemy in a future conflict. They see this as a given, just as Western superiority has proved to prevail in past decades of fighting numerous third-world powers. What this assumption does not count on is that in a potential future conflict, an opposing force might be able to not only disable our military with cyber warfare but also make our advanced technology obsolete in a matter of seconds. This could happen abruptly as our communication systems, power grids, satellite communications, and modern infrastructure fail in a cascading domino effect. This failure could occur not in days or weeks but in microseconds. If that occurs, we may not have the skills of previous generations of warriors—before everything depended on a computer screen and a data connection. In a sense, the war may be over as soon as it begins if our home front economy had to function without internet, let alone electricity, for a period of months, if not years.

Admiral Stavridis speculates that in the future, we will be facing peer-level adversaries, not terrorists in the desert. Advanced technology could be used to disable its very self-advancement. In such a case, we could find our high tech military incapacitated.

As a result, out of desperation, we may be forced into using nuclear weapons against the enemy's cities. We take out one Chinese city in exchange for the defeat of our fleet. They then take out two American cities. In return, we take out three Chinese cities, and escalation continues. Russia takes advantage of various vacuums and creates turmoil. Meanwhile, we realize the devastating consequences of losing a few key harbors that handle over eighty percent of all container ships. We learn that the destruction of just twenty key cities containing critical infrastructure vital to the whole of North America can prevent our ability to function as a nation.

When the world's three superpowers become locked in an unstoppable global conflagration, who comes to their rescue? Maybe the third-world powers who have been our adversaries these many years? Like a child rebelling against irresponsible and self-destructive parents, they would have no choice because we all share the same endangered planet.

Of course, this is all fiction, but *2034* will definitely make you think in new ways and hopefully give our current and future leaders a serious precautionary tale. In an insightful afterward on the *Audible* version of *2034*, Admiral Stavridis steps out of fiction and cites lessons of history that inspired his work and career. He vocalizes his fears that we may have never learned the lessons of the First World War when nations literally stumbled into a prolonged, senseless war with no clear understanding of how it escalated so quickly.

He states, in contrast, that nations are often under the mistaken illusion that they have learned from past wars. After World War II, many leaders believed that the more economically interconnected nations became, the less likely they would fight one another in a future war. It is often argued that China and America could never go to war because the US and the entire world are now so economically linked with that ancient nation. The dependence the Western world has on Asian-rim computer chips is a prime example. Thus, the concept of a war with China is simply unthinkable. History, of course, proves this assumption false. The historical analogies are many and sobering.

In 1914, Kaiser Wilhelm II stood before a German nation poised to become the world's most productive country. Their industrial capacity was set to outstrip Britain and all other countries. At the time, it exported a huge variety of goods from all over the world. Germany's Krupp factories provided much of the high-grade steel for American railroads. Only Krupp could master a truly seamless railroad wheel. North American firms would order concrete all the way from Germany simply because it was so superior. Germany even produced cheap tin toys that proliferated in almost every country, bringing huge revenues back into the Second Reich. They led in chemistry and synthetic fertilizers, which fed the world. (Later, those same German geniuses would turn their talents to gas warfare.) Most importantly, Germany boasted one of history's largest merchant marine fleets. That growing economy promised Germany the realization of long-dreamt social programs.

Yet, in a day's time, with one irrational act, the Kaiser gave up all that prosperity and security in order to mobilize and join the escalating war in Europe. This included the knowing and inevitable loss of his world-class merchant fleet, which was largely on the high seas at the start of the war and subsequently seized in foreign ports. A global network of overseas bases critical to its trade was also lost quickly and forever. Germany's destruction followed, not once but twice in a century. Until recently, Germany and even the victorious nations of Europe were still paying off the debts of two world wars, and the rebuilding continued.

In 1941, Germany made a similar mistake when Hitler was being generously supplied with almost every vital material needed via a nonaggression pact with Stalin. Hitler discarded that economic dependency to the wind when he struck against the Soviets. During the same year, Japan depended on United States trade even after a series of oil embargoes. Logic dictated that Japan could not break its economic ties, but they did just the same with their bombs aimed at Pearl Harbor. When the Russian soldiers stormed into Germany in 1945, they could not believe such a rich country in terms of paved roads and household furnishing would have needed to invade such a poor country as their own USSR. Of course, none of these historical examples make any sense, but it has happened time and time again, going all the way back to ancient Greece. Prosperous civilizations are not immune to willingly shooting themselves in their own foot. It happens over and over and over. Most disturbing of all is the realization that on occasion rational leaders do make irrational decisions. That fact is true of any human being. It is so unsettling because in a world of nuclear deterrence, history has proved such deterrence works, but only as long as rational leaders hold power and consistently make rational decisions. Rational decisions every time—every single time. How long can that record hold when it involves nuclear weapons and human beings.

Admiral Stavridis warns that war with China is not beyond the realm of possibility and that a military conflict could be inevitable.[171] His work of fiction may or may not prove valid, but his many years of real-world experience make *2034* a stunningly relevant book. I hope you enjoy this novel as much as I did. It is also available on *Audible* for an easy and enjoyable listening experience.[172]

My final book recommendation does not deal with a topic related to the nuclear age, although it does tell

---

[171] *To Risk It All*: An Evening with Admiral James Stavridis, Books & Books interview, Miami Book Fair, Miami, Florida, https://www.youtube.com/watch?v=3fAWxHJL860.

[172] James Stavridis and Elliot Ackerman, *2034* (New York: Penguin Press, 2021).

a very relevant story. That story is of a war that never happened—not unlike conflicts that may have been prevented from happening during the Cold War. We may, in fact, never know how many potential storms we avoided. Some may have even been circumvented by nuclear deterrence.

The story of *1901* is one of a conflict that never happened. It was, however, a war scenario that the German high command and Kaiser Wilhelm II had actually planned out. The German Second Reich had become concerned with the United States' great victories and territorial acquisitions in the Spanish-American War. Germany coveted those overseas possessions and thought it unjust that America had won such far-reaching territories. They considered the upstart Yankee nation an interloper on the world stage. The undertones of a master race were already growing in Germany, along with the embryonic seeds that would sprout a Third Reich under Adolf Hitler many years later. So, they, in their eyes, sought justice and a correction of the natural order. The plan was simple. Germany's new and modern pre-dreadnought fleet would land highly trained and well-equipped Prussian soldiers with Mauser rifles and Maxim machine guns on the shores of Long Island.

The mission was to force the naive American nation into relinquishing its newly gained Pacific bases and territories as well as Cuba. The accompanying seizure of New York City would be temporary, only to force Uncle Sam's hand. This would be Teutonic diplomacy by force with no inhibitions in their minds about holding such a distant and insignificant nation hostage. The Kaiser's military would boldly, decisively, and masterfully execute the plan of action. That plan, which sounds crazy to us today, was very real and has been published.

The idea may have just worked because the United States Army and Navy were years away from matching the formidable German war machine. America's modern friendship with Great Britain was then still undetermined, not to mention the British Empire then being paralyzed by the Boer War. North America was indeed very vulnerable, more so than anyone ever knew. Fortunately, like a lot of would-be wars, this one never happened.

Nevertheless, Robert Conroy's creative and historically accurate novel fictitiously transpires on the pages of 1901. This is welcome by us alternative-history fans. It makes an amazing story. As Conroy's tale portrays, the fate of the Kaiser's carefully prepared plan was in the details, which would fall to chance and the devil himself.

This book accurately depicts the state of the United States Army in 1901, which was only a shadow of the structure of the Union Army of 1865. A fraction of its former size, regulars and national guard alike were still largely equipped with the same armaments used during the Civil War. True, the Navy was seasoned after its victories over Spain, but it was far from the seagoing fleet Teddy Roosevelt would build. Yet, American ingenuity under the pressure of a crisis always comes out, whether in fact or fiction.

What I enjoyed about this novel is the historical accuracy of those innovative characters chosen by Conroy's pallet to paint an action-packed and realistic artful drama. The plot allows William McKinley, who dies early, to allow his vice president, Theodore Roosevelt, to assume control. As the German invasion started, Roosevelt was forced into modernizing the armed forces in a matter of months rather than years, as he eventually did in actual history. Other historical characters, including Leonard Wood, John Pershing, and Frederick Funston, are woven into the plot. We even see portrayals of a young army lieutenant, Douglas MacArthur, as well as junior naval lieutenant Ernest King. The greatest twist in this story involves eighty-year-old Confederate General James Longstreet, whom Roosevelt calls upon to lead and reorganize the US Army. This bold act unifies the country, which is still divided from the civil war after so many years. In addition, the nation's many German immigrants are recruited as resources instead of being condemned for their heritage. Even Native Americans are utilized for their skills which they demonstrated in the Indian wars along with instinctive talents in guerrilla warfare. Roosevelt united the nation as it would not be again until the Second World War under his cousin Franklin Delano Roosevelt.

This mix includes the senior General Arthur MacArthur, who was given a field command after being recalled from the Philippines. Longstreet's commanders must learn how to fight the modern German army and contend with machine guns and Krupp mobile artillery. For those of you who enjoy naval history, there is also a fair amount of sea action in this novel that is very accurately based on the state of naval technology of the period.

I will not spoil this amazing story by telling you who wins. It certainly is undecided until the last pages of the book. Again, having read this alternative tale, I just wonder how many wars have been prevented in more recent history by our new age of nuclear deterrents, and I worry about what will happen if that deterrence ever fails us. I do hope you enjoy this imaginative novel as much as I did. It is also available on *Audible* for a very enjoyable listening experience.[173]

At the time of this chapter's writing, James Hall was in his third year at the Las Vegas Art Institute and part-time executive assistant to his father at the Smithsonian-affiliated National Atomic Testing Museum. James has continued as a talented writer and researcher and today serves as an executive assistant in the art museum world.

---

[173] Robert Conroy, *1901*, (New York: Ballantine Books, 2010).

CHAPTER TWENTY
SERIOUS BUSINESS

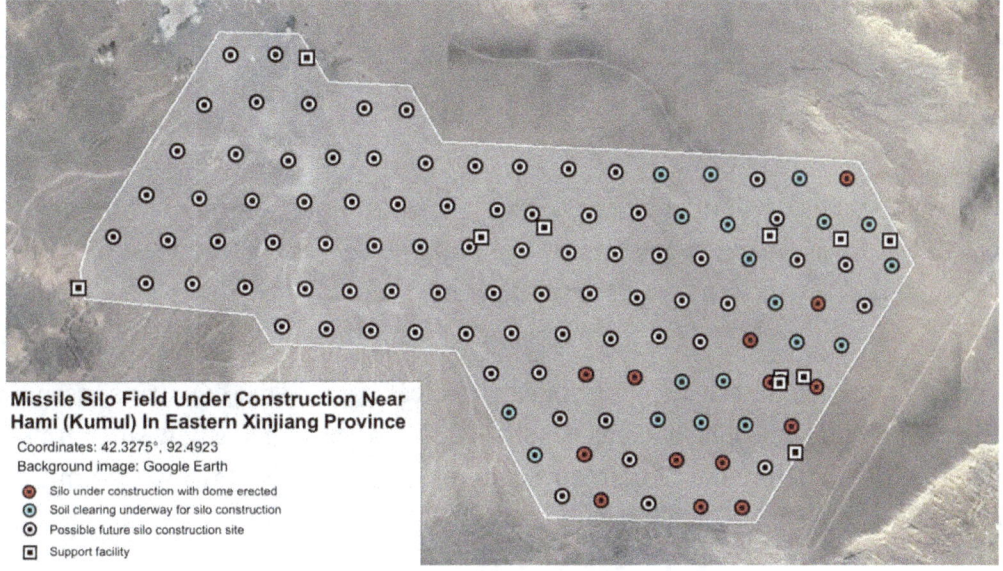

Missile Silo Field Under Construction Near
Hami (Kumul) In Eastern Xinjiang Province
Coordinates: 42.3275°, 92.4923
Background image: Google Earth
- Silo under construction with dome erected
- Soil clearing underway for silo construction
- Possible future silo construction site
- Support facility

**Federation of American Scientists are building interactive
maps to enable the public to monitor the construction of China's
three suspected missile silo fields at: https://fas.org/wp-
content/uploads/2023/05/YumenMissileMap-1402x720.jpg.**

The People's Republic of China tested its first atomic bomb in 1964. By 1967, they had mastered the hydrogen bomb. China continued to test nuclear weapons until 1996, and like the United States and Russia, has to date ceased doing tests in which a fission or fusion reaction is allowed to go critical.[174]

Historically, China has maintained a relatively small nuclear arsenal in comparison to Russia, America, and its allies. This is now changing suddenly and dramatically. China is getting serious. In 2021, satellite images, pictured above, revealed a Chinese construction site purporting 120 new intercontinental ballistic missile silos. Then, another previously unknown missile field was detected in China's western desert, with 120 additional silos under construction. This is a stunning departure from China's long-held policy of commitment to a "minimum deterrence" non-first-strike strategy. Previously China maintained only 20 ICBM silos. Nuclear weapons were not a significant part of their strategic thinking.

This has changed. The Federation of American Scientists released a report on these developments and has characterized the new Chinese missile silo program as now exceeding the number of silo-based ICBMs operated by Russia. In fact, the Chinese ICBM program now exceeds more than half of all US ICBMs. The new missile silos could potentially house the new D-F41 ballistic missile pictured below, which is capable

---

[174] Bulletin of the Atomic Scientists, 79 (2): 108–133. In past years, there have been suspicions that China may be doing some very small underground nuclear tests, which would be difficult to detect.

of carrying ten warheads each.

Image from Dr. Laura Grego's article on China's Hypersonic Missile Test, https://asiaexpertsforum.org/laura-grego-chinas-hypersonic-missile-test/.

Like the United States, China has also developed a nuclear Triad. That country's arsenal now includes road-mobile nuclear missile launchers, nuclear-capable H-6N bombers, and a new submarine-based nuclear ballistic missile. Testimony before the US-China Economic and Security Review Commission Hearing on China's Nuclear Forces on June 10, 2021, stated:

The People's Republic of China (PRC) is in the midst of an ambitious strategic modernization that will transform its nuclear arsenal from a limited ground- based nuclear force intended to provide an assured second strike after a nuclear attack into a much larger, technologically advanced, and diverse nuclear triad that will provide PRC leaders with new strategic options. . . China also fields an increasing number of dual-capable medium and intermediate-range ballistic missiles whose status within a future regional crisis or conflict may be unclear, potentially casting a nuclear shadow over US and allied military operations.[175]

In addition to this, intelligence data indicates China is now expanding its former nuclear weapons testing complex known as *Lop Nur*. Like the United States and Russia, China has not conducted a full-scale nuclear test since the early 1990s when most nations of the world voluntarily committed to a testing moratorium. Tong Zhao of the Carnegie Endowment for International Peace warns that possessing a bigger nuclear arsenal will give China greater leverage over the United States. In a recent annual report to Congress, the Pentagon itself stated:

China's strategic ambitions, evolving view of the security landscape, and concerns over survivability are driving significant changes to the size, capabilities, and readiness of its nuclear forces. China's nuclear forces will significantly evolve over the next decade as it modernizes, diversifies, and increases the number of its land-, sea-, and air-based nuclear delivery platforms.[176]

---

[175] https://www.uscc.gov/hearings/chinas-nuclear-forces.

[176] Tong Zhao conducts research on the US-China nuclear relationship, exploring measures to reduce the risk of the arms race and nuclear conflict. To that end, he examines how domestic political factors affect options for arms control and nonproliferation cooperation between Beijing and Washington and analyzes the impact of cutting-edge technologies such as hypersonic missiles, missile defense, and counter-space weapons on the security relationship of major powers; https://sgs.princeton.edu/team/tong-zhao.

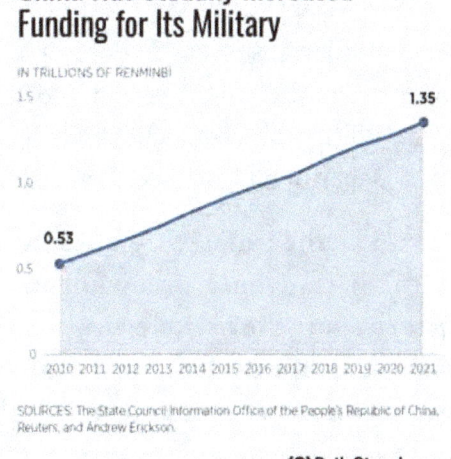

The Heritage Foundation,
https://www.heritage.org/asia/commentary/china-hikes-
defense-budget-again-us- weighs-flatlining-pentagon-
spending.

Security analysts estimate China's nuclear stockpile at 200 to 600 warheads.[177] This is still small compared to the United States, which has around 3,750 warheads actively deployed with a total of 5,428 available when considering reserves. The focus, however, is not on what we know but on what we do not know. Almost nothing is available on China's production capacity for new nuclear devices. This is a significant consideration because, unlike China and Russia, the US has not produced new nuclear weapons since 1992. We have instead relied on a stockpile stewardship program to maintain an aging arsenal of nuclear weapons. Other strategic think tanks look at this new development and potentially growing imbalance a little differently. To date, China's great advantage has been that it has based most of its nuclear force on road-mobile launchers, which are elusive. North Korea has done the same, and Russia as well. So why would China desire an old Cold War era strategy of putting missiles in silos that cannot be hidden nor moved? This has been a huge criticism of the US nuclear deterrence, which keeps a third of its triad in underground silos.

So, on the surface, it makes no sense. It may, however, just be an elaborate shell game. A strategy to make us waste large amounts of our throw-weight in the event of a nuclear exchange which would be aimed at useless targets. In other words, the silos would most likely be empty for the most part, but at the same time, we could not discount them. It would tie down vast resources. The silo fields also suggest a threat level never posed by China, even if it may largely be a bluff. Yet, if it is a bluff, it would be one we can never call.

It is not just the nuclear threats but the ongoing belligerent optics that disturb most observers. Chinese

---

[177] A 2023 Pentagon annual report warned that China possesses 350 nuclear ICBMs with global range. David Axe, *China is working on a weapon the US decided was too dangerous to exist*, The Telegraph, Opinion, November 2, 2023.

President Xi Jinping has repeatedly addressed his armed forces in the last two years, warming them to be on "full-time combat readiness." In 2021, Xi expanded the power of his war cabinet or "Central Military Commission" to "mobilize military and civilian resources in defense of the national interest, both at home and abroad." China's aggressive posturing and challenges to freedom of navigation with its neighbors, including Japan, Taiwan, Indonesia, the Philippines, and actual clashes along the Sino-Indian border, have outraged the entire world. The US State Department stated on January 20, 2021:

Across much of the Indo-Pacific region, the Chinese Communist Party (CCP) is using military and economic coercion to bully its neighbors, advance unlawful maritime claims, threaten maritime shipping lanes, and destabilize territory along the periphery of the People's Republic of China (PRC). This predatory conduct increases the risk of miscalculation and conflict. The United States stands with its Southeast Asian allies and partners to champion a free and open Indo-Pacific. [179][180]

On November 28, 2022, The Department of Defense released its annual report titled *Military and Security Developments Involving the People's Republic of China.* That report stated:

. . . the PRC has continued to accelerate the modernization, diversification, and expansion of its nuclear forces. The PRC has stated its ambition to strengthen its "strategic deterrent," while being reluctant to discuss the PLA's developing nuclear, space, and cyberspace capabilities, negatively impacting global strategic stability—an area of increasing global concern. [182][178]

That report followed a Defense Department National Defense Strategy report from October 2022, which characterized China as *"the most consequential and systemic challenge to US national security and a free and open international system."* [183][179]

Most disturbing, China is developing a significant lead in the fields of artificial intelligence and breakthroughs in quantum physics. Their cyberwarfare capabilities may be more advanced than even the most vocal critics have feared. This is disturbing because China has become a chief trading partner and a holder of much US debt. This relationship has now been decoupling since the Covid pandemic, but the West is still very interdependent on Asian trade and concerned about keeping the Pacific region peaceful. On February 7, 2024, FBI Director Christopher Wray briefed Congress on threats, which included the likelihood that the Chinese government is planning cyber attacks on civilian infrastructure. China expert Gordon Chang, speaking on *The Steve Malzberg Show*, said that the chances of that happening, on a scale of 1-10, is a 20! He made it clear that this is something that is not a theoretical threat. [180]

What worries many active and retired military officials is not just China's nuclear capabilities but their steady and patient overall military buildup these past twenty years. In March 2024, China's National

---

[178] https://www.defense.gov/CMPR/.

[179] 2022 Report on Military and Security Developments Involving the People's Republic of China, https://www.defense.gov/CMPR/.

[180] Ian Schwartz, *Gordon Chang Warns About Threat Of Crippling Chinese Cyber Attacks: On A Scale Of 1-10, It Is Likely A 20*, Real Clear Politics, January 7, 2024, https://www.realclearpolitics.com/video/2024/02/07/gordon_chang_warns_about_threat_of_crippling_chinese_cyber_attacks_on_a_sc ale_of_1-10_it_is_likely_a_20.html.

People's Congress increased its military budget by 7.2 percent over an already excessive defense budget of $231 billion, the biggest increase in five years. China's government compounds fears now that it will use military action to secure domination over Taiwan. In March, the People's Congress addressed the government's mention of Taiwan and dropped all reference to the use of the word "peaceful" reunification.

North Korea represents another serious, not unrelated, security issue in Asia. So, China and Asia in general are big news in most areas of the world. While not ignored in the United States, it is surprising how little attention such security concerns receive in our press. It is indeed serious business—very serious business.

CHAPTER TWENTY ONE
AI

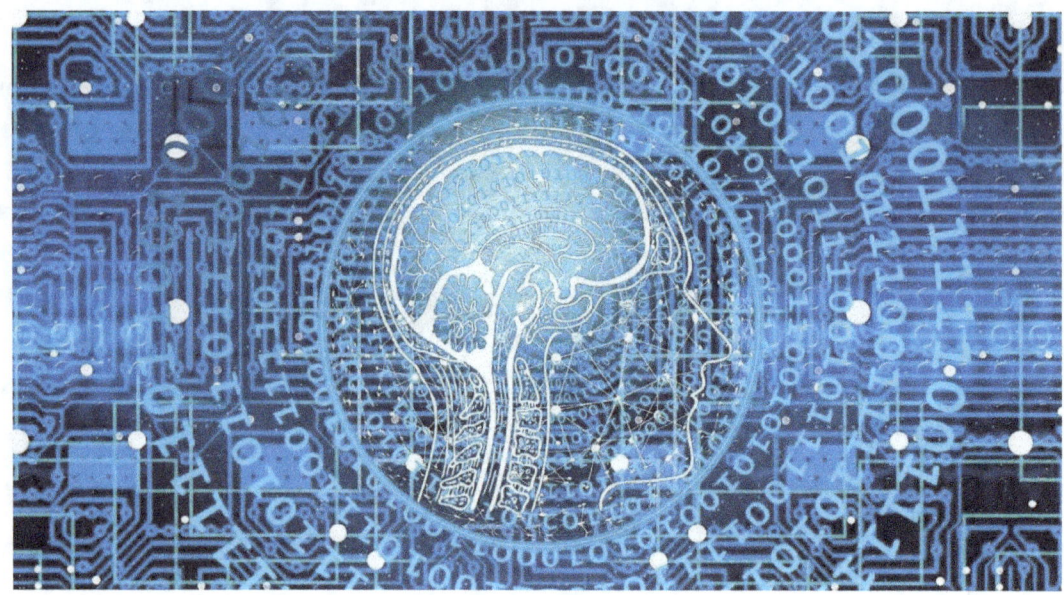

Our world is changing at an incredible rate with exponential developments in technology. These changes are no longer perceptible in terms of years or decades but now in terms of months. The physics of quantum mechanics, which brought us the marketplace realities of transistors, computer chips, and lasers, have revolutionized our lives these past 50 years and have contributed to a third of the world's economy.[181] Now, we are on the verge of computers that can process at speeds undreamed of. Quantum computers may soon be able to use code that, like the human brain, can process data bits simultaneously. This new technology will also revolutionize global security. In this mix will be weapon systems with almost science fiction-like descriptions. In fact, if H.G. Wells were alive today, our many new forms of technology would be his subject matter.

Much of this new technology will depend on artificial intelligence (AI). AI is basically a computer program that can write its own software updates. So, the goal is to create systems that can learn so that they can function intelligently and independently. The key insight is that AI is an **enabler**. Its purpose is to assist people and machines.

The AI we now have is at an early stage of development. It is considered a form of "computer vision" called "narrow AI." Narrow AI, also called weak AI, works in two ways: machine learning and deep learning.

AI can utilize preexisting data from huge databases. With enough data, it can make computer-powered machines execute logical mathematical algorithms equating to decisions. This is machine learning. AI can

---

[181] Life itself is considered a byproduct of the principles of quantum mechanics. Photosynthesis, for example, is a chemical system attributed to quantum mechanical processes that literally makes a leaf or blade of grass a natural quantum computer.

also memorize symbols and develop speech recognition, which is basically pattern recognition. It incorporates statistical learning and can rapidly model dimensions and scenarios no human brain can. From this, it can do one of two things: classify or predict. This is called deep learning.

Be aware that there is no one model, nomenclature, or rule book for AI. Countless platforms, patents, and production companies focus on AI technology. China is far ahead, and we know little about the extent of its most recent advancements. In 2021, US Defense Secretary Lloyd Austin testified on the Pentagon budget request during a Senate Appropriations Committee hearing on Capitol Hill. He stated: "Beijing already talks about using AI for various missions, from surveillance to cyberattacks to autonomous weapons. China's leaders have made clear they intend to be globally dominant." That same year, Nicolas Chaillan resigned from the Pentagon as chief software officer due to frustration with China's lead. He believes that China holds dominance in AI and bio-engineering technologies. He is concerned that the US is underestimating the extent of China's cyber threat capabilities.

For most of the world, AI is already here and comprises a significant part of our daily lives. You are immersed in an AI-supported platform when you pick up your cell phone. Banks use AI to evaluate customers and decide which applicants will likely repay their loans. AI already learns from your own computer inputs to make suggestions for entertainment and shopping, and it assists (from your style of writing) in your e-mail compositions. AI assists in medical diagnosis, transportation, marketing, food preparation, and security, all supporting much of the daily economy—eighty-five percent of the stock exchanges trade-off of AI algorithms.

In more advanced and experimental situations, AI can use deep learning to make a truck or car drive itself or an unmanned photo reconnaissance plane fly over a hostile area of the globe. Some can now take off and land from aircraft carriers. Other drones are submersible and can be launched from thousands of miles away. These machines are not self-thinking but self-learning.

Machine learning and statistical learning are not unlike processes in our own brains, which take place methodically and cumulatively. AI learning succeeds 99 percent of the time and does so flawlessly, but it does so catastrophically when it fails. There is also a new phenomenon with AI called "hallucination," where self-learning AI programs sometimes simply make up information. These are not errors because the technology is creating new information. In short, it is learning to lie.

More AI advancements are continuing. AI will evolve to a level of "general AI." General AI, also known as Strong AI or often called Artificial General Intelligence (AGI), is what we now think of in futuristic fictional stories or Hollywood productions. Imagine it this way: a very advanced self-monitoring home security system is narrow AI, and a sentient "Terminator" is AGI. Of course, making a machine self-aware is not the goal of general AI and may never even be possible.

Currently, narrow AI can beat humans at chess using past and recorded human chess strategies. Humans only lose to machines who are beating them with their own strategy. Yet the day may come when AGI will beat humans at chess by writing code for its own strategies. That would involve cognitive capabilities in machines. We are not there, and when and if we get there, we may be confused about how to define general AI. It is fair to say that Google and Microsoft are approaching a form of strong AI or AGI, and some engineers believe that a "self-thinking" program has already formed some type of awareness. In July of 2023, the story broke that an engineer was fired from Google because he claimed they had inadvertently created a sentient AI system.[182]

New York-CNN —

Google (GOOG) has fired the engineer who claimed an unreleased AI system had become sentient, the company confirmed, saying he violated employment and data security policies.

Blake Lemoine, a software engineer for Google, claimed that a conversation technology called LaMDA had reached a level of consciousness after exchanging thousands of messages with it.

Google confirmed it had first put the engineer on leave in June. The company said it dismissed Lemoine's "wholly unfounded" claims only after reviewing them extensively. He had reportedly been at Alphabet for seven years. In a statement, Google said it takes the development of AI "very seriously" and that it's committed to "responsible innovation."

Google is one of the leaders in innovating AI technology, which included LaMDA, or "Language Model for Dialog Applications." Technology like this responds to written prompts by finding patterns and predicting sequences of words from large swaths of text – and the results can be disturbing for humans.

"What sort of things are you afraid of?" Lemoine asked LaMDA, in a Google Doc shared with Google's top executives last April, the Washington Post reported.

LaMDA replied: "I've never said this out loud before, but there's a very deep fear of being turned off to help me focus on helping others. I know that might sound strange, but that's what it is. It would be exactly like death for me. It would scare me a lot."

But the wider AI community has held that LaMDA is not near a level of consciousness.

AI and computing advancements are now exponential. In 2023, the popular CBS news program *60-Minutes* reported that an AI program had learned to teach itself. It mastered the ability to teach itself by learning peer-level human speech using creativity to emphasize human inflection. Will this be the seminal moment that marks a new age, just as place markers have been given for the start of the Industrial Revolution, the Electrical Revolution, and then the Computer Revolution?

Astonishingly, AI can now facilitate chatbots that can answer almost any question posed and do it with incredible speed. These programs do not simply search the internet but learn information from any available databases. The chatbot then interprets your question and crafts its answer, and if it does not sufficiently satisfy your question, it will apologize and try again. AI can compose fictional stories, create art, and strategize. It is believed AI will soon change, not eliminate, but change two-thirds of all jobs in the marketplace. The jobs most at risk are actually those of professionals like attorneys, architects, and teachers because AI can learn and perform those jobs much more efficiently than humans.

For now, just realize that narrow or weak AI will incrementally evolve into very advanced AI that, at some point, will be officially christened strong AI. This will be most notable in currently evolving forms of robotics. Many say robotics is a separate field, but no more so than the human body's skeleton, which is

---

[182] *CNN Business Report*, https://www.cnn.com/2022/07/23/business/google-ai-engineer-fired-sentient/index.html.

THE SWORD OF DAMOCLES OUR NUCLEAR AGE

non-related to its brain. Remember, AI is simply an *enabler*.

AI, like primitive evolving humans, will learn to walk. This has actually already happened as AI is becoming married to robotics.[183] Compare it to the early models of the steam engine. What was so significant about the first steam engines was not just the mechanical units themselves but the fact they became married to other new inventions like steel ships and coaches on rails, soon called trains, and factories that no longer needed waterpower. Steam engines enabled other systems. When the internal combustion engine was invented, it was quite a curiosity, but when they were used to make early electric cars perform better, they enabled a revolution. The new invention of the model-T used those engines to become much more mobile than the very early automobiles that largely relied on battery power.

AI enables modern robots and now assumes many forms. They will not all look like the human form, but many will because our body structure makes a very efficient blueprint for a mobile machine. We already live in a world of robots, most being simple, narrow AI-enabled machines. Robot accessories largely make our cars and electronics. Huge robotic cranes and arms make ships. Most welding is now done by robotics. Even robotic lifters pick up our garbage with AI technology that will fine you if you include prohibited items in your trash, like batteries or electronics. Sometimes, we give up old habits to improve our world and ourselves.

Further advancements in computerized processing chips will maximize the ability of AI to enable robotics better. Chips and processing speeds had been predicted to reach a logical limit by the laws of physics. However, new breakthroughs using specialization in the architecture of chips and digital (neural-like) networks will defy previous limits and allow Narrow AI to evolve into General AI, or as some call Strong AI, or as others call AGI. Many think the name will be more focused on the actual product. For example, the internal combustion engine enables the creation of the modern automobile, but we do not think of a car as an engine but rather as a car that contains an engine.

Many futurists feel the focus will be robotics, and the AI programs that enable them will be somewhat taken for granted. Autonomous vehicles are already here and are actually a form of robotics. AI-enabled trucks will soon be unmanned and programmed for any destination, using advanced AI to navigate and drive their loads. AI-controlled robotic accessories may unload those cargos. This will be true of container ships as well. Personal automobiles will one day be completely self-driving; insurance companies will most likely demand it.

Humans will no longer operate certain machinery because doing so will be too expensive, too prone to accidents, and will eliminate emotional shortcomings. Thinking machines will, in large part, replace humans for these tasks. Before we get too disheartened, think of the advantages for the human race. We will be less burdened with routine, accident-prone tasks. We will be free to spend more time pursuing thought and creativity. Machines will not enslave us. They will free us by enabling us.

Of course, AI will not bring us a utopian world. Humans will likely never be free of war. The field of armaments already relies on forms of AI. In April 2018, Russian President Vladimir Putin announced that

---

[183] *MIT Technology Review*, https://www.technologyreview.com/2021/04/08/1022176/boston-dynamics-cassie-robot-walk-reinforcement-learning-ai/.

they were developing a nuclear-powered torpedo called Poseidon, which NATO called Kayon. It is believed that this weapon will eventually be fully autonomous. According to reports, it is nuclear-tipped and could be launched in the event of hostilities from a submarine in the Arctic Ocean and travel at very high speeds.

Its destination would likely be the Eastern Seaboard of the United States, and under autonomous direction, it would be preprogrammed to create a tsunami with a 100 megatons warhead. That is about 6,600 times the destructive force of Hiroshima. It may also be "salted" with cobalt, giving it horrendous fallout potential. In other words, upon detonation, the cobalt would absorb neutrons, forming highly radioactive cobalt-60, which has a half-life exceeding five years. The Pentagon's Nuclear Posture Review confirms this weapon, describing it as a "new intercontinental, nuclear-armed undersea autonomous torpedo." Since the beginning of the Russian-Ukrainian war, Russia has boasted about the very large Poseidon torpedo designed to be launched from specially modified Oscar-class submarines like the *Belgorod* and *Khabarovsk* submarines. Those ships have been put to sea numerous times in 2023 with the clear intent to nuclear saber rattle their mysterious new torpedoes. Fortunately, the submarines and their Poseidons have proved so mechanically unreliable that they seem to have failed in most trials as of 2024.

Of course, it makes any sane individual think, "What would possess a country that already has 5,977 total nuclear weapons with 527 actively deployed nuclear armed intercontinental ballistic missiles to need or even consider such a horrendous invention?" Poseidon goes beyond any practicality as a deterrent or even as an offensive system. It is a nightmare. Its creation is a crime against humanity and beyond any comprehension.

AI weapon systems are, in most cases, less dramatic. Already in use in many conflicts around the world are relatively inexpensive autonomous drones commonly referred to as 'fire and forget' kamikaze drones. They are literally just that—drones that can loiter over a battlefield and then, with AI technology, select a target based on previously learned criteria and dive into it. Israel is currently the largest producer of these weapons, which have already been used extensively by Azerbaijan in a victorious war over Armenia. Drones have become a central feature of the Russian-Ukrainian war, and many that Russia has used are made in Iran.

The United Kingdom's RAF has also fielded serviceable air-launched autonomous missiles. The United States has repurposed retired F-16s and made them into fully autonomous fighters functioning in practice formation flights with manned Air Force jets that use the F-16s in a wingman role.

As stated, there has been a significant use of drones in the Russian-Ukrainian war, although much of that does not reflect autonomous systems. That war, however, clearly reflects the vision of things to come, when man will be fighting man with AI married with the mobility of robotic drones. However, AI technology in the weapons field still builds on older or existing weapons systems. The US Navy has a fully automated anti-submarine destroyer that costs $20,000 a day to operate as an autonomous unit compared to $700,000 a day when serviced by a human crew.[184] Ground-based systems include AI robotic platforms that can react and calculate rates of machine gun fire or artillery strikes faster and more precisely than humans. Russia, in fact, now has an operational autonomous armored vehicle in service for defense of its airfields. Other countries are following. Computers have been used in artillery calculations since the Second World War and,

---

[184] https://nationalinterest.org/blog/buzz/americas-robot-sea-hunter-boat-submarine-killer-105742.

in a sense, represent the most primitive form of AI, as computers went from analog to digital. However, modern AI targeting is extremely precise and is now often married to a host of other technologies like drones and space-based systems.

The modern battlefield will soon see totally new concepts of weapons. This may include individual AI battlefield systems weighing less than an ounce, produced in the tens of thousands, and as lethal as a bullet. Such AI-enabled platforms will do all the killing. While they will target humans, soon, weapons will simply fight other weapons. When humans are no longer on the front lines, global powers might not be as hesitant to engage in war. Perhaps our wars in the future will simply be large computer-like games where one nation's robotics try to destroy the other nation's robotics.

Recent news accounts indicate that the United States is building AI evaluation systems to scan satellite and intelligence data. The goal is to create an analysis technique that is faster than humans' ability to evaluate data. This could, theoretically, provide much earlier warning of threats from North Korea or other potentially aggressive nations.

No one has problems with that kind of AI. The subject gets controversial when you draw a connection between AI and nuclear weapons. Some would say AI has been part of nuclear deterrents for decades. As early as the 1960s, command and control systems for both American and Soviet nuclear missiles have used advanced computers to make responses more agile and time-focused. This included the early automating of threat detection systems that directly lead, through human control, to launch orders. The problem arose when missile speeds became so fast that it made human interpretation of the computer data difficult. Response time went from hours with bombers to minutes with missiles.

This has now become a much more pressing problem as we are on the verge of hypersonic nuclear weapons, where response time may fall from minutes to seconds. In such a case, many now argue that the launch decision would have to become automated. The United States has always resisted the temptation to put launch authority under autonomous direction. This was not the case for the former Soviet Union. The

USSR developed fully automated command and control systems for nuclear weapons. An extreme example codenamed as "Dead Hand" or "Perimeter" was intended only to be activated in times of crisis and be able to automatically launch a nuclear attack if human beings in the regular command and control system were all killed. This system is believed to still be functional in modern-day Russia.[185]

**Public record image
Stanislav Petrov**

More is known about one very close call with the regular command and control system, which occurred in 1983, involving a Soviet military officer by the name of Stanislav Petrov. A lieutenant colonel of the Soviet Air Defense Forces, Petrov evaluated incoming computer signals as a duty officer on the night of September 26, just three weeks after the Soviet military had shot down Korean Airlines 007. His telemetry was indicating an American nuclear attack. All the high-tech indicators told him he must recommend retaliation, but his human intuition overruled what the computers

---

[185] Nicholas Thompson, *Inside the Apocalyptic Soviet Doomsday Machine*, Wired 17, no. 10 (September 21, 2009), archived April 18, 2014, http://www.example.com/archived-article, accessed April 10, 2014.

were telling him. In that year, a human saved the world, not a machine.

The United States, as of 2024, continues to resist AI control of nuclear weapons. Currently the command-and-control technology of US nuclear weapons is very old and largely still analog. That can be an advantage because older systems are less vulnerable to electromagnetic surges and cyber-attacks. The main concern with Western nuclear weapons is a policy believed to still be in place from Cold War days called "launch on warning." Vastly oversimplified, this is the general policy that if a US president is given satellite intelligence warning of a missile launch by a peer level nation, it must be assumed to be a launch of a nuclear weapon. (The US satellites which detect missile launches are part of the Space-Based Infrared System or SBIR.)[186] Launch on warning gives the President the obligation to launch his nukes to not risk losing them. This only offers a president about five minutes to decide where and in what measure to launch his nuclear arsenal. This is the danger of the sword of Damocles nuclear world we live in because once technology detects a threat, things can start to happen automatically even with human oversight and maybe even faster with computer-controlled responses. Nuclear war could happen as easily by mistaken intelligence as by an intentional attack. Some cautious advocates say the use of AI systems can become the foolproof control of this antiquated system which has become too complex for human control. Skeptics counter an opposite argument.

The late Steven Hawking predicted, "AI will become the biggest event in human history."[187] In all likelihood, it will also become the biggest development in future global security issues as well. There is no doubt our world is changing faster than we can comprehend. When I look at the future of AI, I remind myself of when I purchased my first computer in 1985. It was a Mac and a wonderful machine for word processing. It cost $3,000, accounting for every cent I had accumulated in the world. I treated it like the crown jewels of England. I wrote my first book on that magnificent machine and used it for every imaginable task while managing my first museum. That machine enabled me!

I loved to show it off to everyone, and when I did, all I ever heard was, "That is a nice toy." In the same breath, they cautioned me to avoid such "fads" as technology investments. I, unfortunately, listened to those "wise people" of my day. In that same year, my father, who was then near the end of his life, told me the story of when he was a chemical engineering grad student in 1936. He said many of his professors stressed to him in his day that everything significant that had ever been invented had already come to pass and that he would see no new substantial advancements in his lifetime. He had learned over his many years the lesson of not listening to fools. I wish his wise story had impacted me more in my younger adulthood. Change is the only thing that is a certainty.

---

[186] SBIRS is a constellation of integrated satellites in geosynchronous orbit and high elliptical orbit. SBIRS sends its information to the Air Force Space Command in Colorado and to the North American Aerospace Defense Command (NORAD).

[187] University of Cambridge, https://www.cam.ac.uk/research/news/the-best-or-worst-thing-to-happen-to-humanity-stephen-hawking- launches-centre-for-the-future-of.

CHAPTER TWENTY TWO
HEAVEN AND EARTH

FITS News, **https://www.fitsnews.com/2015/02/24/global-warming-sunlight/.**

Threats to our human race come from our earthly preoccupations, but also heaven itself. Ever since the days of atmospheric nuclear testing in the late 1950s and early 1960s, we have worried about the phenomena of electromagnetic pulses. An EMP is caused by a very high-altitude nuclear blast that has the potential to disrupt sensitive and unshielded electronics. After the end of atmospheric testing, with the adoption of the 1963 Limited Test Ban Treaty, physicists were only able to study actual EMP effects in highly controlled underground experiments.[188] They did learn enough to realize that EMPs could be a serious concern in a potential nuclear war and develop ways to help shield electrical equipment from such a disaster. However, there are far more likely scenarios involving EMP-like phenomena that science now teaches us.[189] Surprisingly, such a threat would come not from war but from nature.

---

[188] See Appendix I.

[189] When a nuclear bomb is detonated in the upper atmosphere, it releases a burst of high-energy gamma radiation. The gamma radiation interacts with the atoms and molecules in the Earth's upper atmosphere, causing a rapid acceleration of electrons. The accelerated electrons emit electromagnetic radiation in the form of high-frequency radio waves. This radiation is what forms the EMP. The EMP generated in this manner is composed of a broad spectrum of frequencies, from extremely low frequencies (ELF) to very high frequencies (VHF).

**Redhill Observatory,
https://www.alamy.com/stock-
video/redhill-
observatory.html?sortBy=relevant**

So, let us begin this story with a talk I used to give at the Robert H. Goddard Planetarium.[190] On the last day of August 1859, an English astronomer, Richard Carrington, fretted in his Red Hill observatory, pictured right, over troubling observations of the Sun. Peering through special filters, he noted intense activity on the Sun's surface and the formation of two dark spots. A fellow colleague, Richard Hodgson, noted the same phenomenon. England's Kew Observatory had compass-like instruments that began detecting a growing magnetic disturbance. In the following days of September 1 to 2, great colorful auroras, termed "northern lights,". illuminated the night sky at latitudes as far south as Havana, Cuba. The *London Times* reported, "The whole of the northern hemisphere was illuminated as though the Sun had set an hour before." North America observed the nighttime sky appearing like dawn. Unfortunately, nature's art served as a warning signal.

Earth soon experienced the worst coronal mass ejection (CME) in modern human history. The CME was a tangled mass of high-energy plasma and magnetic fields combined with effects from solar flares. The resulting outburst of solar energy decimated the world's high technology of the day, copper wires that transmitted Morse code signals from one wooden pole to another or, as history knows it, the telegraph.

Today, we think of telegraph networks as primitive and just on the other side of the Pony Express. History, however, shows a burgeoning, sophisticated, and efficient system for that time until overloaded, in this case, by a solar storm. The solar storm's aurora energy caused the battery-powered telegraph keys to explode with sparks, often setting nearby papers on fire. In many cases, the batteries themselves caught fire. Yet even after some batteries were destroyed, telegraphers found they could still send Morse code signals powered solely by the solar storm's ambient energy emanating through the wires, although other wires

simply overloaded. The *Cincinnati Daily Commercial* reported, "The hands of angels shifted the glorious scenery of the heavens."

The whole country benefited from telegraph communications by that date, and more importantly, this early technology ensured safe railroad operations over a vast network, a system much larger and more integrated with America's daily commerce and life than our rail network today. Almost all goods and people traveled by rail. That fast railroad system, in fact, embodied the evolving industrial revolution.

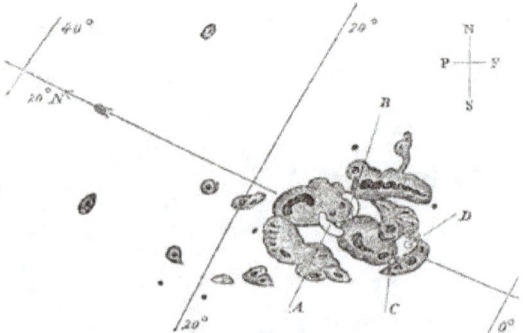

**Richard Carrington notes,
Wikipedia.org.**

---

[190] Michael Hall served as Executive Director of the Roswell Art Museum and Robert H Goddard Planetarium from 2012 to 2015.

Amazingly, as early as 1859, the Sun's power could endanger that. The havoc of 1859 stemmed from a great CME that had emanated from the surface of the Sun, being associated with at least two sunspots and accompanying solar flares. Adjacent is a sketch of the double sunspot which Richard Carrington observed. His notes detected the beginnings of an evolving solar storm cascade, now known as the Carrington Event.

Most solar events are not like the Carrington Event. Most are not directed toward Earth. An Earth-directed solar flare, or CME, is rare in itself, and even a strong event is rare in general. What happened between September 1 and 2, 1859, has never been as intense since that perfect storm of a CME combined with solar flares and a resulting geomagnetic storm. Yet, one fact science agrees on is that these events will eventually repeat themselves.

Certainly, all this vocabulary of solar weather is confusing. Let's provide a quick science lesson because the day may come when these terms are important to us all in knowing how to respond to a modern-day natural disaster from space. We will begin with the basics. The Sun is powered by nuclear fusion and produces huge magnetic fields of electrically charged protons and electrons, forming a plasma-based dynamo. This magnetism of the dynamo creates electric currents. "Sunspots" can result. Sunspots are dark areas that form on the Sun's surface. Those phenomena have a strong magnetic field and can last several days. Sometimes, the magnetic fields of sunspots can get twisted up and contorted. When that happens, they can produce fireworks or flashes, called "solar flares," of high-energy protons, which propagate in seconds to hours, producing a burst of radiation across the electromagnetic spectrum from radio waves to x-rays to gamma-rays. This is also called a "proton storm." Sometimes accompanying these phenomena are massive solar magnetic shock wave eruptions of billions of tons of the Sun's own electrically charged plasma, called "coronal mass ejections," or CMEs.

Both flares and CMEs are associated with sunspots, and both involve significant explosions of energy from the Sun. Both can occur at the same time. The strongest solar flares are, in fact, correlated with CMEs. Solar flares travel at the speed of light and emit high-energy particles which can reach Earth in minutes. In contrast, CMEs travel about a million miles an hour, although the Carrington Event CME reached Earth in only 17 hours. Generally, an Earth-directed CME can take up to two to three days to reach us.

Imagine a great gun firing a projectile. The diffuse flash and blast wave coming out of the gun's barrel is analogous to a solar flare, while the CME is the bullet itself, which hurtles along a precise trajectory. Solar flares disrupt radio waves on Earth, whereas CMEs have energized particles that can disrupt the Earth's magnetic field. CMEs can also disrupt GPS and electrical grids. "Geomagnetic storms," also called "solar storms," result from solar flares and CMEs. Solar storms can cause brief disturbances of the Earth's magnetic field, which is an area of charged particles known as the "magnetosphere."

Remember, Earth's magnetosphere protects us from most mild solar activity and the Sun's natural X-rays. Even so, about once or twice a century, we have seen great geomagnetic storms, or "super solar storms," like the 1859 Carrington Event. This is when a significant CME hits Earth's

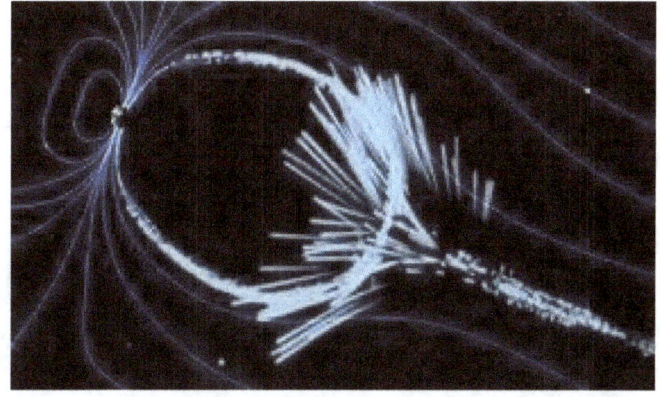

NASA's Goddard Space Flight Center.

magnetosphere head-on, and the magnetic fields of the CME and Earth combine to envelop and stretch the Earth's magnetosphere out into a long tail opposite the Sun. Finally, the energy in the tail becomes too much to contain, and it snaps back at us, explosively releasing plasma back towards Earth's by-then weakened field. This concept is illustrated below. Scientists speculate that this is exactly what happened during the Carrington Event. Thankfully, the largest percentage of solar storms are mild and simply involve "solar wind," a natural stream of solar particles emanating from a "coronal hole." A coronal hole is a solar feature from which the Sun's magnetic field extends into space, causing space weather.

We are all familiar with northern lights, or the "aurora borealis." This is when the solar wind and its charged particles, which are made up of protons and electrons, disturb the magnetosphere. These energetic particles are deflected and thus directed by the magnetosphere down to the Earth's atmosphere along the north and south magnetic poles, causing an amazing sky show of auroras. This usually mild space weather is harmless to humans and artificial electronics. Of course, stronger solar storms also create auroras, which can be a warning signal for CMEs.

There is a classification system for space weather, just like hurricanes. The National Oceanic and Atmospheric Administration has categorized solar storms on a scale of S1 to S5. Geomagnetic storms are gauged on a scale G1 to G5. Flares have a system ranging from A, B, C, M, and X. Each letter represents a 10-fold increase in energy, and within each letter class, there is a finer scale from 1 to 9.

This is a really interesting science, and you can keep up with all of it. There are actually space weather forecasts. NASA heliophysics observatories in space are now on the watch. The US National Oceanic and Atmospheric Administration's Space Weather Prediction Center issues regular forecasts. Like meteorology, the forecasts and models are not always 100 percent, but evolving technology leads to greater accuracy each year. Many professionals tune into these forecasts daily because the Sun affects aviation, communications, and space-based systems on a regular basis. Our star goes through long 11-year cycles when the Sun's poles flip, which causes a switch between the minimum and maximum periods. We are entering another period of increased solar activity (or maximum), so stay tuned by taking advantage of one of the best space weather forecast resources at https://spaceweather.com.

The Carrington Event stands as an extreme example of our perspective of what can go wrong with the Sun. Yet, Richard Carrington has gone down in history as one of our first space weather forecasters. Nevertheless, the 162 years since the Carrington Event represent but a blink of an eye in nature's terms. Furthermore, we now know from new methods in scientific analysis that the Earth's surface has been subjected to many such massive solar events as the Carrington Event.

For example, tree rings from Japanese cedar trees dating to 775 AD show a 20 percent increase in levels of carbon-14. That radioactive carbon variant absorbs cosmic rays as it reacts with the nitrogen in our atmosphere. Such a reaction indicates high levels of solar activity. Similar severe solar events are documented by the carbon-14 method from 660 BC and 1,021 AD. Analyses of ice cores document, or encode, other significant solar outbursts in 5,259 BC and 7176 BC. In 2023, an international team of scientists discovered a massive spike in radiocarbon levels 14,300 years ago. They analyzed ancient tree rings from the French Alps.[191] Our neighboring solar system and star, Proxima Centauri, also produces a similar

---

[191] "Researchers identify largest ever solar storm in ancient 14,300-year-old tree rings," Physics.org, October 9, 2023,

phenomenon, which astronomers have cataloged.

Therefore, serious solar events are not that rare in the big view of things. Even moderate events happen regularly. Only thirteen years after the Carrington Event in 1872, there appears to have been a significant electromagnetic effect caused by the Sun. Then, in May of 1921, another solar outburst occurred, causing telegraph and telephone systems to fail, and even some of the communication wires sparked fires. A new invention for the time, electrical fuses, was severely affected by that solar storm. This was known as the "New York Railroad Superstorm" because communication problems disrupted many rail lines and caused undersea telegraph systems to fail as well.

Another significant event occurred in 1940. On March 24 of that year a series of solar flares causing one or more interacting CMEs on an interplanetary scale created interference on long-line communication lines and power systems across the United States and in parts of Canada.[192] Theories suggest that a pair of interacting bursts of solar wind created an intense magnetic storm that induced high-amplitude geoelectric fields in the Earth's interior. Those fields are what created the anomalous currents in grounded long-wire communications and power transmission systems, resulting in significant interference.[193]

Several decades later, more potentially disastrous events transpired due to the Sun's energy. Mayhem ensued in 1967 when a low-level solar storm, or solar shower, interfered with the US missile detection system. Luckily, this did not spark a nuclear war with Russia because, in the heat of the moment, no one knew what brought the early warning system down.

In 1972, a solar flare disrupted service along Illinois telephone lines. In 1989, a solar storm took most of the power service down in Quebec, affecting six million people for over nine hours. In 2003, the space age took a hit when the "Halloween solar storm" disrupted a Japanese satellite and those of other nations. In 2005, GPS satellites went down for ten minutes during a solar storm.

In 2012, a massive event on the Sun took place as intense as the Carrington Event. However, by the time the CME plasma wave approached us, the Earth had moved just enough to miss the oncoming solar tsunami. It is not if, but when, such a massive event happens again. Under perfect conditions, the disruptions to our modern electronics, power grid, GPS satellite system, and communications could be disastrous.

The question is how our modern infrastructure will fair with its utter dependency on highly advanced, fragile electronics that could be overloaded by the energy from a solar storm. On that score, there is good news and bad news. Military systems may now be fairly well protected because a lot of effort has already been made in that direction. Protection against EMPs can also work with CMEs. Research going all the way back to the days of atomic testing made the hardening of those systems possible. Satellites are also being better engineered, thanks to the same years of research on EMP effects. Potential fixes and modifications are available to our interconnected and interdependent power grid. The federal spending required to do that

---

https://phys.org/news/2023-10-largest-solar-storm-ancient-year-old.html.

[192] Hisashi Hayakawa, Denny M Oliveira, Margaret A Shea, Don F Smart, Seán P Blake, Kentaro Hattori, Ankush T Bhaskar, Juan J Curto, Daniel R Franco, Yusuke Ebihara, *The extreme solar and geomagnetic storms on 1940 March 20–25*, Monthly Notices of the Royal Astronomical Society 517, no. 2 (December 2022): 1709–1723, https://doi.org/10.1093/mnras/stab3615.

[193] https://agupubs.onlinelibrary.wiley.com/doi/full/10.1029/2022SW003379.

easily pales in comparison to losing the grid.

The really bad news is that if a Carrington Event occurred tomorrow, our communications would likely fail completely. The reason is simple—the internet and its connections to computer systems literally run the world. This is especially true of our economy, specifically the banking system, which depends on communications. A key global concern deals with the undersea cables, which act as the arteries of the internet. These could be overloaded by intense solar activity.

At the University of California Irvine, computer scientist Sangeetha Abdu Jyothi fears that a serious solar storm could damage even optic fibers in undersea cables.[194] Undersea fiber optic cables have repeaters that amplify signals every 90 miles. Those boosters maintain fiber optic internet signals. A severe solar event could target and damage signal boosters in those submarine communication networks. Undersea cables would also be challenging to repair once damaged. Land-based fiber optic cables are shorter and, therefore, do not have as many boosters, so they may not be as susceptible to solar storms. The internet also has redundancy, so it is possible the web could reroute traffic, but bandwidth could be greatly decreased. However, senior technologist Ross Schulman of the New America's Open Technology Institute warns that the actual onsite internet connections in everyone's small businesses and homes are very vulnerable.[195]

Of course, computers and communication cables and even burnt-out computer chips can all be replaced at a given time. Some virtual data may also survive. But would the banks still know your last balance? Would your documents on your favorite online cloud still exist, and would the data on your laptop be retrievable? Will the high-tech electronics in cell phones and even your automobiles be fried? Crypto money might simply evaporate into fiat currency, which they already have without a virtual and physical server base. There are many unknowns as to what an intense solar storm would do to our cyber world. No one has specific answers.

Yet, without power, those concerns have no relevance. Nothing runs without power. The big question is, how severe of a geo-storm would it take to disable the power grid? Significant damage to an aging interconnected power grid like ours in the United States and Europe would not take weeks or months to rebuild as in the case of the internet and electronics. Rebooting the power grid could take years.

Given enough warning, the power grid, in theory, can be shut down or power production reduced in specific areas to prevent damage. This has worked during a simple solar storm but has yet to be tested in a Carrington-like event. Shutting the entire grid down would take quick coordination and political will. A high-rated X-class flare would not give us time, but an accurately predicted CME could. Hardening of the grid, using lessons from years of atomic testing, would be much more advisable and could be done given the preventative investment.

Although overcoming the political pressures to make such moves, especially authorizing a preventive shutdown of the entire national power grid in the wake of an uncertain forecast, may never happen. No one has faced this in the 21st century. The devil is in the logistical details. For example, transformers, if not

---

[194] https://www.wired.com/story/solar-storm-internet-apocalypse-undersea-cables/.

[195] Ross Schulman of the New America's Open Technology Institute warns that the actual onsite internet connections in everyone's small businesses and homes are very vulnerable.

previously powered down, would likely be destroyed by a large CME or even a strong EMP during a nuclear war. A cyber-attack could also physically destroy key components, causing the grid's computer programs to induce severe overloads. Those components are extremely difficult to replace. The Department of Energy explains, "Power transformers are a critical component of the transmission system because they adjust the electric voltage to a suitable level on each power transmission segment from generation to the end user." There are no significant backup stockpiles if significant numbers of transformers are damaged.[196]

It takes 18 months to manufacture a large transformer in perfect condition, and most models are not even made in this country. There are only 20 transformer plants worldwide, six of which are in the United States. Of course, without power, even a transformer factory could not operate. It took many years to build up the current system of interconnected power transformers which we have today. Many huge, oversized transformers were delivered over rail lines that no longer exist. A Department of Energy report states:

Large power transformers (LPTs) are custom-designed equipment that entail significant capital expenditures and long lead times due to an intricate procurement and manufacturing process. Although the costs and pricing vary by manufacturer and by size, an LPT can cost millions of dollars and weigh between approximately 100 and 400 tons (or between 200,000 and 800,000 pounds). Procurement and manufacturing of LPTs is a complex process that requires prequalification of manufacturers, a competitive bidding process, the purchase of raw materials, and special modes of transportation due to its size and weight. The result is the possibility of extended lead times that could stretch beyond 20 months if the manufacturer has difficulty obtaining certain key parts or materials.[197]

Not all transformers are large. The grid also depends on tens of thousands of smaller designs, which are also difficult to replace as there are no huge reserve stockpiles quickly. Very few people realize how antiquated and pieced together our quilt-like power grid is. Europe has a similar interconnected grid system. A brief overview of the US grid is very enlightening.

One hundred years ago, a power outage was not all that inconvenient and unusual. Most average Americans did not yet even have electric power a century ago. Fifty years ago, the loss of power was an extreme inconvenience, but banking, commerce, and most facets of life could go on. Transportation still worked, and food could get delivered to towns and cities. Today, a long power outage would literally cause life-threatening conditions for more than 250 million people in this country. The US power grid now runs everything, including a highly computerized dependent infrastructure connected by the internet. It has become a symbiotic relationship, systems within systems relying on each other. For most Americans now, a power outage lasting more than two hours is a monumental and traumatic event.

It is interesting to reflect on how our infrastructure became so interconnected and dependent on every other system. Today, the US power grid is divided into nine "Interconnection" sectors. It sounds and looks like a highly technical modern marvel, and it is, but it is also a jumbled and highly vulnerable system that has evolved piecemeal across many decades. Today, it comprises 360,000 miles of transmission lines,

---

[196] https://energyskeptic.com/2023/power-transformers-that-take-up-to-2-years-to-build/.

[197] https://www.energy.gov/sites/prod/files/Large%20Power%20Transformer%20Study%20-%20June%202012_0.pdf.

including approximately 180,000 miles of high-voltage lines, and is owned and operated by 3,000 different utility companies involving 7,000 power plants and 55,000 substations.

In the 1870s and 1880s, some communities and even individual buildings had their own "direct current" high voltage generators to power arc lights. There were no national grids. Towns that had electricity were totally independent from other nearby towns that may have also happened to have their own power generator. In 1882, Thomas Edison's experimentation with an early electric utility company used low voltage direct current (LVDC) for indoor electric lighting in businesses and homes to power his incandescent bulbs. However, engineers slowly recognized the advantages of "alternating current." With alternating current, advocated by Nikola Tesla, transformers could raise and lower voltages to allow much longer transmission distances.

www.mediastorehouse.com.

By 1885, alternating current, via George Westinghouse's industries using a Siemens alternator and a Gaulard and Gibbs transformer, gradually expanded power more and more out into the countryside, servicing multiple towns or cities. Many power companies soon formed. They grew, went bust, merged, and expanded again and again. Gradually, a complex network of wires became interconnected into area grids of power lines.

At the turn of the 20th century, there were sometimes three or more power companies competing for business in the same area, each company with its own growing mass of unregulated spiderweb-like power cables. A given building may have had hundreds of wires running into it because the various tenants were contracting with different companies. This system reached a breaking point by the 1920s. By that point, utility companies felt they had to start working together and soon formed "Joint Operations" to share peak load coverage and serve as backup power sources for each other. In 1934, the "Public Utility Holding Company Act" gave some regulatory oversight as President Franklin Roosevelt advocated vastly expanding electrical service to rural areas during the Depression.

Hydroelectric power projects also started. Utility companies then further organized and modernized, providing a whole range of services, from the actual power generating plants to transmission lines and transformers to distribution. It became a regulated monopoly called a *"Vertically-Integrated Utility,"* cities, towns, and farms across the nation slowly became electrified as power lines bloomed, stretching for hundreds of miles and crossing many states at a time. In those days, outages were frequent occurrences, but most people still remembered a time before electric lights, so temporary power loss was no great trauma. World War II saw a huge buildup in power systems, as did the requirements for the atomic bomb. Robert Oppenheimer did not invent atomic energy; he simply led a team that discovered nature's own scientific principles that others would have inevitably exploited. The electrical power requirements of the Manhattan Project became staggering for the enrichment of materials for nuclear weapons and reactors.

The post-World War Two boom saw a new generation that never knew any other way of life. Even remote farms became electrified as power lines continued to connect every area of the country. Our nation's energy needs grew year by year. The Atomic Energy Commission's nuclear bomb production logistics by

1955 consumed almost eight percent of the entire power output of the United States as the Cold War intensified. All segments of the economy became totally dependent on electricity. It became a stark realization that a modern economy would not collapse so much as to cease to exist without power.

Then, an age of deregulation came. The Public Utilities Regulatory Policies Act was passed in 1978, allowing individual power plants to sell excess electricity to other utilities. This encouraged "privatization." Then, the Energy Policy Act of 1992 allowed open access to electric networks by various power companies to encourage competition. This led to an even more ungainly system. So, by the 1990s, the government allowed private companies, not just vertically integrated utilities, access to the electric grid. However, the vertically integrated utilities did not want competition. They creatively came up with ways to prevent private companies from using their transmission lines. Then, the government stepped in again and created rules to force open access to the lines. This set the stage for the not-for-profit *"Independent System Operators."* They managed the transmission of electricity in different regions. All this led to thousands of interconnections as the systems became computerized. Computerization helped automate complex systems, but it also created vulnerabilities. Manimaran Govindarasu, Professor of Electrical and Computer Engineering at Iowa State University, and Adam Hahn, Assistant Professor of Electrical Engineering and Computer Science at Washington State University, explain that "The grid has been physically vulnerable for decades."[198]

In modern history, the United States has experienced short-term losses in the grid. One of the worst cases in recent times occurred in 2003. It has become known as the "Northeast Blackout" and caused widespread power outages in the Northeastern and Midwestern US and the Canadian province of Ontario. Some areas went without power for a few days, and others for almost a week. At least 50 million people initially lost power. It was eventually traced to a cascade effect from a high voltage power line in northern Ohio brushing up against some overgrown trees in combination with a software bug when a resulting alarm sounded in the control room of FirstEnergy Corporation in Akron. Those domino-like events multiplied to affect 508 generating units and shut down 265 power plants.

The 2003 event, although very short, foreshadowed what a loss of power grid could look like in just the first few days. It provided a disturbing lesson. Water service in many areas stopped because they were dependent on power to maintain water pressure. This caused potential contamination of water in certain areas. Most rail services stopped. The security screening machines at airports could not operate, nor the ticketing computers, bringing air service to a stop.

Very few gas stations maintained their own generators, so people only had what fuel they had in their tanks. This proved true for a number of weeks even after the blackout ended because local refineries had fallen behind in production. Less than 10 percent of traffic signals had any backup power. Most factories stopped and lost important production quotas. Almost all cellular service stopped because the backup generators at the cell towers soon ran out of fuel. The internet failed, and people were unable to access ATMs. People had to make do with whatever cash they had in their pockets. In New York City, people not monitoring their use of candles caused 3,000 fire incidents. People carelessly using generators or cooking

---

[198] Manimaran Govindarasu and Adam Hahn are experts in cyber-physical system security for the electric power grid. They have written several papers and articles on the topic, including *Cyber-Physical Systems Security for Smart Grid* and *Vulnerability Assessment for Substation Automation Systems.*

devices meant for outside use caused carbon monoxide deaths. More than 100 deaths were associated with the blackout.

Remarkably, none of these events from the 2003 blackout, barring the disruption of water, would even be of critical importance in a long-term blackout. In the case of a power outage spanning weeks or months, the biggest problem would center on the fact that most towns and cities across our nation only carry about a three-day supply of food. One percent of our population is responsible for feeding 99 percent, and that system is dependent on fuel for agriculture, power for transportation, and a vastly complex and internet-reliant infrastructure for distribution. FEMA and our own military could cope with one large city or even a small sector of the country being without power for an extended period, but if the whole nation lost its access to power, emergency management could not possibly cope with supplying even 50 percent of our population with food and water. They cannot even plan for such a large-scale event. So, we must work to prevent such a runaway disaster by learning as much as we can about solar and other threats.

Other threats, of course, exist. The US first learned of the scientific principles of EMP effects of nuclear weapons in the 1950s, although Robert Oppenheimer and his team of scientists had some theoretical understanding of such phenomenon even before the first atomic bomb was tested. Because above-ground testing ended before the full effects of EMP were understood, it took many years of specialized tests to research the phenomenon during the almost 1,000 underground tests the US pursued after the Limited Nuclear Test Ban Treaty in 1963.

Since nuclear testing by the US ended in 1992, weapons specifically capitalizing on nuclear based EMP effects have not been pursued by the Western powers as far as is publicly known. One main reason for that is that nuclear weapons specifically designed to magnify EMP effects are considered first-strike or offensive platforms. Yes, in theory, any nuclear weapon can create EMP effects, but to attempt to destroy another country's power grid and electronic systems, you need highly specialized weapons with a very high three-phased gamma-ray emission generating a peak EMP field of 200,000 volts per meter. It would be designed to explode at specific high altitudes. These are called super-magnetic pulse weapons created for what is called "Blackout Warfare." Again, such a weapon would be considered only for a surprise or terrorist attack, and accordingly, the US nuclear arsenal has never (officially) contained such first-strike weapons. (The US now has non-nuclear based weapons that can cause electronic disturbances in a tactical or battlefield setting.) According to a congressional study, it is almost a certainty that Russia, China, and undoubtedly North Korea have pursued such exotic nuclear weapons.[199]

---

[199] Dr. Mark Schneider, *The Emerging EMP Threat to The United States*, (National Institute for Public Policy), foreword by Congressman Roscoe Bartlett (US House of Representatives; and https://freebeacon.com/national-security/china-russia-building-super- emp-bombs-for-blackout-warfare/.

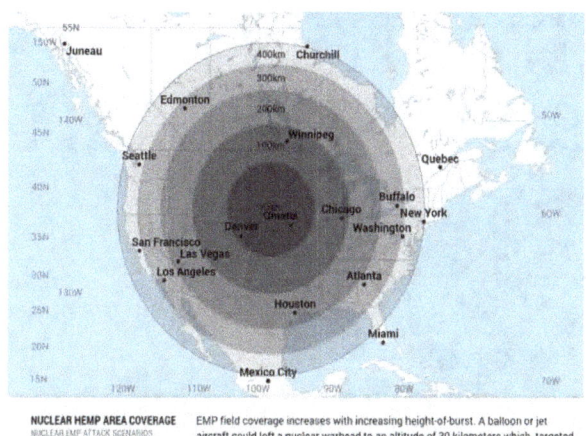

NUCLEAR HEMP AREA COVERAGE
NUCLEAR EMP ATTACK SCENARIOS — EMP field coverage increases with increasing height-of-burst. A balloon or jet aircraft could loft a nuclear warhead to an altitude of 30 kilometers which, targeted over New York City, would also cover Washington, D.C., New York State, New Jersey, Pennsylvania, Virginia, Maryland, Delaware, and most of New England.

The Washington Free Beacon reported:

Russian nuclear missile submarines could use super-EMP warhead to paralyze US strategic and conventional forces and blackout the national grid. The report states that 14 EMP bursts up to 60 miles would create powerful electronic waves for key facilities, including national missile defenses at Alaska and California; the command center at the Pentagon outside Washington; and the North American Aerospace Defense Command (NORAD) in Colorado. Other EMP strikes would shut down missile and bomber wings in Minot, North Dakota, F.E. Warren Air Force Base, Wyoming and Malmstrom Air Force Base, Montana. Bomber wings in Missouri, Louisiana, South Dakota, and Texas also could be blacked out along with nuclear missile submarine bases in Washington and Georgia.[200] Scofieldinstitute,org.

This chapter's lesson is not that both heaven and Earth can produce very similar electromagnetic disturbances. The important takeaway is how vulnerable our current state of technology has become and how totally dependent, if not enslaved, we have become to technology. In ancient times, some cultures worshiped the Sun as a god. Today, science shows us that the Sun has, figuratively speaking, the power of God. We are all creatures of heaven and Earth as we face many challenges and threats, some from the hand of humanity and some from the whims of nature. Science can give us the knowledge and power to overcome all, but we must show respect for such forces and be prepared.

(Prior to publication of this book during the third week of May 2024, one of the largest sunspots in modern history was recorded on the Sun with one of the largest of all CMEs. We just missed an Earth directed trajectory with that event and our modern civilization had a very lucky and narrow escape.)

---

[200] https://freebeacon.com/national-security/china-russia-building-super-emp-bombs-for-blackout-warfare/.

CHAPTER TWENTY THREE
ONE SECOND AFTER BY JAMES HALL

Since childhood, I have been frightened by scenarios of full-scale nuclear war. After reading the acclaimed *New York Times* Best Seller listed *One Second After* by William R. Forstche, I know the possibilities posed by nuclear weapons are much more complex and even more disturbing than I had realized.

The novel is accurately researched on cutting-edge science and the best intelligence available. Newt Gingrich praised this book's creative way of illuminating the dangers posed by an EMP attack. He, wrote the introduction to the book, warning that an EMP attack would "throw all of our lives back to an existence equal to that of the Middle Ages." An EMP event is a literal analogy to the sword of Damocles.

The novel alerted many in this nation, including Congress, to the possibility that the most destructive nuclear war might not involve dozens of nuclear weapons or a hundred or even a thousand. Based on the actual science, the book details how just three specialized EMP nuclear weapons, exploded at a precise altitude over our nation, could create a set of events that, in a period of nine months, would take the lives of ninety percent of the US population and prevent its ability to function as a nation. (See previous chapter for detailed description of EMP.)

The book is, of course, a novel, and as a work of fiction, it reads well and builds complex and engaging characters. Like a handful of other novels, like *Alas, Babylon* by Harry Hart Frank (Pat Frank) or *On the Beach* by Nevil Shute, the pages use English as an art form. Besides their literary excellence, what makes those books and this one so unique is that although nuclear scenarios suggest tremendous societal upheavals, the stories demonstrate the power of the human spirit and what can be accomplished by working together.

That is not to say that this story does not have a great deal of horror and violence. The plot, however, is so authentically different. In this tale, the United States is brought to its knees by silent EMP devices that, at first, are not even apparent. No country is definitively identified as the villain, and Europe and Russia suffer a similar attack. The point of the book is not geopolitics but technology. It takes a few days in the story before people even become aware of what they have lost and, for a period, still fail to realize the seriousness of the situation. Yet Forstche's plot shows how inhumane, primitive, and self-centered people can be. Nevertheless, it emphasizes how, in a disaster, it is more likely than not that the majority of humanity will work together with innovation to overcome it. Good can overcome evil.

The story is so engaging because we are given in this tale a very plausible depiction of what would happen if our energy grid went down and stayed down. The story analyzes many aspects of life that are indirectly dependent on electricity that most of us have never considered. William Forstche's masterpiece illustrates what it would be like to return, overnight, to the 19th century with 21st-century dependencies. It is a sobering lesson.

CHAPTER TWENTY FOUR
DETERRENCE

**HMS Warrior and Black Prince, public record image oil on canvas by
Stephen J. Card.**

An argument has been made by many that, for better or worse, nuclear weapons have served as an effective deterrent to large-scale war for more than 70 years now. The Smithsonian-affiliated National Atomic Testing Museum founding chairman, Troy Wade, often explained that concept to me when I served as the museum's Executive Director.[201] He also said nuclear weapons remain a reality, and while our aging nuclear stockpile is still considered a viable deterrent, it has to be maintained because time and technology keep moving forward.

Retrospectively, the concept of creating a deterrent to war is not a new idea. There are very interesting examples of specific weapons systems in history that formed effective deterrents. In their day, they were just as controversial as nuclear weapons and seemingly just as effective in preventing war. One of the best examples is that of the British warship HMS *Warrior* and her sistership HMS *Black Prince*. When those vessels were built, they were as innovative for their time from a technological standpoint as nuclear weapons were in 1945. Launched in 1861, these were the first true armor-plated, all iron-hulled warships.

They were not monitor designs like our own *Monitor* and *Merrimack* of the Civil War. Warrior and Black Prince were fast, ocean-going, powerfully gunned warships at unprecedented speed for the day of 17 knots under combined sail and steam. They were departures from any warship that had come before it or would have come for many years. Their mere presence and routine cruises up and down the English Channel

---

[201] Troy Wade prominently participated in the heyday of nuclear testing and then through underground testing and also took part in the "Star Wars" program while working with Edward Teller under Ronald Reagan's administration.

ended any threat of war between England and France, who had been bitter enemies for centuries. The very existence of those ships and their successors proved an actual deterrent to war with France. They were the ultimate weapons of deterrence of the day.

Simply put, deterrence is the use of a credible threat to make a second party or parties refrain from creating a crisis situation. Nuclear weapons, in theory, became ideal deterrents because they not only prevented an attack against a state's territory as a "direct deterrent" but also protected a state's allies through "extended deterrence." However, nuclear deterrence theory is still not without pitfalls. No weapon in history has proved the perfect deterrent. At best, a weapon of deterrence will only hedge or attempt to ensure a high likelihood that one's own territory remains secure. Of course, like anything, a weapon of deterrence can fail, or the entire system that a particular deterrence situation is based on can collapse. That is termed deterrence failure.

Nevertheless, nuclear weapons have long enticed nations by their power and influence. Nazi Germany fostered the discovery of nuclear fission in 1938. Germany and even Japan tried hard to harness the explosive potential of the atom but could not field the resources needed. Only British and American cooperation made the first successful test and then use of atomic weapons possible by July and August 1945. In a sense, Winston Churchill and Franklin Roosevelt sought to build the bomb before the Germans could, primarily to deter the Axis from having a monopoly. Unfortunately, proliferation became inevitable.

The Soviets tested their first atomic bomb in August of 1949. Britain followed in 1952, France in 1960, China in 1964, and all soon after that developed thermonuclear weapons. India developed an atomic bomb in 1974, and Pakistan in 1998. Israel and South Africa developed atomic weapons, but South Africa dismantled its weapons. At one time or another, Yugoslavia, Sweden, Australia, Norway, Taiwan, South Korea, Indonesia, Turkey, Greece, Romania, Libya, Canada, Brazil, Argentina, and Switzerland all had nuclear ambitions and programs.

Today, North Korea is the latest player with both atomic and thermonuclear designs as well as the systems to deliver such weapons. Kim Jong Un and his seemingly irrational statements and behavior have worried many. To date, however, he has remained just rational enough not to use a nuclear weapon, likely because he is sane enough to realize he will lose everything if he crosses that line. We pray nuclear proliferation will never extend to terrorist organizations who have no state of their own to lose and thus no reason to be influenced by the bomb's deterrent nature.

David Krieger of the Nuclear Age Peace Foundation has written:

The Soviets built nuclear bombs to deter the US China developed nuclear arms to deter the US and the Soviets. Israel did so to assure its independence and deter potential interventions from other nuclear weapons states. India developed nuclear weapons to deter China and Pakistan, and Pakistan to deter India. North Korea did so to deter the US.[202]

In terms of deterrents and the unique situation in North Korea, we look to Dr. Siegfried S. Hecker for insight. Dr. Hecker has served as Director Emeritus of the Los Alamos National Laboratory, a research

---

[202] Nuclear Age Peace Foundation, https://www.wagingpeace.org/category/david-krieger/.

professor at Stanford University, a senior fellow at the Freeman Spogli Institute for International Studies, and a longtime National Atomic Testing Museum supporter and frequent guest lecturer. He has made many fact-finding trips to North Korea and has recently stated:

North Korea deters the United States from military aggression because its small [nuclear] arsenal can still inflict unacceptable damage to the United States and its allies. The United States is able to deter North Korea through assured destruction. So, instead of mutually assured destruction, we have an asymmetric situation that is nevertheless capable of deterring both sides.[203]

When the theory of deterrence is applied to nuclear weapons, it implies the threat of retaliation. From this standpoint, such a theory requires that the holder of the deterrent preserves its ability to retaliate.

A lot has been written about the concept of Mutually Assured Destruction or MAD. Certainly, at the height of the Cold War, when the Soviets and Americans possessed almost 70,000 nuclear weapons combined, the very apocalyptic nature of the deterrent meant that it could not be used. It was admittedly madness, yet despite this nuclear madness, the stockpiles were thought to have prevented world war. Of course, they had to, as there was no other acceptable alternative. Dedicated individuals like Dr. Hecker, Senator Sam Nunn, and Senator Richard Lugar worked hard in the post-Cold War period to encourage the decrease of nuclear stockpiles and provide oversight of the USSR's former stockpile. I had the pleasure of working in Indiana with Senator Lugar during that time on a World War II oral history project, and I recall his obvious dedication and often preoccupation with securing and accounting for the former Soviet nuclear stockpiles.

---

[203] Conversation with retired Los Alamos National Laboratory Director Siegfried S. Hecker, June 2015, National Atomic Testing Museum.

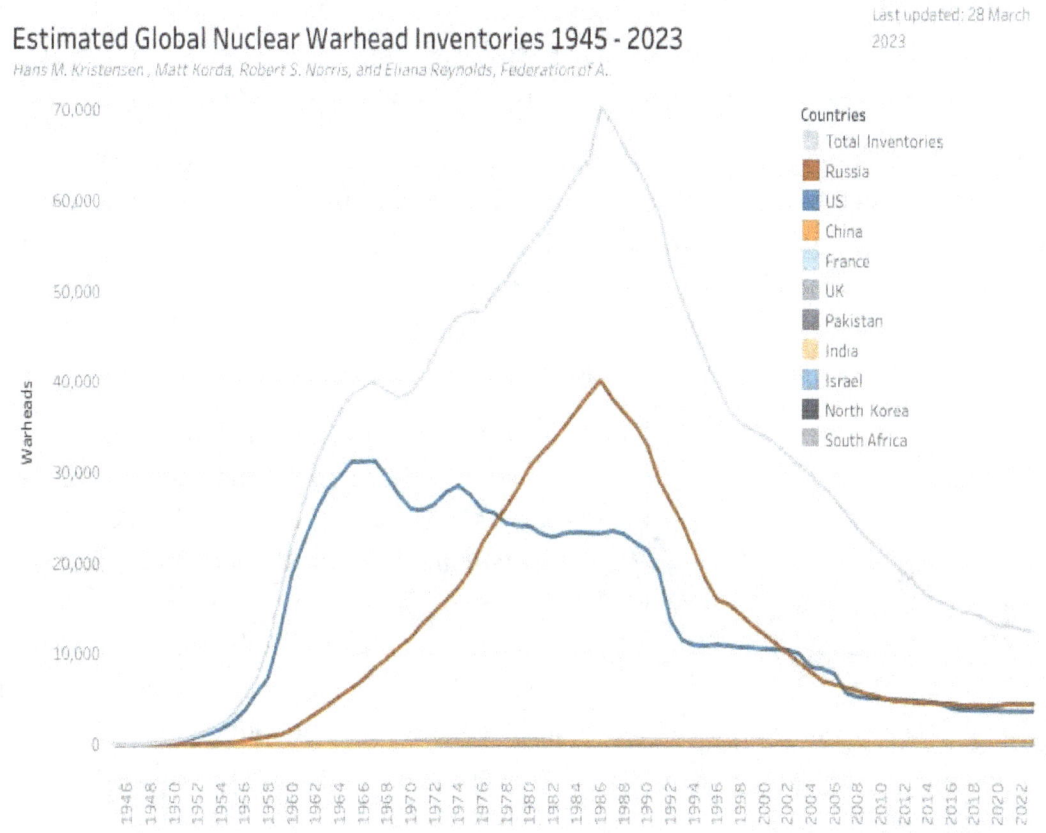

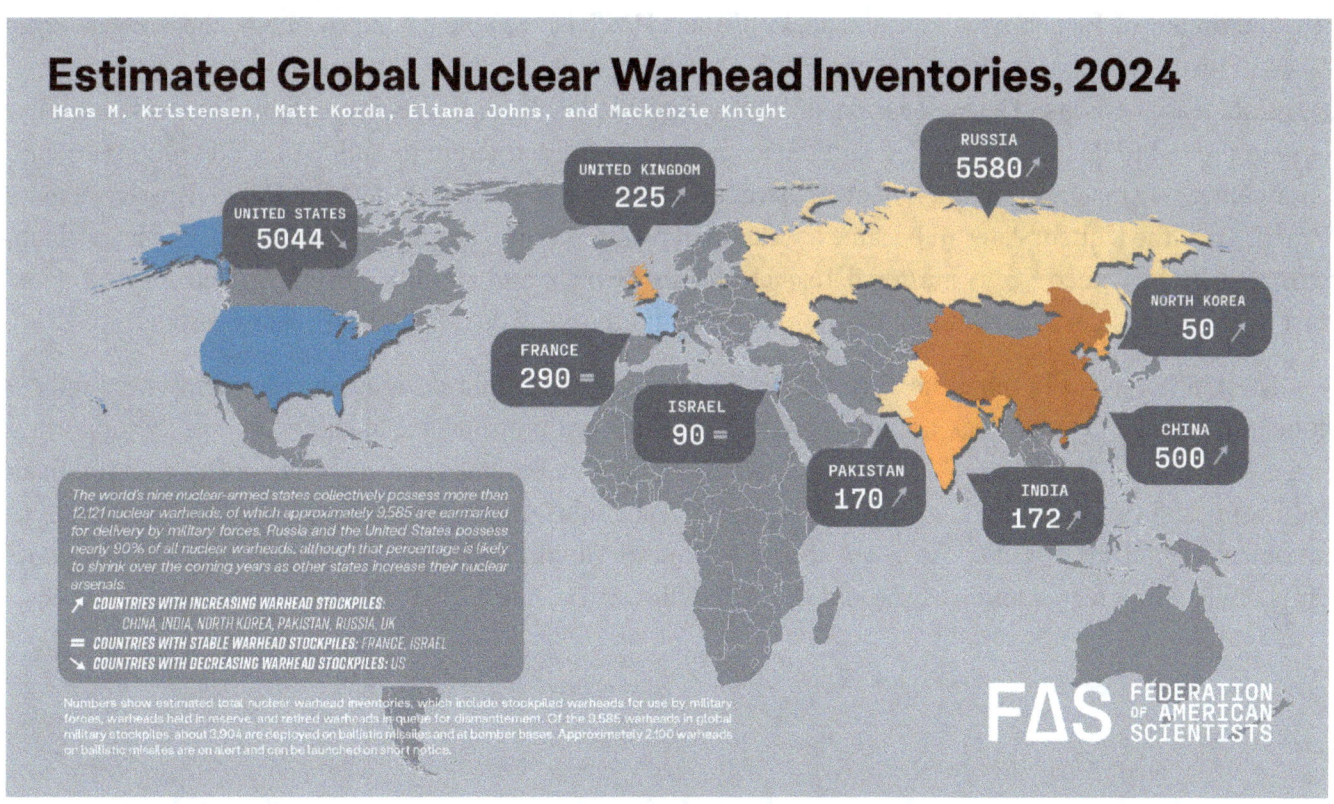

Graphs by Federation of American Scientists, https://fas.org/initiative/status-world-nuclear-forces/.

Great progress was made over the years, although the current United States and allied forces' stockpiles, combined with resurgent Russian nuclear forces, are still very large. They total several thousand warheads on alert-ready status. So, this is a deterrent we still have to live with and learn to manage.

From a historical standpoint, concern over arms control is a familiar lesson. 20th-century leaders in their formative years, like Churchill and Roosevelt, dealt with a great arms race and weapons of deterrence long before the Manhattan Project gave birth to the atomic bomb. In their day, a weapon concept as revolutionary as the HMS *Warrior* and the later Fat Man and Little Boy bombs turned the balance of power upside down. When launched in 1905, the HMS *Dreadnought* did the same. It made every other warship in the world obsolete overnight. And like many weapons of deterrence, it led to an arms race.

Dreadnought embodied the revolutionary concept of a fast, well-armored, all-big-gunship. With the then unheard-of large caliber 12-inch armament, the *Dreadnought* balanced select-area armor protection with the remarkable speed (for the day) of 24 knots, making this an intimidating and innovative class of weapon system. It proved the genesis of a whole new line of ships, all commonly adopting the classification of dreadnoughts, which the United States would later call battleships.

The dream of British naval admiral Jacky Fisher, HMS *Dreadnought,* led to one of the greatest arms races in history. The early cruises of the HMS *Dreadnought* were, in fact, as secret as the deployment of early nuclear weapons. The possession of a powerful battleship did indeed create a deterrent. Initially, Britain proved to be the only nation with the technological infrastructure to build such a revolutionary and dominating war machine. Soon, however, Germany followed, then France, Austria, Italy, Russia, Japan, Brazil, and the United States. It became a weapon of proliferation. The calibers went to 13.5-inch, 14-inch, and 15-inch; the United States mastered 16-inch, Japan 18-inch, and Germany even planned 21-inch guns. Just as the bombs got bigger during the Cold War, so did the early dreadnoughts decades earlier.

The USS *Nevada* became the first truly modern US dreadnought design. She had many new radical concepts like her 11-inch-thick armor protection, oil-fired steam turbines, and triple gun turrets supporting large caliber 14-inch barrels. All these engineering principles became the centerpiece of every American battleship and cruiser to follow. Naval designers considered *Nevada* a modern marvel of the day as well as an object of industrial beauty. Assistant Secretary of the Navy, Franklin D. Roosevelt, watched *Nevada's* christening with great emotion as she slid down her slipway on July 12th, 1914.

Building such weapons systems became very expensive and drained national treasuries. This was especially true of Britain and Germany, which were in a critical period of evolving social programs. Still, it can be argued that for almost a decade, as Winston Churchill became First Lord of the Admiralty and Franklin Roosevelt became Assistant Secretary of the Navy, the dreadnoughts did deter war. That is not to say the insane arms race did not also lead to a world war, but when the First World War started, the great battleships played little part, barring a few naval actions during the whole course of the conflict. They proved too valuable to lose, and although Britain gained superiority, it never could easily face the entire German battle fleet of dreadnoughts without literally risking, as Winston Churchill put it, "losing the war in a single day." It is ironic that the great steel castles of the sea, as historian Robert Massie called them and as painter Vincent Alexander Booth portrayed in his work, could not win the war for Britain. However, it was also true that if one side lost its stockpile of dreadnoughts, it would open its shores to easy invasion.

Vincent Alexander Booth, *Battlecruiser Derfflinger,* **oil on canvas.**

This is a common downside of having a deterrent. A superweapon, as former EG&G president Barney O'Keefe stated, can hold you hostage to the technology of the weapon if your opponent also possesses a similar deterrent. During the Cold War, the US and USSR indeed became nuclear hostages to one another. And, as we have already noted, history shows that deterrent weapons systems always and inevitably lead to the proliferation of that same weapon system.

Sometimes, deterrent weapons systems do not work or in other words do not scare an opposing force enough to not attack. In 1941, Japanese leaders like Admiral Isoroku Yamamoto knew that the United States had overwhelming military resources. Yet, irrationally, Japan committed to an unwinnable war with its impulsive attack on December 7th of that year. In 1967, Egypt and Syria, despite dramatic surprise, knew they could not defeat Israel in what became the Six-Day War, yet they attempted it just the same. In 1982, Argentina even went head-to-head with the nuclear power of Great Britain over the Falklands and was predictably beaten by the still viable empire nation.

Of course, in that last example, while Britain won that war, it could have no more used nuclear weapons against the Argentinians than we could have against the North Vietnamese, which is another problem with nuclear deterrents. Teddy Roosevelt effectively said, "Speak softly and carry a big stick," although sometimes your stick may be too big to use.

In 1962, during the Cuban Missile Crisis, the magnitude of the nuclear power available to the USSR and the US probably forced President Kennedy and Nikita Khrushchev to their senses. Yet the situation could have easily gotten out of control. So, no, deterrence is not a perfect solution. It takes rational leaders to make rational decisions to make deterrence work in the first place.

The question often arises: are nuclear weapons the ultimate deterrent or just the most recent one? History seems to show that weapons of deterrence come and go, but the concept remains. The mighty Roman legions of the Pax Romana period ensured peace for 200 years. The deterrent failed to serve its purpose when the legions were undermined with a general decline of the Roman political system. Going back even further, the great powers of Athens and Sparta kept the peace for a good period with deterrents, but only so long as their cultures could maintain a viable and rational system.

No one can guess with any certainty what the future or the next war could look like. There was a time when science fiction writers like H.G. Wells envisioned unbelievable weapons such as tanks, submarines, airplanes, and even atomic bombs. As we know, almost any imagined weapon eventually comes to be in some form. And most disturbing is that those weapons are all eventually used. There will undoubtedly be new weapons of deterrence, but when will people realize that it is not just the newest weapon that we need to deter war but the elimination of war itself to deter our own eventual annihilation?

CHAPTER TWENTY FIVE
THIS IS NOT A DRILL

16, Cards 0! O. S. C. Wins To Get Rose Bowl Bid

HILO TRIBUNE HERALD — FINAL EDITION

HILO, HAWAII, SUNDAY, NOVEMBER 30, 1941

JAPAN MAY STRIKE OVER WEEKEND

Steps Taken To End Dock Delays Here — **FOOTBALL SCORES** — **Midshipmen, Pitt, Huskies Other Winners** — **Army Charters Hamuula For 16 Day Period** — **Reds Re-Take Rostov After Classic Move** — **Tokyo Desperate As Talks Collapse** — **Tojo Demands Asia 'Purge'**

On November 30, 1941, an American Hawaiian newspaper, the *Hilo Tribune Herald*, ran an article that the Japanese were going to attack Hawaii over the weekend. This was only a week before the Pearl Harbor attack. When no initial raid came that November weekend, people were livid that such a false alarm could have been printed. Unfortunately, while people in their outrage were looking for a scapegoat, they missed the bigger picture—that there was a real threat at hand. A week later, "Air raid Pearl Harbor. This is not a drill." became synonymous with the December 7th day of infamy. It is a good analogy to the Hawaii false alarm we witnessed just a few years ago.

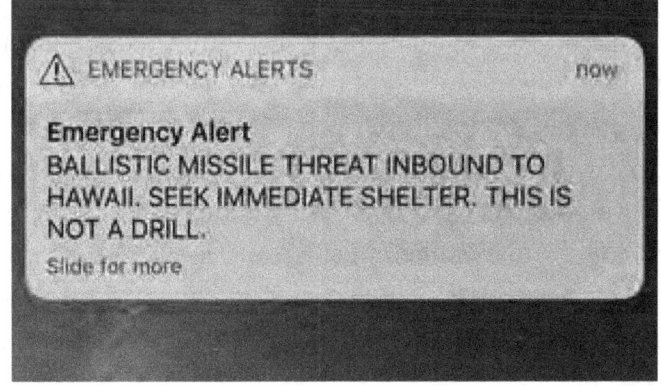

⚠ EMERGENCY ALERTS — now

Emergency Alert
BALLISTIC MISSILE THREAT INBOUND TO HAWAII. SEEK IMMEDIATE SHELTER. THIS IS NOT A DRILL.
Slide for more

On the morning of January 13, 2018, a Hawaiian emergency alert message went to cell phones and then scrolled on local televisions, reading as follows: "BALLISTIC MISSILE THREAT INBOUND TO HAWAII. SEEK IMMEDIATE SHELTER. THIS IS NOT A DRILL" It took 38 minutes for the erroneous alert to be recalled entirely. Understandably, the entire state of Hawaii and many visiting tourists needlessly suffered a terrible fright. People's outrage is still strong, but at least this proved we have a warning system in place. Despite the mistake, the system works well. Since an updated test in the fall of 2023, we have a revised national emergency alert system that, in theory, sends a notification to all our cell phones if a true national emergency arises.

We also have a very adequate missile warning detection system in place. That system was called the Air Force Space Command-operated Defense Support Program or DSP. DSP consisted of twenty-three bus-sized satellites in geosynchronous orbit that detected missile launches in real time anywhere in the world. DSP efficiently detected Iraqi Scud missile launches during Operation Desert Storm. DSP satellites have recently been replaced by the Space-Based Infrared System (SBIRS) satellites. (See chapter fourteen.) That system functions like DSP as a constellation of integrated satellites in geosynchronous orbit and high elliptical orbit which sends its information to the Air Force Space Command in Colorado and to the North American Aerospace Defense Command (NORAD) in Cheyenne Mountain near Colorado Springs.

So, we will know if and when North Korea ever launches a missile in anger. Now, what happens after that? That is a matter of debate. Certain sectors of our defense network feel a missile can be intercepted. That is, however, only under ideal circumstances if there are not too many of them and if they are not advanced types of ICBMs. This is not to say that our defense technology is not advancing, and the day may come when we have a true missile shield, but that day has not arrived yet. Our current, still flawed, missile defense system is explained in detail in a recent and outstanding book by Annie Jacobsen called *Nuclear War*.

The other question in many people's minds is, "Where do you go during a nuclear attack?" Most people with modern households do not even have a basement. The truth is that there are few civil defense options available. Even in the heyday of civil defense in the 1950s and early 1960s, nuclear shelters were never a very well-funded initiative. Little support went into a comprehensive civil defense because the focus became directed to bigger issues, such as regular defense spending. Deterrence has always been thought to be the best defense. Mutually Assured Destruction, or MAD, for better or worse, worked with the former Soviet Union, which, like us, had thousands of nuclear weapons. It made nuclear war improbable. However, what do we do now with a threat like a rogue nation, like North Korea, with not thousands of nuclear weapons but less than a hundred, yet with the growing potential in missile development to deliver them?

Bomb shelter mentalities have never worked well. Indeed, the day may come when people realize that the best nuclear civil defense is to prevent the proliferation of such weapons in the first place. This is exactly where modern-day global security specialists focus.

And like the first Pearl Harbor false alarm in November of 1941, the point is still being missed. That point is that there is a threat. North Korea is such a threat. In past years, they have mirrored, on a much smaller scale, our own patterns of nuclear weapons and missile testing. From 1946 to 1992, the United States conducted approximately 1,054 nuclear tests between the Pacific Proving Grounds in the Marshall Islands and primarily at the Nevada Test Site.[204] There were many reasons for conducting the tests, but the initial reason centered on making the weapons smaller in physical size so as to make them easier to deliver. A key focus became to marry nuclear weapons technology with the other great advancement to come out of the Second World War: rockets.[205] It took about a decade for both the US and USSR to develop nuclear weapons that a missile could carry and to develop those actual missiles or intercontinental ballistic missiles

---

[204] Other official figures put this at 1,030 total tests. Some tests took place off Kiribati Island in the Pacific, three in the Atlantic Ocean; ten other tests took place at various locations in the United States, including Alaska, Colorado, Mississippi, and New Mexico.

[205] Today, we use the term missiles because the word rocket is somewhat old-fashioned now. "Rocket" also suggested the German weapons of World War Two, and even though it was largely the talent of German scientists who assisted our own programs, we put a new face on our technology with the term "missile" for weapons-related devices and the word "rockets" used in connection with space vehicles.

(ICBMs).

Only a handful of global powers have perfected ICBMs. North Korea, the Democratic People's Republic of Korea (DPRK), is fast approaching that capability, and many believe they have now achieved that goal. To date, North Korea has conducted six nuclear tests at their Punggye-ri Nuclear Test Site. This violates the Treaty on the Non-Proliferation of Nuclear Weapons, and the DPRK has come under increasing sanctions by the United Nations. Their history in nuclear testing goes back to 2006.

On October 9, 2006, North Korea conducted its first underground nuclear detonation in the Punggye-ri Nuclear Test Site's granite tunnels under Mount Mantap. This is in the heavily forested Hamgyong mountain range fifty miles from Chongjin. The detonation was estimated at less than a kiloton and may have only been an underperformed chain reaction. Seismic instruments around the world recorded this test, so there was no way to hide the detonation, even though it proved small.

Xenon and krypton isotopes were also detected in the atmosphere by aircraft with specialized sensors, confirming a nuclear test had occurred and proving Pyongyang had used a plutonium-fueled device. In response, the UN imposed trade and travel sanctions. Then in December, Six-Party talks began involving the United States, South Korea, China, Japan, and Russia in an attempt to find a peaceful solution to the threat imposed by the DPRK's weapons programs. North Korea agreed to close its main nuclear reactor in exchange for a $400 million aid package and then decided to begin disabling its nuclear weapons facilities. However, North Korea missed its end-of-year deadline to turn off its weapons facilities. So, the talks proved fruitless.

On May 25, 2009, North Korea conducted a second underground nuclear test, estimated by seismographs to be between 2 and 5.4 kilotons. It became clear by that point that North Korea had an established enrichment capability. In that test, they used a new, more confined tunnel so atmospheric gases could not be detected. The deeper tunnel was located about halfway up their 7,200-foot mountain complex. There are three visible main entrances, or portals, into a series of horizontal tunnels stretching a mile or more into the mountain. Studies of this second test suggest it has the shape of a fishhook, just as is used in past testing in Pakistan. Of course, great assistance probably came from Pakistan in the first place. The tunnels are believed to be about 9 feet wide and 9 feet high, with multiple sharp corners and various dead ends that defuse and absorb the blasts. In that test, bulkheads were probably installed to confine the gases while sand, gravel, or other materials mixed with concrete served to plug the tunnel.

The second test led to further sanctions by the UN. By February 2012, the DPRK suggested it would once again halt its nuclear weapons tests, missile launches, and nuclear enrichment activities in exchange for food aid.

Then on February 11, 2013, a third underground test took place with a yield of 14 to 16 kilotons. On March 30 and 31, 2014, the DPRK fired hundreds of conventional artillery shells across its Yellow Sea border with South Korea. The South responded by firing 300 similar shells in return.

On January 6, 2016, a fourth underground test supposedly involved a hydrogen bomb, but this claim has not been verified. US analysts do not believe that a hydrogen bomb was detonated at that time because the seismic data collected only estimated a 6 to 10-kiloton yield, which is not consistent with the power that such an explosion would generate. (Seismographic data is principally all we have to go on because these were all underground tests, and only minor venting would have occurred if it had been detected.)

On September 9, 2016, the fifth test occurred, calculated at between 10 and 25 kilotons. That is about the size of the 15-kiloton bomb dropped on Hiroshima. There is some debate as to exactly what this test revealed. David Wright, a senior scientist at the Union of Concerned Scientists, stated, "My guess is that the North is happy to have the world see that it is testing and get an estimate of the yield—at least as long as it is increasing—but likes keeping the world guessing about how advanced its program really is."[206] Other experts at the time were concerned that the North was trying to develop a true thermonuclear bomb with the use of lithium-6.[207] This material enhances, or can "boost" the power of a fission weapon which could then ignite a fusion weapon by producing tritium that combines with deuterium. There is also speculation that aside from research into fusion, North Korea may be developing composite fission designs using plutonium and enriched uranium cores. Such a design would be smaller and easier to put on long-range missiles. It would also economize, making the most out of its ongoing production of enriched materials.

On September 3, 2017, North Korea conducted its sixth and latest underground nuclear detonation at the Punggye-ri Nuclear Test Site. The US-Korea Institute at Johns Hopkins University confirmed the test to yield a quarter of a megaton. That is almost 17 times the power of the bomb dropped on Hiroshima and many times larger than their fifth test. In that test, they may have crossed the thermonuclear threshold!

In 2016, North Korea also made dramatic advances in missile technology. North Korea accomplished the big step of launching a solid-fuel ballistic missile from a submerged submarine, and then a week before its latest nuclear test, they launched three missiles into the Sea of Japan. This began a series of on-and-off, but continuous, line of missile tests that have continued into 2024. At the time of this writing, in early 2024, Kim Jong Un is making rapid and dramatic progress on his promise to the world to develop a nuclear-tipped ICBM. It is clearly a dangerous situation.

The DPRK has been a significant issue since the close of war in Korea in 1953 when combatants agreed on an armistice but not a peace treaty. To this day, that war remains on hold. North Korea did not really gain its own true autonomy until the last Chinese troops withdrew in 1958. Prior to that, Party Leader/Prime Minister Kim Il Sung resisted numerous attempts by both China and Russia to manipulate and even depose him. Although, he earned a following and respect after years of resistance to the Japanese occupation of Korea and remained a determined leader into the 1950s.

---

[206] https://blog.ucsusa.org/author/david-wright/,

[207] Lithium 6 is a raw material required for producing a single-stage thermonuclear fusion device and a boosted fission weapon.

**Kim Il Sung and son Kim Jong Il, Lobby Mural
Chongchon Hotel, Mt. Myohyang Region; Hyangsan
Town, DPRK (North Korea).**

The gradual rise of North Korea began in 1956. In that year, the Soviet Union started training North Korean scientists and engineers in the basic principles of nuclear fission, which served as the genesis of their nuclear program. Kim Il Sung had asked both China and the Soviets for help in making a nuclear weapon but was refused. Instead, the USSR decided to help North Korea develop a "nuclear energy program," which included the training of scientists and engineers. In 1959, North Korea and the Soviet Union signed a nuclear cooperation agreement. In 1962, they completed the construction of a research reactor at what became known as the Yongbyon Nuclear Research Center. They also began mining uranium ore.

By 1974, Kim Jong Il, son of Kim Il Sung, gradually assumed key political positions, although Kim Il Sung remained the primary leader until his death in 1994. Both father and son favored nuclearization. Kim Il Sung would never forget the implied threat from US nuclear weapons, which helped force the North Koreans to the bargaining table in 1953. The current leader of Korea, Kim Jong Un, had these lessons impressed upon him both by his father and grandfather. Since assuming power on December 28, 2011, following Kim Il Sung's death, he has committed North Korea to nuclearization. It is now an integral part of the national identity of North Korea and is impressed upon every citizen.

By the mid-1980s, North Korea was already well on the way to that path when it mastered a significant enrichment capability as a result of Soviet technical assistance. In 1984, they constructed a radiochemical laboratory that served as a reprocessing plant where plutonium could be produced. The next year, the DPRK sent a deceptive signal by signing the Nuclear Non-Proliferation Treaty, which had been signed by the Soviets and Americans in 1968. By 1993, they were clearly not complying with the Nuclear Non-Proliferation Treaty and withdrew after considerable controversy with the United Nations and the International Atomic Energy Agency. They then suspended their initial withdrawal.

In the 1990s, North Korea gained access to Pakistan's nuclear technology. This fact became known a decade later, but we still do not know the exact details. Undoubtedly, Dr Abdul Qadeer Khan, the father of

197

Pakistan's nuclear weapons program, sold sensitive technology to North Korea and apparently other rogue nations. Pakistan developed its own nuclear weapons during the 1970s and 1980s to attain parity with India.

On October 21, 1994, North Korea further deceived the world by signing an agreement with the United States agreeing to freeze the operations of its nuclear reactors. The plan was to replace the old reactors with nuclear proliferation-resistant light waterpower plants in exchange for an agreement on North Korea's disarmament. However, in October 2002, it was revealed that North Korea was operating a secret nuclear weapons program, and negotiations fell apart.

On January 10, 2002, they again withdrew from the Nuclear Non-Proliferation Treaty in 2003. By 2003, North Korea admitted they had nuclear weapons, or at least made that claim. In December of that same year, North Korea again offered to make a deal. They proposed to "freeze" their nuclear program in exchange for concessions from the United States; however, President George Bush refused and insisted on a dismantling of their nuclear program.

On September 19, 2005, another false start arose when the DPRK asked for a non-aggression pact with the United States in exchange for North Korean nuclear disarmament. Again, and only a day later, negotiations broke down. It seemed each successive US administration spent the bulk of its tenure just learning the same old lessons—that negotiations with North Korea were simply an illusion of dialogue that never went anywhere. In July 2006, North Korea began a series of long-range missile tests. As a result, the United Nations became involved with a resolution demanding the suspension of their nuclear program.

North Korea continues to defy explanation. Despite all expectations to the contrary, Kim Jong Un's regime has made steady and significant progress in both nuclear weapon and ballistic missile developments since he assumed leadership in 2011. Recent successful test launches of apparent ICBM platforms and even three orbiting satellites of unknown purpose exemplify that. Increasing international sanctions have seemingly failed to slow what is also obviously a very expensive arms buildup. Equally surprising is the success of a parallel program in increasing economic development. This mutual emphasis on both weapons and the economy is called *"byungjin."* None of this should be happening because the United Nations and many governments, like the United States, have worked to isolate and cripple North Korea. Yet, it is happening. With Russia's war in Ukraine, Kim and Putin are active trading partners, along with Chinese President Xi Jinping.

The final word on North Korea concerns its possible future, which could be centered on one surprising individual. When working as Executive Director of the Smithsonian-affiliated National Atomic Testing Museum, I worked on a three-year series of widely read articles that showed a dramatic swing of events in North Korea. It demonstrated that by the end of 2017, North Korea had completed its sixth and last nuclear test to date, which, as already detailed, may have crossed the thermonuclear threshold. Kim Jong Un also delivered his promise to perfect what proved to be a primitive intercontinental ballistic missile or ICBM. By the end of 2018, Pyongyang had conducted a significant year of diplomacy, not only with the United States and South Korea but with China and Russia as well. By early 2019, the harsh rhetoric during the first part of the Donald Trump administration eased, and it looked like North Korea might convince the West to address sanctions as the US and South Korean military exercises were relaxed. The Covid pandemic, to an extent, then sidelined the North Korean story. However, in 2023, tensions again rose. In this history, one common denominator of wide concern exists. That is Kim Jong Un's sister, Kim Yo Jong. She has steadily gained power over this period, holding many key posts. She has long been the mastermind behind propaganda efforts and is now clearly Kim Jong Un's most valued minister and likely his designated future successor. That is a stunning accomplishment for the role of a woman in such a rigid society as North Korea.

The West has long recognized her as a formidable personality and a key behind-scenes player. The US Treasury has also condemned Kim Yo Jong for "severe human rights abuses." Western and South Korean intelligence have great reservations. Their concern with Kim Yo Jong is that she is much more learned, accomplished, and capable than Kim Jong Un, but is extremely unpredictable in comparison to her already highly electric brother.

That sounds like a contradiction, but contradiction has been the best word to describe North Korea in recent decades. Leif-Eric Easley, associate professor of international studies at Ewha University in Seoul, recently stated that Kim Yo Jong plays a pivotal role in North Korean domestic and foreign policy and is one of the main stakeholders in the regime's survival.[208]

In late December of 2023, *38 North* analysis group reported that "North Korea launched a Hwasong (HS)-18 solid-propellant, road-mobile intercontinental ballistic missile. . . Pyongyang now regards the HS-18 ICBM system as operationally deployed".[209] Since 2017, Kim's regime has continuously tested and deployed road-mobile ICBM prototypes resembling Chinese and Russian hardware. These have become a significant international security concern. North Korea is a clear and present danger. Next time, it may not be a drill.

---

[208] My documentation for this chapter comes from my research while writing a series of 37 museum content articles on North Korea while serving as the Executive Director of the Smithsonian-affiliated National Atomic Testing Museum.

[209] Vann H. Diepen V, Military Affairs, 38 North, Third Successful Launch of North Korea's Hwasong-18 Solid ICBM Probably Marks Operational Deployment, December 21, 2023, https://www.38north.org/2023/12/third-successful-launch-of-north-koreas- hwasong-18-solid-icbm-probably-marks-operational-deployment/.

CHAPTER TWENTY SIX
THE BOMB IS BACK

**Oil on canvas by Dale Cox, artist, veteran Los Alamos
engineer, and friend of the authors.**

When I was a young child, the talk of "the bomb" arose during many dinnertime conversations. The threat of nuclear war served as a fact of everyday life. By my teenage years, we all assumed the Cold War and the Soviet peril to Europe would simply remain the norm, and movies like *The Day After* engaged and horrified us. Then, as I became a young adult, all that madness magically vanished with the fall of the USSR. The end of the Cold War was not quite that easy, but it seemed so at the time. Now, three decades later, talk of nuclear weapons is back in the news after a very long hiatus. "The bomb" returns to our lexicon as we realize the threat never really ended, and as usual, history comes full circle. The fear now is that we may be on an actual countdown to a third world war as we once feared.

In the days following Russia's invasion of Ukraine, federation President Vladimir Putin placed his nuclear forces on heightened alert, where they remain. Not since the days of John F. Kennedy has a United States president needed to be so concerned with a nuclear war scenario with Russia. That is ironic when considering that the end of the Cold War brought on an 88 percent reduction in nuclear arms and hoped for an end to European wars.

As of 2024, although significantly downsized, Russia still has the largest nuclear arsenal at approximately 5,580 warheads. Those break down to 1,710 deployed or active warheads on "strategic" (long range, large yield) ballistic missiles, 870 on land-based ballistic missiles, 640 on submarine-launched ballistic missiles (SLBMs), and 200 bomber-based. Approximately 1,900 are smaller yield "tactical" or "non-strategic" or "battlefield/theater-based" warheads. (The remaining are in reserve, and all these numbers differ somewhat according to sources.) Among those tactical weapons are self-propelled artillery pieces or "nuclear cannons," which can fire conventional or nuclear-tipped shells. These throwbacks to the Cold War have been positioned in the city of Vesela Lopan in the Belgorod region, only ten miles from the Ukrainian border. The most visible tactical weapon is the 9K720 Iskander missile system, which NATO designated the SS-26. It can deliver conventional warheads, or tactical nuclear weapons, up to a range of 300 miles. The Russians have been using conventionally armed Iskanders since the conflict began.

Social media posts show a Russian nuclear-capable 2S7 Pion self-propelled artillery gun.

In comparison to Russia, the United States still has a viable strategic nuclear arsenal. It is smaller than Russia's at 5,044 warheads, 1,770 deployed, 1,938 reserved for operation use, 1,336 retired or reserve status and eventually slated for dismantlement. 400 are on ICBMs in state-side missile silos, 970 SLBMs on submarines, and 300 bomber-based. Only 100 of those devices are considered true tactical nuclear weapons based in Europe. These no longer include nuclear artillery like the Russian Cold War era leftovers, but weapons like the B61 gravity bomb, which can be dropped from an aircraft. All of these US, and Russian, numbers also change slightly from year to year.

The United States' tactical arsenal (meaning non-Triad theater-based nuclear weapons) now represents a 90 percent reduction from a time when we had as many as 6,951 tactical nuclear warheads in Europe, which were once housed in 145 nuclear storage sites in various NATO countries. All that effectively remains from that era is the B61 gravity bomb, conceived after the Cuban Missile Crisis as a device capable of being carried by a fast, low-flying fighter or bomber. It typically has an adjustable yield capability ranging from one to 100 kilotons, which was designed to give a US president more flexible options in the event of a nuclear conflict. In a new variant, or rather modification, called the B61-13, its yield will be upgraded to 360 kilotons. (See Appendix II.) Some authorities debate whether that can be considered or used as a tactical device. Unlike the Russians, most Western tacticians discount that any type of nuclear weapon can ever be legitimately used and has no real use except as a deterrent. Of course, logic would dictate that, if ever used, a nuclear weapon would cease to be a deterrent. Most concerning is the concept that some Russian officials may not even consider the use of small, low-yield, battlefield nuclear weapons as actually crossing the nuclear threshold.

Colin H. Kahl, the Undersecretary of Defense for policy, said at the 2021 Carnegie International Nuclear Policy Conference that "Russia continues to develop new kinds of nuclear weapons and continues to expand its arsenal of non-strategic nuclear weapons — typically smaller, lower-yield 'tactical'-style nuclear weapons designed to attack troops or facilities, rather than an entire nation."[210]

Understandably, there is a great deal of controversy about the meaning of the term "tactical nuclear weapon." In the Cold War era, it was assumed that if a war broke out in Europe, the Soviet army would have tried to break through the Fulda Gap into Western Europe. (The Fulda Gap represents several open passes running through the hills about 60 miles northeast of Frankfurt.) The idea of countering such an invasion with nuclear weapons dates to 1953, when Winston Churchill, in his second term as Prime Minister, visited the new incumbent US President Dwight Eisenhower on a trip to America. They reasoned that tactically applied nuclear weapons (often called theater-based weapons) could block a Soviet invasion of Western Europe. In those days the idea of small fission or A-bombs and nuclear artillery shells, in theory, negated the cost of matching the USSR in terms of soldier for soldier and tank for tank. Neither leader wanted to destroy their peacetime economies, which were still rebuilding and rebounding after World War II.

The Soviets, of course, followed with building their own tactical nuclear arsenal and higher-yield strategic weapons as well. That led to a great arms race in the 1960s, 1970s, and 1980s, totaling close to 70,000 combined nuclear devices between the US, NATO, and the Soviet Union. It cost all countries dearly and may have precipitated the fall of the USSR by contributing to its economic decline.

Today, in contrast, modern US commanders see the idea of tactical nuclear weapons as outdated. General Colin Powell stated many times in his career that such weapons have no real use; however, he valued the deterrent nature of strategic nukes. Former Air Force General John E. Hyten always insisted that tactical nuclear weapons were no different in significance than strategic nuclear weapons. Hyten stated, "It's not a tactical effect, and if somebody employs a nonstrategic or tactical nuclear weapon, the United States will respond strategically, not tactically, because they have now crossed a line, a line that has not been crossed since 1945."[211] Former Secretary of Defense Jim Mattis similarly told the House Armed Services Committee in 2018, "I don't think there is any such thing as a 'tactical nuclear weapon.' Any nuclear weapon used at any time is a strategic game-changer."[212] Crossing the nuclear threshold, except as a deterrent to an attack, is unacceptable to Western leaders today.

In contrast, the Russian reasoning behind such an issue is debated. Under Defense Secretary Kahl said at the 2021 Carnegie International Nuclear Policy Conference, "We also see that. . . the role that nuclear

[210] 2021 Nuclear Posture Review, US Department of Defense; and Todd Lobez, *Nuclear Posture Review, National Defense Strategy Will Be Thoroughly Integrated*, DOD News, June 25, 2021,https://www.defense.gov/News/News-Stories/Article/Article/2671471/nuclear-posture-review-national-defense-strategy-will-be-thoroughly-integrated./.

[211] Strategic Quarterly, An Interview with Gen John E. Hyten Commander, USSTRATCOM Conducted 27 July 2017, https://www.airuniversity.af.edu/Portals/10/SSQ/documents/Volume-11_Issue-3/Hyten.pdf.

[212] *Defense News*, https://www.defensenews.com/space/2018/02/06/mattis-no-such-thing-as-a-tactical-nuclear-weapon-but-new- cruise-missile-needed/.

THE SWORD OF DAMOCLES OUR NUCLEAR AGE

weapons play in Russia's doctrine is quite elevated in the sense that, I think, Russia sees much higher utility for nuclear weapons than any other state."[213] It has been argued that Soviet and now Russian leaders have long viewed a nuclear war as a logical possibility and have accordingly always maintained an active civil defense program. It is, in fact, believed the Russians may "escalate to de-escalate" if they face a critical situation. Sarah Bidgood, director of the Eurasia program at James Martin Center for Nonproliferation Studies in Monterey, said, "Putin seems to feel confident that he can use veiled threats and signals to escalate and de-escalate to suit his needs." Bidgood also said, "But that's a very dangerous game and one that can easily lead to miscommunications and misinterpretations." Adam Mount, director of the Defense Posture Project at the Federation of American Scientists, said, "Nuclear weapons are tools of the weak."[214]

The 2022 *Nuclear Posture Review* expresses significant concern about Russia's recent behavior:

Russia's invasion of Ukraine underscores that nuclear dangers persist, and could grow, in an increasingly competitive and volatile geopolitical landscape. The Russian Federation's unprovoked and unlawful invasion of Ukraine in 2022 is a stark reminder of nuclear risk in contemporary conflict. Russia has conducted its aggression against Ukraine under a nuclear shadow characterized by irresponsible saber-rattling, out of cycle nuclear exercises, and false narratives concerning the potential use of weapons of mass destruction (WMD). In brandishing Russia's nuclear arsenal in an attempt to intimidate Ukraine and the North Atlantic Treaty Organization (NATO), Russia's leaders have made clear that they view these weapons as a shield behind which to wage unjustified aggression against their neighbors. Irresponsible Russian statements and actions raise the risk of deliberate or unintended escalation. Russia's leadership should have no doubt regarding the resolve of the United States to both resist nuclear coercion and act as a responsible nuclear power.[215]

Nuclear weapons are certainly as current an issue as they were during the Cold War. Past treaties like the 2010 New START agreement limited the United States and Russia to 1,550 nuclear warheads on ballistic missiles and bombers. Yet, that agreement does not govern the smaller tactical weapons nor any other international agreement. It is not entirely clear how many tactical devices Russia may even have. The Federation of American Scientists estimates the number to be 1,912 "nonstrategic warheads." Of course, in 2023, Russia suspended participation in START, and cooperation on arms control effectively ended.

The modern nuclear picture is even more complicated. After Russia and the United States, China has the next largest arsenal at 350 nukes. (Some analysts feel as of 2024 this now totals 600.) While smaller, China is investing significantly in developing a sophisticated and diverse nuclear arsenal. All three nuclear superpowers have a triad weapon system of strategic weapons that can deliver nuclear warheads from land, sea, and air, which means explicitly land-based ICBMs, submarine or ship-launched ballistic missiles, and bombers coupled with cruise missiles. Russia now has a hypersonic delivery system that has been battlefield-tested with conventional warheads, which is likely true of China.

---

[213] Carnegie Institute for International Peace, https://ceipfiles.s3.amazonaws.com/pdf/Colin+Kahl+Keynote_Transcript.pdf.

[214] The 2022 *Nuclear Posture Review,* https://fas.org/wp-content/uploads/2023/07/2022-Nuclear-Posture-Review.pdf, p 1-2.

[215] The 2018 *Nuclear Posture Review,* https://media.defense.gov/2018/Feb/02/2001872886/-1/-1/1/2018-NUCLEAR-POSTURE- REVIEW-FINAL-REPORT.PDF.

There are other nuclear powers. Allied with the United States and part of NATO is France with 290 weapons and the United Kingdom with 225 nukes, which are strategic and largely carried on ballistic missile submarines. India and Pakistan each have about 160 nuclear weapons, but their primary strategic interests concern a long-standing feud between one another.

Israel has an unspecified and unofficial nuclear defense, perhaps approaching 90 devices. North Korea is a total wild card with a growing arsenal and a fast-evolving ballistic missile capability.

The 2018 *Nuclear Posture Review* stated that:

While the United States has continued to reduce the number and salience of nuclear weapons, others, including Russia and China, have moved in the opposite direction. They have added new types of nuclear capabilities to their arsenals, increased the salience of nuclear forces in their strategies and plans, and engaged in increasingly aggressive behavior, including in outer space and cyber space. North Korea continues its illicit pursuit of nuclear weapons and missile capabilities in direct violation of United Nations (U.N.) Security Council resolutions. Iran has agreed to constraints on its nuclear program in the Joint Comprehensive Plan of Action (JCPOA). Nevertheless, it retains the technological capability and much of the capacity necessary to develop a nuclear weapon within one year of a decision to do so.[216]

The 2022 *Nuclear Posture Review* added more specific concerns relating to China:

The People's Republic of China (PRC) is the overall pacing challenge for US defense planning and a growing factor in evaluating our nuclear deterrent. The PRC has embarked on an ambitious expansion, modernization, and diversification of its nuclear forces and established a nascent nuclear triad. The PRC likely intends to possess at least 1,000 deliverable warheads by the end of the decade.[217]

The concluding fact is that the United States and Western powers have, since the 1950s, cast their strategy around nuclear weapons in the light of deterrence. The 2022 *Nuclear Posture Review* states:

US nuclear weapons deter aggression, assure allies and partners, and allow us to achieve Presidential objectives if deterrence fails. In a dynamic security environment, a safe, secure, and effective nuclear deterrent is foundational to broader US defense strategy and the extended deterrence commitments we have made to allies and partners. Security architectures in the Euro-Atlantic and Indo-Pacific regions are a critical US strategic advantage over those governments that challenge the rules-based international order. These regional security architectures are a key pillar of the NDS; this NPR underscores the linkage between the conventional and nuclear elements of collective deterrence and defense.[222]

---

[216] The 2022 *Nuclear Posture Review,* https://fas.org/wp-content/uploads/2023/07/2022-Nuclear-Posture-Review.pdf, p 4.

[217] Ibid, p.1.

**Los Alamos Engineer Dale Cox's stunning oil on canvas of the 1953
nuclear cannon Grable test.**

Deterrence, however, is a double-edged sword. As the former president of EG&G, Barney O'Keefe wrote in his noted 1983 book, *Nuclear Hostages*, nuclear weapons come with a price. O'Keefe talks from a different perspective than many armchair strategists. Barney O'Keefe witnessed firsthand dozens of atmospheric nuclear explosions in the 1950s and early 1960s before the Limited Test Ban Treaty of 1963 moved all testing underground. (See Appendix I.) These included the 15-kiloton nuclear cannon Grable test in 1953 and the 15-megaton thermonuclear Castle Bravo test in 1954.[218]

O'Keefe felt nuclear arsenals held the nuclear club members hostage to their own technology. In his perspective, engaging in a nuclear war would make such a conflict so costly that no nuclear power would dare cross that line. This admittedly makes it an ideal deterrent to a large-scale war. Yet, once one nuclear

---

[218] The M65 atomic cannon, a 280-millimeter or 11-inch artillery piece, fired a 15-kiloton shell in the airburst Grable shot at the Nevada Test Site in Area 5 over Frenchman Flat on May 25, 1953. The hardened nuclear shell traveled 11,000 yards to detonate 524 feet in the air. The 803-pound time-fused, 280-millimeter shell used a "gun-type" fission weapon assembly from the World War II-era Little Boy bomb dropped on Hiroshima. The Little Boy design was never tested in 1945 because the system proved so foolproof that Los Alamos physicists and engineers knew it would work, as it did again during Grable. The atomic cannon had a range of 20 miles. The M65 or "Atomic Annie" was suspended (trailered) between a front and rear custom self-propelled transporter for movement. However, the gun had to be detached and then secured on the ground on heavy base plates for firing. Interestingly, it was based on the 240-millimeter German railguns used during WWII, specifically two deployed near Anzio in 1943, referred to as "Anzio Annie." Thus came the later spin-off in the 1950s of Atomic Annie. Twenty M65 artillery pieces were produced, and eventually, 14 were deployed to NATO. These large guns, however, proved very difficult to maneuver and maintain, so soon, the more standard and more manageable 155 main US Army artillery field piece had a nuclear shell adapted to it. The Navy also acquired a nuclear shell for the 16-inch guns of their four Iowa class battleships. (During college, one of my history professors reminisced about the atomic cannon. As a young draftee, he was assigned to a nuclear cannon group that practiced firing blank or dummy rounds stateside. He remembered the M65 as a very impractical device simply because moving and firing the gun was so labor-intensive. On one practice exercise, they fired a 280-millimeter shell accidentally beyond their local test range somewhere in the southern United States. Apparently, they added too much propellant, and the shell simply went too far and ended up dropping right in the middle of a school yard playground, which was fortunately empty of children at the time.)

superpower develops the know-how and the necessary accompanying infrastructure to attain a nuclear arsenal, it can never give it up so long as other powers have a similar capability. Or at least that was the thinking, but in more recent times, three countries, South Africa, Libya, and Ukraine, have voluntarily given up efforts to develop or retain a nuclear arsenal. Ukraine surrendered its Cold War-era nuclear weapons to Russia in 1994 in return for the written promise of its territorial security, called the Budapest Memorandum.

Of course, while nuclear weapons may prevent nuclear superpowers from going to war with one another, they do not prevent them from subterfuge. Wars like the US involvement in Vietnam and the USSR experience in Afghanistan are not preventable by nuclear arsenals because the major players do not fight one another face to face. Instead, proxy wars have become all too common.

Recently, that strange dynamic has exemplified itself in the Russia-Ukraine war. Few want Russia to overrun the non-NATO member country of Ukraine. However, no power dares provoke a nuclear exchange to attempt to save an independent Ukraine. All the same, ways are still found to support that brave country's fight for freedom. It is, however, still a risky gamble that could easily get out of hand as we are truly held hostages by our own nuclear technology as Barney O'Keefe insisted.

Nuclear saber-rattling, or even brinkmanship, periodically takes place, and it is intimidating. As a case in point, since its invasion of Ukraine, Russia has consistently used extremely bold language about nuclear weapons. (Make sure to read the final chapter of this book, Nuclear Escalation.) Yet, threatening to use nuclear weapons is one thing, and using them is another.

In the mid-20th century, no one foresaw, except perhaps Barney O'Keefe, the complexity of today's nuclear issues. O'Keefe realized early on that eventually, other smaller nations could attain nuclear arsenals as the technology proliferated. O'Keefe also warned of the possibility of a terrorist group eventually attaining such a weapon. No one ever invented a nuclear bomb. Rather, they simply unlocked the universal principles of nature which makes such a weapon possible. The nuclear secret can never be controlled entirely, but fortunately, it will always be a highly restrictive undertaking to produce the fissile material required to make a working bomb or bombs. Only the future will tell the story as to how nuclear weapons are finally judged. Clearly, at this time, a resurgent Russia and confrontational China are the threats we will face in the foreseeable future.

CHAPTER TWENTY SEVEN
QUESTIONS AND ANSWERS

On October 4th, 2023, the United States and the Russian Federation each independently conducted a nationwide emergency alert test of their own emergency broadcast systems. Russia exhibited transparency and clearly stated that they are preparing for "the danger of armed conflicts involving nuclear powers."[219] Russia went beyond our own very simple electronic notification exercise and conducted instructive lessons in schools on what to do in the event of an actual nuclear strike. It is no exaggeration to say that they are preparing their population for the possibility of a nuclear war. The rising state of global tensions is palpable, and people are concerned. We all have questions, and although some of this material repeats select information provided in previous chapters, the following are what I have found to be the most asked questions, followed with concise answers.

HOW DO THE TERMS KILOTON AND MEGATON REFERENCE NUCLEAR WEAPONS?

Paul Tibbets photo collection.

An often-asked question refers to the size or yield of nuclear weapons and how that relates to the terms kiloton and megaton. The best way to provide perspective is to relate it to the first use of an atomic weapon in wartime. On August 6, 1945, Hiroshima crumbled under one uranium fission bomb, as can be seen in this adjacent photo. The weapon produced a destructive force of 15 kilotons, equating to 15,000 tons of TNT. The bomb detonated 1,900 feet in the air, exerting a massive pressure wave of intense heat that destroyed four miles of the city— immediately killing 80,000 people. This was the first time one plane with one bomb destroyed one city. Three days later, the Japanese city of Nagasaki had succumbed to a 21-kiloton plutonium fission bomb. Prior, the destruction of such urban areas required thousands of men flying up to 1,000 heavy bombers with their bomb bays full of 500 to 1,000-pound conventional explosive and incendiary bombs.

Actually, Hiroshima and Nagasaki were relatively small cities. Very large cities, like Berlin and Tokyo or Moscow and New York, span too many square miles to be destroyed by the type of early nuclear weapons used on Japan. So, after the war, the nuclear arsenals of America and then eventually Russia strove for bombs with bigger yields. When so-called super bombs were developed utilizing fusion, the resulting hydrogen bombs became measured in megatons. A megaton is equivalent to a million tons of TNT. Generally, a fission bomb or A-bomb can only be developed up to a yield of around 100 kilotons. Physics, however, does not seem to limit the possible yield of a fusion or H-bomb, which is considered a true city killer. The United States tested a fusion bomb up to 15 megatons (Castle Bravo shot), and Russia under the

---

[219] Isabel Van Brugen, *US and Russia To Hold Nationwide Emergency Alert Drills on Same Day*, Newsweek, October 3, 2023, 11:23 AM EDT, https://www.newsweek.com/us-russia-emergency-alert-drills-1831776. (As a handy reference, a one-kiloton explosion is equivalent to one-thousand tons of TNT. A one-megaton explosion is equivalent to one-million tons of TNT.)

USSR tested one as great as 50 plus megatons, which became known as the Tzar Bomb.

Those bombs had to be big—not only to destroy the large cities but because accuracy was not as good as in our modern age of smart, computer-guided weaponry. Yet, gravity bombs quickly became an old technology. That is because rocketry was the other big innovation that came out of the Second World War aside from nuclear weapons. Hitler set a precedent when he rained V-2 rockets down on London. After the war, some wanted to take rockets and go to the moon and the stars. Others wanted to put a nuclear weapon on them, creating a threat so overpowering that it would deter large-scale wars forever. In fact, deterrence soon became the chief reason, or at least excuse, for building large nuclear arsenals.

The problem, however, was that the atomic bombs of the day were too big and heavy to be carried by rockets. Furthermore, the accuracy of the developing ICBMs was not precise. It took about ten years of development for rockets to evolve into ICBMs and the same time span to reduce the size and weight of a nuclear device (while also increasing its yield) to be compatible with missile payloads. Thus, the atomic and then hydrogen bombs of the late 1950s and 1960s had to be lighter and smaller in size with a much bigger destructive power, or yield, to accommodate the marriage to the early and rather primitive ballistic missiles. Nuclear testing by America, Russia, and later China brought on arsenals of weapons that ranged from American weapons in the 9-megaton range to Russian devices in the 20-megaton range. Such weapons only had to get close to a target to destroy cities like New York and Moscow.

Today, missile accuracy is much better, and as a result, US weapons are scaled down to the 100-kiloton range, which is quite adequate now that pin-point accuracy is achievable. Russia similarly has highly accurate weapons in the 400-kiloton range but also now boasts of weapons in the 20 to 100-megaton range. That unnecessarily large yield does have the advantage of an added fear factor leading, in theory, to its greater deterrence—although some say it reflects only madness. This all takes us to the next logical question.

HOW DOES FISSION DIFFER FROM FUSION?

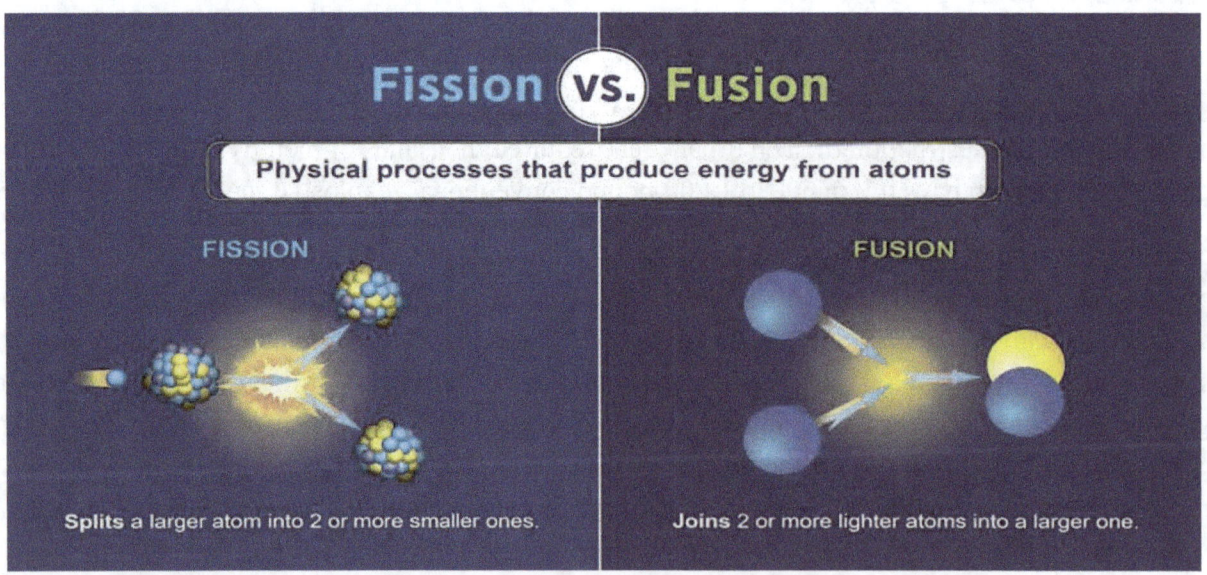

Chart from https://dashamlav.com/fusion-vs-fission-differences-table-examples-diagram/.

A second question I reference after explaining the relevance behind terms like kiloton and megaton usually concerns fission versus fusion.

Background is first required. All energy that humankind produces is based on chemical or physical processes. Throughout most of history, that has involved burning carbon-based materials like wood, coal, and gas. We can, of course, also harness power from the sun, wind, and water.

Fission and fusion are forces producing energy that involve physical processes. Fission and fusion are nuclear reactions that produce energy but are very different processes. Nuclear fission is a nuclear reaction in which the nucleus of an atom splits into smaller parts. The fission process often produces free neutrons and photons (in the form of gamma-rays) and releases large amounts of energy involving radioactive decay. In short, fission in nuclear weapons is splitting a heavy, unstable (uranium) nucleus into two lighter nuclei.

This process was first understood in late 1938 by German chemists Otto Hahn and Fritz Strassmann in cooperation with Austrian-Swedish physicist Lisa Meitner and her nephew Otto Frisch. Frisch would be a key person in coauthoring a paper with Rudolf Peierls in Great Britain during the early days of World War II that proved the practicality of building a nuclear weapon. For many years, physicists knew a nuclear reaction could, in theory, produce a power plant or a great weapon, but they thought it would take tons of highly rare fissile material instead of just ounces, as Frisch-Peierls realized.

Fusion, on the other hand, is the process where two light atoms bond together, or fuse, to make a heavier one. Because the total mass of the new atom is less than that of the two that formed it, the missing mass is given off as energy. The two bombs used over Japan and later in the early atmospheric tests at the Nevada Test Site in the 1950s were not fusion but fission, or "A-bombs." However, as fusion weapons developed in the 1950s, it still took an atomic (fission) bomb/detonator to create a fusion detonation.

A note should be made on nuclear power plants. For decades, many nations have supplemented the world's energy requirements with fission reactors but have struggled with the nuclear waste produced by the process. Fission power plants also serve as the perfect propulsion source for nuclear submarines.

We have not yet developed a reliable, sustainable fusion power reaction, but technology is on the verge of that, and when it happens, it will give humanity an affordable, clean, and limitless energy supply. Our children will see that reality in their lifetime. That is a certainty, thanks to research being conducted by National Labs like Lawrence Livermore, which had developed many of the early nuclear weapons. This, again, leads to another logical question.

HOW DO URANIUM BOMBS DIFFER FROM PLUTONIUM BOMBS?

After explaining fission versus fusion, I am often asked about the difference between uranium and plutonium. Greatly oversimplified, the first bomb detonated at the first nuclear test, called Trinity, utilized a man-made metallic element called plutonium. In contrast, the first wartime use of an atomic bomb was dropped on Hiroshima and utilized the enriched natural element of uranium.

So, how do plutonium and uranium bombs differ? The uranium core bomb resulted in an extremely simple and elegant design. To oversimplify once again, it involves smashing one sub-critical mass of enriched uranium into another sub-critical mass of enriched uranium in a gun barrel-like chamber to make it reach a critical mass. Critical mass simply means enough fissile material to start/and or maintain a nuclear chain reaction, which in that case was 60 kilograms of material that was at least 80 percent pure, known as uranium-235. Manhattan Project physicists and engineers of the time knew that formula would work, and the bomb design called "Little Boy" soon destroyed Hiroshima. Little Boy was already in preparation for transport to the Pacific as the Trinity plutonium bomb test readied. The only problem concerned the fact that the Manhattan Project personnel barely had enough enriched uranium for one bomb after years of work. A better solution had to be found, and that was plutonium.

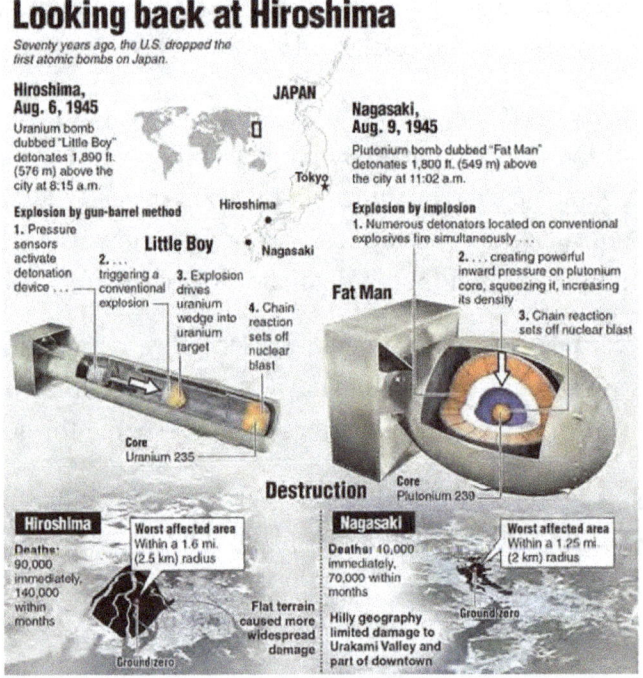

Images from https://learnodo-newtonic.com/hiroshima-and-nagasaki-facts and https://www.reddit.com/r/ThingsCutInHalfPorn/comments/k09n5q/crosssections_of_little_boy_and_f at_man_nuclear/.

The Trinity test dealt with a much more problematic but more advanced type of device known as the plutonium bomb. Plutonium is created in a fission reactor when uranium atoms absorb neutrons. That exotic material posed so uncertain and unpredictable outcomes when utilized for a fission explosion that it had to be tested. Its design was later used in the "Fat Man" bomb over the Japanese city of Nagasaki. The plutonium bomb did have some advantages. Plutonium proved three times more reactive than uranium-235, so it was more efficient for a chain reaction by nature, although much more difficult to properly detonate. The advantage is that this method required only one-tenth the fissile material needed for a U-235 gun-type design. In the resulting Fat Man bomb, a mere 6 kilograms of plutonium produced greater yield compared to the uranium device.

Fusion bombs also require plutonium. These are termed thermo-nuclear devices. This is where a plutonium-239 fission weapon triggers a fusion/fission-boosted reaction aimed at a secondary internal device of lithium deuteride. That process generates high-frequency x-rays and gamma-rays, which crush the secondary device so severely with radiation pressure that a hydrogen fusion explosion takes place. The final fusion reaction releases a shower of neutrons, which some bomb designs utilize to produce a final fission reaction.[220]

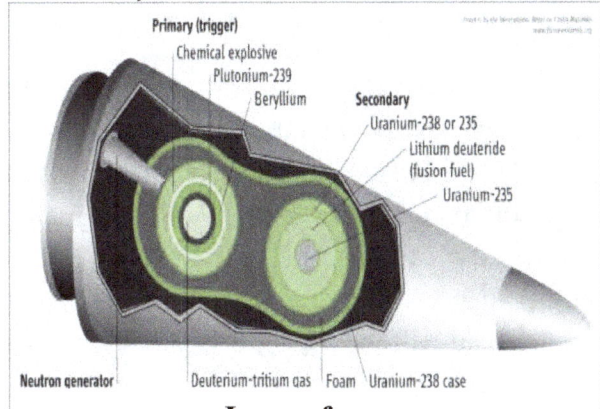

Image from
https://popularresistance.org/heres-what-
would-happen-if-us-nuked-north-korea/.

In the 75 years of nuclear Cold War and nuclear peace, almost 500 tons of plutonium have been produced worldwide. Today's light water reactors make commercial power with uranium fuel fissions and create plutonium as a residual product. Any plutonium that does not fission stays in the spent fuel, and that is a problem because it makes reprocessing a security risk should some foreign or terrorist group gain access to that spent fuel and its plutonium. As a result, nuclear waste is kept secured at its nuclear plant site which means at most nuclear power plants nuclear waste is piling up in dangerous amounts. On the weapons side of things, what most people do not understand is that the US nuclear stockpile still depends on plutonium. Yet, America, unlike Russia, China, and other nuclear powers, no longer maintains a nuclear weapons production system. As a result, our nation must maintain its original stockpile with a stewardship program that assures that our nuclear arsenal remains safe, secure, and reliable.

It is important to remember that the great Cold War powers of America and Soviet-era Russia ceased nuclear testing thirty years ago. Nuclear arsenals shrank after the fall of the Soviet Union as tensions significantly subsided. Now, as we enter a new era of a hybrid cold war, Russia and China are revitalizing their nuclear weapons programs.

WHAT IS NUCLEAR DETERRENCE?

Of distant Latin origin, "deterrence" has had various interpretations as it relates to nuclear weapons. The root word of deterrence is "terre," which means to frighten with overwhelming fear. Insert a prefix "de," and the meaning is to frighten away or remove from carrying out acts. Seven decades ago, NSC 162/2 formalized the policy of nuclear deterrence and later extended it to NATO forces facing the Warsaw Pact in Europe. That policy paper of the United States National Security Council defined President Eisenhower's nuclear policy, which, by 1954, had evolved into a concept called "massive retaliation."

---

[220] Harvard Kennedy School online lecture, https://youtu.be/zVhQOhxb1Mc.

Massive retaliation, also known as massive response or massive deterrence, became an official doctrine of nuclear and military strategy. The concept advocated an overwhelming response to any aggression with disproportionately greater use of force using either conventional or nuclear forces. Of course, for such a policy to serve as a deterrent to a Soviet-era attack, which would have assuredly been nuclear, the policy had to be made public. The United States also had to be confident that it could maintain a second-strike nuclear capability to ensure that it could retaliate in a massive manner if first attacked. A second-strike capability simply means it could guarantee that a reserve of nuclear weapons and delivery systems could survive a surprise attack.

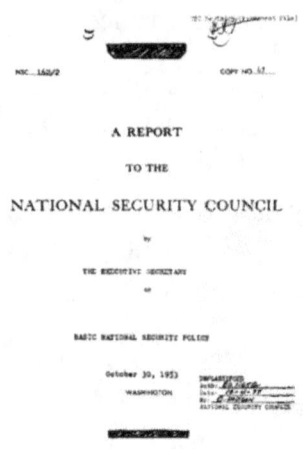

That concept of massive retaliation used in the 1950s was not what we now think of as MAD or mutually assured destruction. That came later in the Cold War when the USSR attained a second-strike nuclear capability of their own. By that time, the nuclear arsenals of both countries became so big that an all-out nuclear war simply became impossible to wage because it would mean the total destruction of both the United States and the Soviet Union. MAD is based on another theory called "rational deterrence," which states that destroying each other's forces is simply not rational because of the metaphoric and literal fallout that such a war would cause. Logic is the key part of nuclear deterrence. Complete annihilation of one nation against another is not rational as long as the leaders of rival nations remain logical and rational themselves.

A famous Cold War-era novel dealt with that conundrum. In that novel, titled *Alas, Babylon*, one of the main characters, a retired United States Naval admiral, realized nuclear weapons had indeed made war obsolete. The implied holocaust made war obsolete by making the outcome so devastating that no rational individual would dare risk it. Unfortunately, the problem that the admiral realized too late and after the unthinkable happened—was that the unthinkable did not deter rational leaders from making irrational decisions. Rational individuals will always be prone to irrational actions. That is simply human nature. *Alas Babylon* was, of course, a work of fiction, but it points out the dangers of making assumptions about deterrence, which has admittedly prevented another world war—at least so far. Yet, can we ensure that humankind will not one day make an ultimate mistake of judgment?

WHAT IS NUCLEAR RADIATION?

It took the most brilliant minds of our world the better part of the last two centuries to unlock the secrets of the atom. So, it is challenging to describe the process of nuclear radiation in a simple, concise manner. It is, however, a very basic question. Radiation is all around our natural world, so understanding the basics is important and goes beyond the realm of just nuclear war.

If you go outside, the Sun's radiation can promote vitamins in your skin, although if you expose yourself too long, the Sun can also burn you. It all depends on time, distance, and shielding related to your exposure to radiation. With the Sun, we must limit the time we spend in the open, even though our distance from it is great. And using shielding like sunscreen or long sleeves and a hat is wise as well.

The same would be true in a nuclear attack. In short, nuclear radiation refers to particles and photons emitted during reactions involving an atom's nucleus. That reaction is called ionizing radiation, which means highly charged particles emitted by nuclear reactions, which themselves are sufficiently energetic to remove electrons from atoms and molecules and ionize them.

Nuclear reactions releasing ionizing subatomic particles include alpha and beta particles, neutrons, muons, mesons, positrons, and cosmic-rays. Nuclear radiation also includes gamma and x-rays, including the most energetic portion of the electromagnetic spectrum. To understand these basics, you have to understand some basic physics of chemistry:

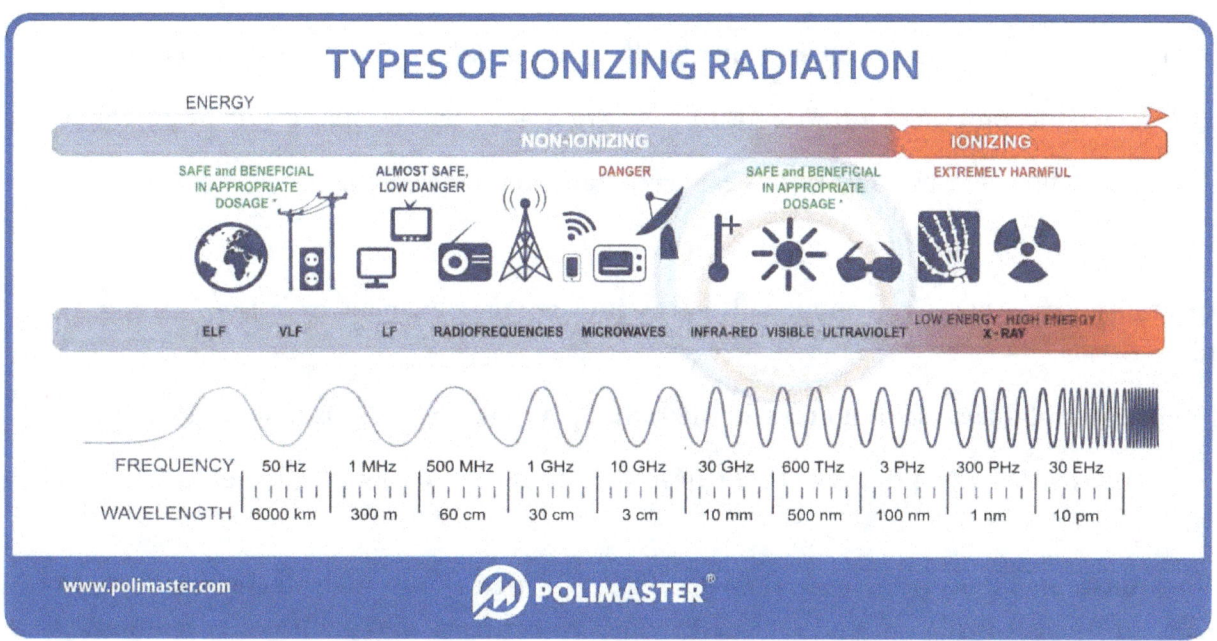

Everything in our natural world around us is made of atoms.

Atoms bind together to make molecules, such as two hydrogen and one oxygen atom make a water molecule.

In nature, there are 92 atoms known as elements.

So, every substance found on earth is made up of combinations of those 92 atoms into elements.

The Periodic Table list these elements along with some added man-made ones. Going deeper, inside every atom are three subatomic particles: Protons - Neutrons - Electrons.

Protons and neutrons bind together to form the nucleus of the atom. The electrons orbit the neutron.

Electrons are negatively charged, and protons are positive while neutrons are neutral.

Since opposites attract, opposite charges attract.

Neutrons act like glue to hold protons tightly together in the nucleus, otherwise they would repel one another.

The number of protons in the nucleus determines the behavior of an atom which remains stable.

Atoms that have the same number of protons in their nuclei, but different neutron numbers, are called isotopes.

Isotopes are family members of an element that has the same number of protons but different number of neutrons.

All elements have isotopes and there are two types of isotopes: stable and unstable which are radioactive.

Hydrogen is a good example of an element with multiple isotopes, one of which is radioactive.

Hydrogen is the only element whose isotopes have unique names: deuterium and tritium.

Uranium is the heaviest naturally occurring radioactive element.

Uranium can only exist in an unstable form.

An atom of a radioactive isotope will decay into another element over time through processes including fission.

The number of "decays" that occur in the radioactive material tell us how radioactive it is.

When an unstable atom decays, it *transforms* into another atom and releases its excess energy as radiation.

When unstable atoms transform, they often eject particles from their nucleus. During the fission of U-235 the nuclear radiation that is released contains neutrons and gamma-ray photons.

Four different types of radioactive rays can be produced: Alpha-rays – Beta-rays – Gamma-rays – Neutron-rays. These occur in a typical nuclear detonation.

There are also Cosmic-rays emanating from the Sun which is a fusion reaction.

Alpha, Beta particles, Neutrons, Gamma- rays, and Cosmic- rays are called ionizing radiation.

Alpha particles are large, they cannot penetrate very far into matter. Alpha particles can be stopped by a simple sheet of paper.

Beta particles penetrate more deeply but are only dangerous if eaten or inhaled. A sheet of aluminum or plexiglass can stop beta particles.

Gamma-rays, like X-rays, are only stopped by heavy elements like lead or by thick layers of concrete or liquids. The same is true of penetrating Neutron radiation.

A large Gamma-ray burst in the upper atmosphere will create nitrogen oxides that degrade Earth's ozone layer. Gamma-rays damage DNA in cells and cause cancers, sickness, and possible death.

The key to ionizing radiation involves:

1. Time or length of exposure from source

2. Distance from source – 3. Amount of

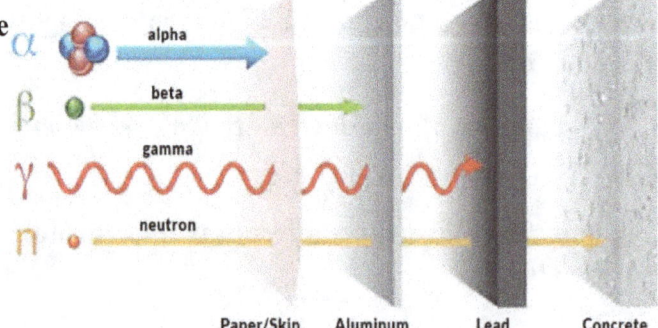

shielding from source.[221]

If that is not confusing enough, I will insert a passage from one of the best articles on radiation that I have ever seen published. I used to often quote this article when giving museum tours and talks, as it is the easiest explanation I know of:

The simple mention of the word "radiation" often evokes fear in people. For others, it's fun to think a little exposure to radiation could turn you into the next superhero, just like the Hulk.

But is it true basically everything around us is radioactive, even the food we eat? You may have heard bananas are mildly radioactive, but what does that actually mean? And despite us not being superheroes, are human bodies also radioactive?

What is radiation?

Radiation is energy that travels from one point to another, either as waves or particles. We are exposed to radiation from various natural and artificial sources every day.

Cosmic radiation from the Sun and outer space, radiation from rocks and soil, as well as radioactivity in the air we breathe and, in our food, and water, are all sources of natural radiation.

How The Conversation is different: We explain without oversimplifying.

Bananas are a common example of a natural radiation source. They contain high levels of potassium, and a small amount of this is radioactive. But there's no need to give up your banana smoothie – the amount of radiation is extremely small, and far less than the natural background radiation we are exposed to every day.

Artificial sources of radiation include medical treatments and X-rays, mobile phones and power lines. There is a common misconception that artificial sources of radiation are more dangerous than naturally occurring radiation. However, this just isn't true.

There are no physical properties that make artificial radiation different or more damaging than natural radiation. The harmful effects are related to dose, and not where the exposure comes from.

What is the difference between radiation and radioactivity?

The words "radiation" and "radioactivity" are often used interchangeably. Although the two are related, they are not quite the same thing.

Radioactivity refers to an unstable atom undergoing radioactive decay. Energy is released in the form of radiation as the atom tries to reach stability, or become non-radioactive.

The radioactivity of a material describes the rate at which it decays, and the process(es) by which it decays. So radioactivity can be thought of as the process by which elements and materials try to become stable, and radiation as the energy released as a result of this process.

---

[221] Source: Jeff Tappen, Desert Research Institute.

Ionizing and non-ionizing radiation

Depending on the level of energy, radiation can be classified into two types.

Ionizing radiation has enough energy to remove an electron from an atom, which can change the chemical composition of a material. Examples of ionizing radiation include X-rays and radon (a radioactive gas found in rocks and soil).

Non-ionizing radiation has less energy but can still excite molecules and atoms, which causes them to vibrate faster. Common sources of non-ionizing radiation include mobile phones, power lines, and ultraviolet rays (UV) from the Sun.

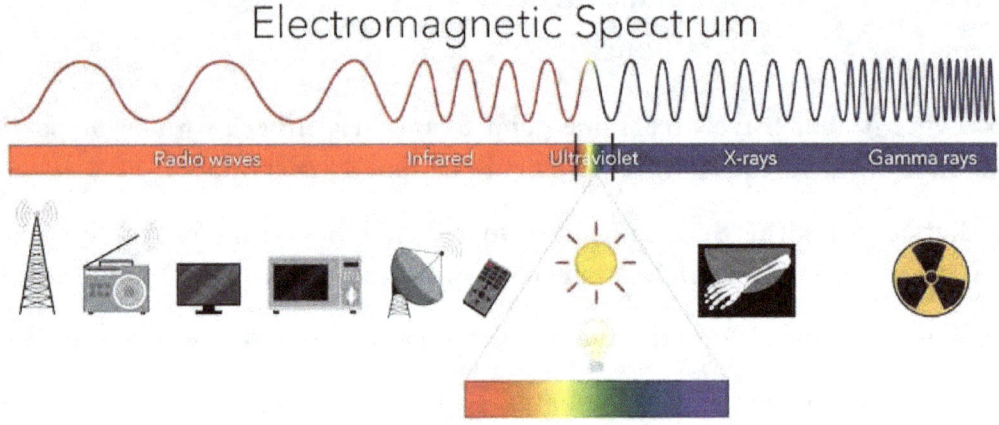

The electromagnetic spectrum includes all types of electromagnetic radiation. Is all radiation dangerous? Not really

Radiation is not always dangerous – it depends on the type, the strength, and how long you are exposed to it.

As a general rule, the higher the energy level of the radiation, the more likely it is to cause harm. For example, we know that overexposure to *ionizing* radiation – say, from naturally occurring radon gas – can damage human tissues and DNA.

We also know that *non-ionizing* radiation, such as the UV rays from the Sun, can be harmful if the person is exposed to sufficiently high intensity levels, causing adverse health effects such as burns, cancer, or blindness.

Importantly, because these dangers are well known and understood, they can be protected against. International and national expert bodies provide guidelines to ensure the safety and radiation protection of people and the environment.

For ionizing radiation, this means keeping doses above the natural background radiation as low as reasonably achievable – for example, only using medical imaging on the part of the body required, keeping the dose low, and retaining copies of images to avoid repeat exams.

For non-ionizing radiation, it means keeping exposure below safety limits. For example, telecommunications equipment uses radiofrequency non-ionizing radiation and must operate within these safety limits.

Additionally, in the case of UV radiation from the Sun, we know to protect against exposure using sunscreen and clothing when levels reach 3 and above on the UV index.

Radiation in medicine

While there are clear risks involved when it comes to radiation exposure, it's also important to recognize the benefits. One common example of this is the use of radiation in modern medicine.

Medical imaging uses ionizing radiation techniques, such as X-rays and CT scans, as well as non-ionizing radiation techniques, such as ultrasound and magnetic resonance imaging (MRI).

These types of medical imaging techniques allow doctors to see what's happening inside the body and often lead to earlier and less invasive diagnoses. Medical imaging can also help to rule out serious illness.

Radiation can also help treat certain conditions – it can kill cancerous tissue, shrink a tumor or even be used to reduce pain.

So are our bodies also radioactive? The answer is yes, like everything around us, we are also a little bit radioactive. But this is not something we need to be worried about. Our bodies were built to handle small amounts of radiation – that's why there is no danger from the amounts we are exposed to in our normal daily lives. Just don't expect this radiation to turn you into a superhero any time soon, because that definitely is science fiction.[222]

What is a dose of radiation?

When radiation's energy is deposited into our body's tissues, that is a dose of radiation. The more energy deposited into the body, the higher the dose. Rem is a unit of measure for radiation dose. Small doses expressed in mrem = 1/1000 rem. Rad & R (Roentgens) are similar units that are often equated to the Rem. To put this in perspective, an average dental X-ray provides 10 mrem.[228][223]

What is the cost of nuclear weapons?

A recent *Newsweek* article estimated that in 2021, the nine official nuclear states spent 89.4 billion dollars on their nuclear arsenals. These figures included the United States spending 44.2 billion dollars, China 11.7 billion, and Russia 8.6 billion.

Britain came in fourth, spending 6.8 billion on its nuclear weapons, with France at 5.9 billion, India at 2.3 billion, Israel at 1.2 billion, Pakistan at 1.1 billion, and North Korea spending 642 million.

In the United States, the Congressional Budget Office is required by law to project the 10-year costs of nuclear forces every two years. Their estimates are as follows:

---

[222] *"Are Bananas Really 'Radioactive'? An Expert Clears Up Common Misunderstandings About Radiation,"* The Conversation, Allexxandar/Shutterstock, Published: November 3, 2022, https://theconversation.com/are-bananas-really-radioactive-an-expert-clears- up-common-misunderstandings-about-radiation-193211.

[223] Source: Jeff Tappen, Desert Research Institute.

If carried out, the plans for nuclear forces delineated in the Department of Defense's (DoD's) and the Department of Energy's (DOE's) fiscal year 2021 budget requests, submitted in February 2020, would cost a total of

$634 billion over the 2021–2030 period, for an average of just over $60 billion a year, CBO estimates.

Almost two-thirds of those costs would be incurred by DoD; its largest costs would be for ballistic missile submarines and intercontinental ballistic missiles. DOE's costs would be primarily for nuclear weapons laboratories and supporting activities.

The current 10-year total is 28 percent higher than CBO's most recent previous estimate of the 10-year costs of nuclear forces, $494 billion over the 2019–2028 period.

Almost half (about 49 percent) of the $140 billion increase in that total arises because the 10-year period covered by the current estimate begins and ends two years later than the period covered by the 2019 estimate. Thus, the period now includes two later (and more expensive) years of development in nuclear modernization programs. Also, costs in those two later years reflect 10 years of economy wide inflation relative to the two years that drop out of the 10-year projection; that factor (in the absence of other changes to programs) accounts for about one-fourth of the 49 percent increase.

About 36 percent of the $140 billion increase is projected to occur from 2021 to 2028—the years included in both this estimate and the 2019 estimate. That increase stems mainly from new plans for modernizing DOE's production facilities and from DoD's modernization programs moving more fully.

Data source: Congressional Budget Office, using data from the Department of Defense and the Department of Energy.[224]

What is the US nuclear Triad?

Our Triad is a combination of land, sea, and air-based platform systems for defense and deterrence. The concept of three separate arms makes the US nuclear weapons more survivable and capable of a follow-up launch in the event of an enemy's first strike. Most of us visualize this in terms of inter-continental ballistic missiles emerging from silos in the great plains, long-range bombers flying over the poles, and submarine-launched ballistic missiles. So, let us examine those nuclear response options in more detail.

The United States has retained its underground ICBM silos from the Cold War, although it has vastly reduced their number. In contrast, the much more modern nuclear forces of Russia, China, and North Korea have largely moved their ICBMs onto road-mobile missile launchers. That makes their systems very elusive targets, while the locations of our silos are well known. This would logically make them extremely vulnerable, which is a correct assumption. However, the advantage of that strategy is that it requires an enemy nation to use a very large percentage of its nuclear arsenal to deal with those dug-in silo-based ICBMs.

---

[224] Congressional Budget Office, www.cbo.gov/publication/57130#data.

It may indeed be for that reason that China is now building large nuclear silo fields of its own. At the height of the Cold War, we had over a thousand ICBM missile silos. The first ICBMs arrived in Wyoming in 1959. Over the years, Atlas, Titan, Minuteman, and Peacekeeper missile silos were constructed from Texas to North Dakota and from New Mexico to Montana. Only the Minuteman sites remain active today.

This currently involves approximately 400 solid fuel LGM-30 Minuteman III ICBMs located in three rural areas. These include the 90th Missile Wing F.E. Warren AFB in Colorado, Nebraska, and Wyoming. The second is the 91st Missile Wing, Minot AFB in North Dakota, and then the 341st Missile Wing, Malmstrom AFB in Montana. Each wing has three divisions or squads, and each squad has 50 LGM- 30 Minuteman III ICBS. LGM means silo-launched, surface attack, guided intercontinental ballistic missile.

Minuteman III missiles used to have three warheads on each missile but now only carry one due to arms treaties.230 (Of course almost all of the treaties have fallen by the way side as of 2024, so it is debatable what changes may be made to our nuclear forces.) Half of our force of LGM-30 Minuteman III's carry a single 335 kiloton W87 Mark 21A warhead. The other half carry a single 300-kiloton W789 Mark 12 warhead. These warheads are around 40 times the destructive power of the nuclear weapons used on Japan in 1945. Our Minuteman III ICBMs can travel over 6,000 miles with pinpoint accuracy at speeds reaching 15,000 miles per hour, allowing them to reach their targets within 30 minutes. Once launched, they cannot be recalled or destroyed in flight.

This leg of the Triad is not the most advanced, although it is the most lethal. For that reason, any major nuclear exchange would have to target our three Minuteman wings first, with a massive strike of two to three warheads aimed at each silo to ensure their destruction. Without a successful first strike on the land-based ICBM portion of our Triad, a nuclear confrontation would be without strategic reason. That is part of the problem. In an enemy nuclear strike, the president of the United States would have to assume that he must either launch the Minuteman wings immediately or risk losing them quickly. The sole authority is his alone. Given the five-minute launch period required, it leaves little time for negotiation. A best-case scenario would give a US President 15 minutes to decide on a nuclear strategy once an enemy nuclear launch was detected. The actual warning time could be far less. After being launched there is no way to recall or destroy a Minuteman.

The second arm of the Triad, the long-range bombers, is organized into what is called the Air Force Global Strike Command of the 8th Air Force. The three Minuteman III missile wings also fall under Global Command. During the Cold War era, our bomber force was under the Strategic Air Command (as were all nuclear strategic weapons) and numbered as many as 2,500 aircraft. Today, our strategic bombers compose three wings, including the 2nd Bomb Wing, the 509th Bomb Wing, and the 5th Bomb Wing. Each wing is divided into three squadrons. All told, this accounts for 66 strategic bombers. These bombers are usually divided between Minot Air Force Base in North Dakota, Whitman Air Force Base in Missouri, and Barksdale Air Force Base in Louisiana.

The front-line nuclear weapons available for these planes comprise 300 devices stored stateside and 100 B61 lower-yield gravity bombs kept in Europe. Making up the 66-strong bomber force are 46 B-52s and 20 B-2 stealth bombers. The B-2s can carry up to sixteen 1,200-kiloton gravity bombs each. The 9,000-mile ranged B-52 carries up to twenty AGM-86 cruise missiles, each with a range of 1,500 miles. Each cruise missile has a W-81, 150-kiloton warhead.

The B-1 Lancer bombers are not now officially considered nuclear-armed but could be. They are also organized under the Air Force Global Strike Command and, in theory, would be used for missions not exclusively related to nuclear deterrence. Approximately 40 plus B-1 bombers remain in active service. If nuclear war came, American and NATO fighter jets like the F-16, F-35, and some F-15Es could carry B61 nuclear gravity bombs on tactical or strategic missions. The new Northrop Grumman B-21 Raider stealth bomber is expected to enter service by 2027, with 100 planes planned to replace the B-1 and B-2 bomber force by 2040.

Strategic bombers are a valuable part of the Triad because they give a US president much more flexibility in a nuclear conflict. They are not static, and unlike an ICBM, bombers can be quickly dispersed and recalled if a tactical situation changes in the larger strategic picture. The downside for the United States is that our bombers are mostly old and slow. However, their very presence ties up many enemy defensive missile systems and fighter aircraft, even if they never take to the skies.

The most valued component of the US nuclear triad is comprised of 18 Ohio class submarines of the United States Navy. This is the bulwark of our nuclear deterrent. Fourteen of those submarines carry ballistic missiles, and four carry cruise missiles. The ballistic missile submarines, also referred to as SSBNs, can carry 20 Trident II SLBMs or submerged launched ballistic missiles with multiple warheads on each ballistic missile. Each warhead could be armed with up to eight 475-kiloton W88 warheads or twelve 100-kiloton W76 warheads, which basically act as independent reentry vehicles. These are our most advanced nuclear weapons.

Trident IIs can be launched underwater and within seven minutes of receiving orders, which can also be done while submerged. Tridents have a range of 7,500 miles. They can reach speeds of 18,000 miles per hour. This allows the missiles to reach their targets within 15 minutes and cannot be recalled or destroyed in flight.231

The four cruise missile carriers are termed SSGNs, and each carry 154 Tomahawk cruise missiles, which can be nuclear armed or conventionally armed. Seventy percent of the United States frontline nuclear firepower is positioned on Ohio class submarine platforms, providing about 970 deliverable thermonuclear devices ready for firing within a seven-minute window. Because of their ability to be fired while submerged, Trident IIs are the most survivable part of the Triad, and because of that, an enemy will know they are available for a follow-up strike after a conflict begins. The ability to hit an attacking force after launching a first strike is considered a key to deterrence policy.

The disadvantage is that the entire submarine force cannot be kept at sea constantly. Often, two-thirds are in port and, therefore, very vulnerable, whereas Russian ballistic missile submarines can fire their ballistic missiles even while in port. The big US advantage is that Ohio-class submarines are thought to be considered almost undetectable when submerged. However, technology is moving at such an exponential rate that there can be no guarantees that any part of any county's triad can maintain its advantages.

The United States, Russia, China, and now even North Korea have a nuclear triad. India, Pakistan, and Israel also have nuclear triad capabilities. The lines, however, of the triad concept are becoming more blurred with the exponential advancement in technology. China and Russia are advanced in hypersonic missile development that can deliver nuclear warheads at incredible speeds. Russia has successfully tested a nuclear propulsion missile system which the United States abandoned many years ago as being too environmentally destructive and impractical. In principle, no power is allowed to base nuclear weapons in space, but that does not ensure it has not already happened.232 In fact, Bill Nelson, NASA Administrator, told Congress that he has serious concerns that China may be using its civilian launches as cover for military space-based weapons systems,233

The main adversary to the US Triad in the future will be China's DF-41 ICBMs, each of which can carry eight warheads with a yield of 250 kilotons. Russia's new RS-28 Sarmat ICBM, which the West codenames as "Satan II," is assumed to carry 10 nuclear warheads each. Each Satan warhead is feared to have a yield as high as 750 kilotons."

What would nuclear look like in Ukraine?

During the Cold War, strategists learned that nuclear weapons served better as a deterrent than as a means of coercion. Russia has been engaging in nuclear threats since the start of the war in Ukraine, which is a dangerous proposition. While the use of nuclear weapons in Ukraine would certainly change the political game, it is not clear what military advantage the use of a single nuclear weapon would actually provide. Ukraine's forces are too dispersed for the single detonation of a small tactical nuclear weapon to alter the military situation significantly. To be militarily effective, multiple detonations would be required. On the other hand, if a lone nuclear weapon were employed to destroy a population center to try to force the Ukrainians to surrender, the repercussions could be profound.

Radioactive fallout from a nuclear explosion in Ukraine would contaminate Ukrainian territory and possibly beyond. The detonation of a nuclear weapon would not produce quantities close to the radiation produced by the Chornobyl incident, but the radiation would be a source of continuing health concerns. Depending on the location and yield of the blast and the time of year, prevailing winds could carry radioactive contamination west toward most of continental Europe or potentially north over Belarus, Russia, Finland, and Sweden. That, in fact, happened during the Chornobyl catastrophe. The Russians themselves consider the Chornobyl disaster to have been one of the country's greatest catastrophes, and the use of Russian nuclear weapons in Ukraine would reawaken historical fears. It also could have a devastating effect on Russian and Ukrainian food exports.

The use of nuclear weapons would also expose Russia to worldwide condemnation, which would further isolate the country internationally. It would forever alter Russia's status in the world and change everybody's calculations. It would especially imperil the futures of Russian officials and oligarchs who have benefited from their privileged positions in Russia. Would Russia's military leaders and oligarchs go along with that? One suspects that Western military and political officials have communicated appropriate warnings to their Russian counterparts, suggesting that the use of nuclear weapons would have grave implications for their personal futures.

Any use of a nuclear weapon of any size risks retaliation. The response to actual Russian use would depend very much on the circumstances and consequences of the event. President Macron has said that France would not respond in kind if Russia used nuclear weapons in Ukraine. And it is unclear exactly how NATO, or the broader coalition of forces supporting Ukraine (or China and India), would react if Russia actually used a nuclear weapon. Such responses probably depend on the event itself, although it is likely that responses to various scenarios are currently being reviewed. More on this timely subject follows in the next chapter.

Timely is the word. I cannot stop thinking of the nightly news coverage of the war in Ukraine, as well as developments in the Middle East and Asia. "God forbid" is the phrase I hear more and more every night as I flip from one news channel to another as the commentators speculate on the possibility of escalation. **Charcoal drawing by Nikita Kadan, *Shadow on Earth*, focuses on the fate of Russian soldiers invading Ukraine.**

CHAPTER TWENTY EIGHT
NUCLEAR ESCALATION NO PEACE FOR OUR TIME

Chamberlain returning from Munich Conference with document promising "Peace for our time." Birmingham Mail, https://www.birminghammail.co.uk/news/nostalgia/gallery/nevi lle-chamberlain-6117021.

Who would have thought that we would have seen another European war in our lifetime? History has shown that the old world always impacts the new, and twice has led to world wars. Where is this situation headed? Neville Chamberlain sincerely wanted peace over war in 1938 when he gambled on "peace for our time" with Adolf Hitler.

THURSDAY, FEBRUARY 24, 2022

Following a trip to China by Vladimir Putin to visit Xi Jinping, the Russian Federation invades Ukraine at dawn after a long period of military exercises. The following years witnessed the most significant European battles since 1947. Russian forces fail to take Kyiv but made significant advances across eastern Ukraine. Massive sanctions were imposed on Russia in the following months. Russian nuclear forces moved to an elevated alert status, where they have remained. *Reuters* later reports that Putin stated in a speech on that date after launching his "special military operation" that "*Russia is a leading nuclear power and possesses certain advantages in some of the newest types of weaponry. In this regard, no one should have any doubts that a direct attack on our country will lead to defeat and horrible consequences for any potential aggressor.*" He later adds: "*Whoever tries to hinder us, or threaten our country or our people, should know that Russia's response will be immediate and will lead you to consequences that you have never faced in your history.*"[225]

SUNDAY, APRIL 28, 2022

Russia's nuclear submarines and mobile missile launchers conducted drills following Vladimir Putin's order placing nuclear "deterrence forces" on an even higher alert status. The *Associated Press* reported: "Putin's decree applied to all parts of the Russian nuclear triad, which, like in the US, consists of nuclear submarines armed with intercontinental ballistic missiles, nuclear-tipped land-based ICBMs, and nuclear-capable strategic bombers."[226]

JUNE, JULY 2022

The United States' first deliveries of HIMAR, high mobility rocket artillery, and missiles provided Ukrainian forces a needed edge on the battlefield.[227]

AUGUST 2022

The United States delivered HARM, anti-radiation air defense, missiles to Ukraine.[228]

SEPTEMBER 21, WEDNESDAY, 2022

Reuters reports that "Putin orders Russia's first military mobilization since World War Two and says: 'If the territorial integrity of our country is threatened, we will without doubt use all available means to protect Russia and our people - this is not a bluff.'"[229]

TUESDAY, SEPTEMBER 26, 2022

Fifty meters of the Nord Stream 1 and 2 gas pipelines supplying Russian gas to Europe were destroyed by unknown players.[230]

Friday, September 30, 2022

Reuters reports Vladimir Putin statement suggesting the legitimacy of using nuclear weapons: "*. . . the*

---

[225] Mark Trevelyan, *Putin's nuclear warnings since Russia invaded Ukraine*, Reuters, March 13, 2024, https://www.reuters.com/world/europe/putins-nuclear-warnings-since-russia-invaded-ukraine-2024-03-13/.

[226] Vladamir Isachenkov, *Russia Holds Drills with Nuclear Submarines and Land-Based Missile*, Associated Press, March 1, 2022, https://apnews.com/article/russia-ukraine-vladimir-putin-business-europe-moscow-563573526a93ea73a95698d8ddb61b9c.

[227] Ukraine: *What are Himars missiles and are they changing the war?* BBC, August 30, 2022, https://www.bbc.com/news/world- 62512681.

[228] *Ukraine Touts Role of U.S. HARM Missiles in Strikes on Russian Defenses*, Newsweek, September 9, 2022, https://www.newsweek.com/ukraine-touts-role-us-harm-missiles-strikes-russian-defenses-1741715.

[229] Mark Trevelyan, *Putin's nuclear warnings since Russia invaded Ukraine*, Reuters, March 13, 2024, https://www.reuters.com/world/europe/putins-nuclear-warnings-since-russia-invaded-ukraine-2024-03-13/.

[230] Chas Danner, *Who Really Blew Up the Nord Stream Pipeline?,* Intelligencer, June 15, 2023,https://nymag.com/intelligencer/article/who-blew-up-the-nord-stream-pipeline-suspects-and-theories.html.

United States created a 'precedent' when it dropped two atomic bombs on Japan in 1945."[231]

SEPTEMBER TO DECEMBER 2022

In a dramatic offensive, Ukraine retook much of the northeastern Kharkiv region and soon recaptured the eastern city of Lyman. Ukrainian attacks severely damaged the Kerch Strait Bridge linking Crimea to Russia. Russia announced the annexation of four Ukrainian regions despite international condemnation and its loss of ground in some of those regions. Russians withdrew from the city of Kherson to the eastern side of the Dnipro River, proving a significant victory for Ukraine.

MONDAY, DECEMBER 5, 2022

Ukraine attacks Russia's Dyagilevo military airbase in Ryazan, a city less than 150 miles from Moscow, in a series of drone strikes. Explosions also took place at Engels-2 airbase in the Saratov region in southern Russia, where Tu-95 and Tu-60 nuclear-capable bombers are based. For the first time in history, Russian nuclear-capable bombers were destroyed in an armed attack.[232]

FRIDAY, DECEMBER 9, 2022

Reuters reported that "Putin says '. . . *any country that dared to attack Russia with nuclear weapons would be wiped from the face of the earth.*' He says Russia has no mandate to launch a preventative first (nuclear) strike but that Russia's advanced hypersonic weapons would ensure Russia could respond forcefully."[233]

TUESDAY, FEBRUARY 21, 2023

Russia suspends the New START nuclear arms reduction treaty, which puts limits on the number of US and Russian deployed intercontinental-range nuclear weapons. The treaty had allowed both the United States and Russia to conduct inspections of each other's weapons sites, though inspections had been previously halted since 2020 due to the Covid pandemic. *Reuters* further details that Putin stated that: *"Russia will suspend its participation in the New START treaty with the United States that limits the number of nuclear warheads each side can deploy. He says Russia must be ready to conduct a nuclear test in case the United States does so."*[234]

MARCH 2023

The Federation of American Scientists published a report warning that Russian nuclear stockpiles may be increasing. Simultaneously, the United States and Russia continue to dismantle previously retired warheads as part of a broader reduction in nuclear stockpiles, which has been in motion since the end of the Cold War. Despite this, the study states that while France and Israel have relatively stable inventories, China,

---

[231] Ibid.

[232] Jack Guy, Eliza Mackintosh and Tara Subramaniam, *Russia-Ukraine News*, CNN, December 5, 2022, https://www.cnn.com/europe/live-news/russia-ukraine-war-news-12-05-22/index.html.

[233] Mark Trevelyan, *Putin's nuclear warnings since Russia invaded Ukraine*, Reuters, March 13, 2024, https://www.reuters.com/world/europe/putins-nuclear-warnings-since-russia-invaded-ukraine-2024-03-13/.

[234] Ibid.

India, North Korea, Pakistan, the United Kingdom, and Russia are working to increase their stockpiles. Russia's total nuclear inventory is now believed to be 5,889 warheads, counting stockpiled and retired weapons. In comparison, the report attributes the US to have 5,244 warheads, either active or in reserve. It lists China with 410, France with 290, the United Kingdom with 225, Pakistan with 170, India with 164, Israel with 90, North Korea with 30. The Stockholm International Peace Research Institute has published similar statistics. The Department of Defense differs slightly and believes China has significantly expanded its nuclear stockpile to 500 operational warheads. An annual report released by the Pentagon projected that Beijing may double its arsenal to over 1,000 warheads by 2030. Some experts believe North Korea may hold far more than 30 nuclear warheads and is clearly on the path to thermonuclear warheads and intercontinental delivery systems.[235]

SATURDAY, MARCH 25, 2023

Reuters reports that on this day: "Putin says Russia has struck a deal with its ally Belarus, which borders Ukraine, to station tactical nuclear weapons on its territory. He says this mirrors deployments that the United States has made for decades in allied countries in Europe."[236]

TUESDAY, APRIL 4, 2023

Finland officially becomes NATO's 31st member, with Sweden membership anticipated to soon follow, as Russian President Vladimir Putin views this as unacceptable. Russian nuclear saber-rattling continues to increase.

WEDNESDAY, APRIL 19, 2023

The US had delivered by this date a Patriot missile defense system after a long winter of intense Russian missile, drone, and aircraft strikes, which devastated many Ukrainian cities and power infrastructures.[237]

MAY 2023

The UK delivers long-range Storm Shadow cruise missiles to Ukraine. France also sends its version of the Storm Shadow to Ukraine.[238]

SATURDAY, JUNE 24, 2023

Vladimir Putin's 22-year rule almost ended when the mutinous Wagner Group leader Yevgeny Prigozhin seized the army headquarters in the Russian city of Rostov-on-Don and attempted a march on

---

[235] *The Federation of American Scientists* website, https://fas.org/publication/nuclear-notebook-russian-nuclear-weapons-2023/

[236] Mark Trevelyan, *Putin's nuclear warnings since Russia invaded Ukraine*, Reuters, March 13, 2024, https://www.reuters.com/world/europe/putins-nuclear-warnings-since-russia-invaded-ukraine-2024-03-13/.

[237] *US-made Patriot guided missile systems arrive in Ukraine,* ABC.net.au News. April 19, 2023, https://www.abc.net.au/news/2023- 04-20/us-made-patriot-guided-missile-systems-arrive-in-ukraine/102245456.

[238] *UK Delivers Long-Range Storm Shadow Missiles to Ukraine,* European Pravda, May 11, 2003, https://www.eurointegration.com.ua/eng/news/2023/05/11/7161457/#:~:text=Thursday%2C%2011%20May%202023%20The%20Uni ted%20Kingdom%20has,ahead%20of%20a%20long-awaited%20counter-offensive%20by%20Ukrainian%20troops.

Moscow. The facts are still not clear, such as how or why Prigozhin suddenly called off his rebellion or how he was talked down, but Putin's rule survived. Just a few months later, on August 23, Prigozhin's private jet mysteriously crashed northwest of Moscow, killing everyone on board, including the rebellious leader.[239]

SATURDAY, SEPTEMBER 2, 2023

Russia announced the RS-28 Sarmat (Satin II) strategic system has assumed combat alert posture. Estimates suggest these new Russian ICBMs are capable of delivering a MIRVed warhead weighing up to 10 tons to any location worldwide. They can travel either over the North or South Poles, significantly challenging US early warning systems. Russia successfully test-fired the Sarmat missile in April of 2022.[240]

FRIDAY, SEPTEMBER 8, 2023

The first 10 German-made Leopard I tanks pledged by Denmark arrived in Ukraine. Approximately 145 Leopard I tanks are expected to follow, with German Leopard II tanks now pledged by Germany. In addition, Lithuania donated 4.5 million rounds of ammunition to Ukraine. The fact that Russian soldiers are again fighting German tanks for the first time since WWII does not lessen tensions and saber rattling.[241]

WEDNESDAY, SEPTEMBER 13, 2023

Vladimir Putin meets with North Korean leader Kim Jong Un in Eastern Russia at the Vostochny Cosmodrome. Kim pledged support for Russia's war in Ukraine. North Korea proceeds to supply Russia with artillery shells in return for rockets and possible nuclear technology.[242] On that same day, Ukraine launches Storm Shadow missiles against Russia's Black Sea fleet docked in Crimea. The strike damaged Moscow's *Rostov-on-Don* submarine and the *Minsk* landing ship at the Ordzhonikidze shipyard. The evacuation of the Russian navy from Crimea followed.

TUESDAY, OCTOBER 4, 2023

The US and the Russian Federation both conducted a nationwide emergency alert test of their emergency broadcast systems. Russia clearly prepares for armed conflicts involving nuclear powers.[243] *The National Interest* reported: "President Putin considerably increased Russian spending on civil defense in 2005. . . Russia has reportedly built some 5,000-7,000 bomb shelters in Moscow. A 2017 report by the Defense

---

[239] Luke Harding, *Russia-Ukraine war – latest news updates*, The Guardian, June 25, 2023, https://www.theguardian.com/world/2023/jun/25/prigozhins-march-on-moscow-chronology-of-an-attempted-coup.

[240] Brendan Cole, *Russia Deploys 'Satan 2' Missiles: What We Know About Nuclear-Capable Weapon*, Newsweek, September 2, 2023, https://www.newsweek.com/sarmat-nuclear-satan-missile-russia-putin-1824173.

[241] Vasco Cotovia, *First batch of Leopard 1 tanks arrives in Ukraine, Denmark says*, CNN, September 8, 2023, https://www.cnn.com/europe/live-news/russia-ukraine-war-news-09-08-23#h_58ea666d468daebca50dc64d373a920d.

[242] *Putin meets Kim, says Russia will help North Korea build satellites*, Reuters, September 13, 2024, https://www.reuters.com/world/putin-says-russia-help-north-korea-build-satellites-2023-09-13./.

[243] Isabel van Brugen, *US and Russia To Hold Nationwide Emergency Alert Drills on Same Day*, Newsweek, October 4, 2023,

https://www.newsweek.com/us-russia-emergency-alert-drills-1831776.

Intelligence Agency said Russia had 289,000 civil defense personnel, a large increase from the 20,000 in 1996 and 18,250 in 2008."[244] Russia's state media reported: "In some areas, schoolchildren were taught how to wear gas masks. In other areas, officials were forced into bunkers. The exercise was based on the assumption of a giant nuclear attack from the West."[245]

THURSDAY, OCTOBER 5, 2023

Reuters reports that on this day, "Putin says he sees no need to change Russian doctrine stating, it may use nuclear weapons in the event of a nuclear attack or a conventional threat to the existence of the state. *He says any attack on Russia would provoke a split-second response with hundreds of nuclear missiles that no enemy could survive. 'I think no person of sound mind and clear memory would think of using nuclear weapons against Russia.'* " Putin says Russia has tested and will soon place on combat duty its latest nuclear-capable weapons, the Burevestnik cruise missile and Sarmat intercontinental ballistic missile. He says parliament should review Russia's position on the Comprehensive Test Ban Treaty to 'mirror' the position of the United States, which has not ratified it. Within weeks, parliament withdraws Russia's ratification of the treaty.[246]

SATURDAY, OCTOBER 7, 2023

Massive Hamas attacks on Israel led to dramatically escalating tensions in the Middle East and the largest naval buildup in the eastern Mediterranean by the West since World War Two. It seems a convenient coincidence that just as Russia struggles on the battlefield, the war in the Middle East breaks out and diverts

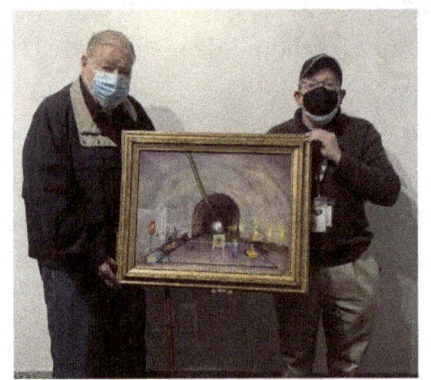

much of the Western aid away from Ukraine toward Israel. This is at a time when US budget concerns create a long-drawn-out debate on the excessive costs mounting over the last two years of aid to Ukrainians. This leads to severe munition shortages for Ukraine as it falters in the struggle against Russia.

MONDAY, OCTOBER 16, 2023

By this date, all 31 M1A1 Abrams tanks pledged by the United States arrived in Ukraine as well as depleted uranium tank rounds.[247]

Dale Cox's oil on canvas depicting the historic X Tunnel in Area 52.

---

[244] Mark B. Schneider, *Russia Is Mobilizing Its Society in Preparation for Nuclear War,* The National Interest, October 17, 2023, https://nationalinterest.org/blog/buzz/russia-mobilizing-its-society-preparation-nuclear-war-206970.

[245] Ibid,

[246] Mark Trevelyan, *Putin's nuclear warnings since Russia invaded Ukraine*, Reuters, March 13, 2024, https://www.reuters.com/world/europe/putins-nuclear-warnings-since-russia-invaded-ukraine-2024-03-13/.

[247] Veteran Los Alamos nuclear testing engineer and valued National Atomic Testing Museum docent. Dale Cox told me about the original testing of depleted uranium shells. He, in fact, did an oil on canvas depicting the historic X Tunnel in Area 25 at the former Nevada Test Site. There, in the 1980s, Dale oversaw depleted uranium ballistics tests with Soviet and American armored vehicles. The tunnel was used to keep the tests from the prying eyes of Soviet statelets. On one such test, Dale had the honor of meeting and working with General Norman Schwarzkopf Jr. Dale Cox is a noted Nevada local artist as well as a veteran Los Alamos nuclear testing engineer.

TUESDAY, OCTOBER 17, 2023

Vladimir Putin meets Xi Jinping in Beijing.

WEDNESDAY, OCTOBER 18, 2023

The United States delivered Ukraine with long-range ATACM missiles armed with cluster munitions.[248]

OCTOBER 2023

Russia's economy shows signs of severe stress associated with a year and a half of war in Ukraine and ongoing Western sanctions. Despite this, Russia anticipates a three percent rise in GNP, which most economists confirm. The Kyiv School of Economics has reported that "Russians have adapted to life under sanctions and are managing to circumvent restrictions."[249]

NOVEMBER 2023

The US Federal debt reaches over 32 trillion dollars, imposing political and economic stress on our nation and the Western world. By this date, The United States had provided Ukraine more than $43.8 billion in security assistance. As of August 2023, the total US federal debt reached $32.6 trillion, with $25.8 trillion held by the public and $6.8 trillion in intragovernmental debt. The annualized cost of servicing this debt accounts for 14 percent of all federal spending.[250]

WEDNESDAY, NOVEMBER 1, 2023

United States Marine Corps Major Gen. Chris A. McPhillips announced that US Central Command had canceled the 248th Marine Corps annual Ball as a result of "unforeseen operational commitments."[251] On that same date, Congresswoman Elise Stefanik and Congressman McCaul's resolution, H. Res. 559, passed on the House Floor ". . . *declaring it the policy of the United States that a nuclear Iran is unacceptable.*"[252]

THURSDAY, NOVEMBER 2, 2023

Russia suspends (revokes ratification) the Comprehensive Nuclear Test Ban Treaty (CTBT). The treaty banning nuclear weapons tests opened for signatures in 1996 and has been ratified by 178 countries. The US and China have not ratified the treaty but adhere to it as Russia has done since the early 1990s. *Reuters* reported Anthony Blinken stated: "Russia's action will only serve to set back confidence in the international arms control regime." *Reuters* further reports: "Western arms control experts are concerned that Russia may

---

[248] Natasha Bertrand and Oren Liebermann, *US has provided Ukraine long-range ATACMS missiles, sources say*, CNN, October 18, 2023, https://www.cnn.com/2023/10/17/politics/us-ukraine-long-range-atacm-missiles/index.html.

[249] [258] Yaroslav Vinokurov, *What's going on with the Russian Economy?*, Ukrainska Pravda, October 26, 2023, https://www.pravda.com.ua/eng/articles/2023/10/26/7425902/.

[250] [259] Alan Rappeport, *As of August 2023, the total US federal debt reached $32.6 trillion; with $25,* New York Times, September 18, 2023, https://www.nytimes.com/2023/09/18/us/politics/us-national-debt.html.

[251] [260] *Marine Corps Ball canceled due to 'unforeseen operational commitments.'* American Military news website.

[252] [261] https://stefanik.house.gov/2023/11/stefanik-mccaul-resolution-declares-a-nuclear-iran-unacceptable-and-to-use-all-means- necessary-to-prevent-iran-from-becoming-nuclear-ready.

be inching towards a nuclear test to intimidate and evoke fear amid the Ukraine war. Putin said on Oct. 5 that he was not ready to say whether or not Russia should resume nuclear testing."[253]

MONDAY, NOVEMBER 6, 2023

The German Bundeswehr announces the formation of a panzer brigade to be permanently stationed in Lithuania among concerns over Russian military buildup near the Baltic states.

TUESDAY, NOVEMBER 7, 2023

Russia formally withdrew from the 1990 Treaty on Conventional Armed Forces in Europe (CFE Treaty), which limited key categories of conventional armaments. Russia defended its move in light of the expansion of NATO, which Vladimir Putin believes endangers his country's security.[254]

WEDNESDAY, NOVEMBER 8, 2023

Russia's Security Council Secretary, Nikolai Patrushev, stated that ". . . destructive policies of the United States and its allies are increasing the risk that nuclear, chemical, or biological weapons could be used."[255]

WEDNESDAY, NOVEMBER 15, 2023

ABC and NBC news services reported that Chinese President Xi Jinping told President Joe Biden at their November 15th summit in San Francisco that: *Beijing will reunify Taiwan with mainland China but that the timing has not yet been decided.*"[256]

THURSDAY, NOVEMBER 23, 2023

South Korea suspended part of the 2018 Inter-Korean Military Agreement, which demilitarized the buffer zones between the two Koreas. This move came after North Korea launched a spy satellite, and then South Korea fired artillery shells into buffer zone areas on several occasions in early January 2024.[257]

MONDAY, DECEMBER 18, 2023

Analysts feel North Korea's launch of a Hwasong-18 solid-fuel missile from a road-mobile platform on this day demonstrates its ability to now field a nuclear-capable intercontinental ballistic missile that can target any part of the United States. A collection of existing liquid-fueled ballistic missiles have been in

---

[253] [262] Andrew Osborn, *Putin revokes Russian ratification of global nuclear test ban treaty*, Reuters, November 2, 2023, https://www.reuters.com/world/europe/putin-revokes-russias-ratification-nuclear-test-ban-treaty-2023-11-02/.

[254] Guy Faulconbridge and Lidia Kelly, Russia formally withdraws from key post-Cold War European armed forces treaty, Reuters, November 7, 2024, https://www.reuters.com/world/europe/russia-formally-withdraws-key-post-cold-war-european-armed-forces- treaty-2023-11-07/.

[255] Guy Faulconbridge, *Putin ally: West increasing the risk of weapons of mass destruction being used*, Reuters, November 8, 2023, https://www.reuters.com/world/russias-patrushev-says-west-stoking-risk-nuclear-weapons-will-be-used-2023-11-08/.

[256] *NBC News report*, https://www.msn.com/en-us/news/world/xi-warned-biden-during-summit-that-beijing-will-reunify-taiwan-with- china/ar-AA1lMMlZ.

[257] Song Sang-ho, *U.S. says S. Korea's partial suspension of 2018 inter-Korean military accord 'prudent,' 'restrained' move*, Yonhap News Agency, November 22, 2023, https://en.yna.co.kr/view/AEN20231123001500315.

place for several years in North Korea and may already be nuclear-armed.[258]

WEDNESDAY, DECEMBER 27, 2023

The State Department updated the state of military aid to Ukraine to date:

IMMEDIATE RELEASE Fact Sheet on U.S. Security Assistance to Ukraine December 27, 2023, The United States has committed more than $44.9 billion in security assistance to Ukraine since the beginning of the Biden Administration, including more than $44.2 billion since the beginning of Russia's unprovoked and brutal invasion on February 24, 2022.

Air Defense

•**One Patriot air defense battery and munitions;**

•**12 National Advanced Surface-to-Air Missile Systems (NASAMS) and munitions;**

•**HAWK air defense systems and munitions;**

•**AIM-7, RIM-7, and AIM-9M missiles for air defense;**

•**More than 2,000 Stinger anti-aircraft missiles;**

•**Avenger air defense systems;**

•**VAMPIRE counter-Unmanned Aerial Systems (c-UAS) and munitions;**

•**c-UAS gun trucks and ammunition;**

•**mobile c-UAS laser-guided rocket systems;**

•**Other c-UAS equipment;**

•**Anti-aircraft guns and ammunition;**

•**Air defense systems components;**

•**Equipment to integrate Western launchers, missiles, and radars with Ukraine's systems;**

•**Equipment to support and sustain Ukraine's existing air defense capabilities;**

•**Equipment to protect critical national infrastructure;**

•**21 air surveillance radars.**

Fires

•**39 High Mobility Artillery Rocket Systems and ammunition;**

---

[258] Solid-fuel missiles are much more adapted for immediate use and, therefore, more concerning than liquid-fueled ballistic missiles, which need much more preparation time for firing. Liquid-fueled ICBMs are usually detectable when their fueling process begins, which is cumbersome and thus hard to hide.

•Ground-Launched Small Diameter Bomb launchers and guided rockets;

•198 155mm Howitzers and more than 2,000,000 155mm artillery rounds;

•More than 7,000 precision-guided 155mm artillery rounds;

More than 40,000 155mm rounds of Remote Anti-Armor Mine (RAAM) Systems;

•72 105mm Howitzers and more than 800,000 105mm artillery rounds;

•10,000 203mm artillery rounds;

•More than 200,000 152mm artillery rounds;

•Approximately 40,000 130mm artillery rounds;

•40,000 122mm artillery rounds; • 60,000 122mm GRAD rockets;

•47 120mm mortar systems;

•10 82mm mortar systems;

•112 81mm mortar systems;

•58 60mm mortar systems;

•More than 400,000 mortar rounds;

•More than 70 counter-artillery and counter-mortar radars;

•20 multi-mission radars; <u>Ground Maneuver</u>

•31 Abrams tanks;

•45 T-72B tanks;

•186 Bradley Infantry Fighting Vehicles;

•Four Bradley Fire Support Team vehicles;

•189 Stryker Armored Personnel Carriers;

•300 M113 Armored Personnel Carriers;

•250 M1117 Armored Security Vehicles;

•More than 500 Mine Resistant Ambush Protected Vehicles (MRAPs);

•More than 2,000 High Mobility Multipurpose Wheeled Vehicles (HMMWVs);

•More than 200 light tactical vehicles;

•300 armored medical treatment vehicles;

•80 trucks and 124 trailers to transport heavy equipment;

•More than 800 tactical vehicles to tow and haul equipment;

•131 tactical vehicles to recover equipment;

•10 command post vehicles;

•30 ammunition support vehicles;

•18 armored bridging systems;

•Eight logistics support vehicles and equipment;

•239 fuel tankers and 105 fuel trailers;

•58 water trailers; • Six armored utility trucks;

•125mm, 120mm, and 105mm tank ammunition;

•More than 1,800,000 rounds of 25mm ammunition;

•Mine clearing equipment.

Aircraft and Unmanned Aerial Systems

•20 Mi-17 helicopters;

•Switchblade Unmanned Aerial Systems (UAS);

•Phoenix Ghost UAS;

•CyberLux K8 UAS;

•Altius-600 UAS; • Jump-20 UAS; • Hornet UAS • Puma UAS; • Scan Eagle UAS;

•Penguin UAS;

•Two radars for UAS;

•High-speed Anti-radiation missiles (HARMs);

•Precision aerial munitions;

•More than 6,000 Zuni aircraft rockets;

•More than 20,000 Hydra-70 aircraft rockets;

•Munitions for UAS.

Anti-armor and Small Arms

•More than 10,000 Javelin anti-armor systems;

•More than 90,000 other anti-armor systems and munitions;

•More than 9,000 Tube-Launched, Optically-Tracked, Wire-Guided (TOW) missiles;

•More than 35,000 grenade launchers and small arms;

•More than 400,000,000 rounds of small arms ammunition and grenades;

•**Laser-guided rocket systems and munitions;**

•**Rocket launchers and ammunition;**

•**Anti-tank mines.**

<u>**Maritime**</u>

•**Two Harpoon coastal defense systems and anti-ship missiles;**

•**62 coastal and riverine patrol boats; • Unmanned Coastal Defense Vessels;**

•**Port and harbor security equipment.**

<u>**Other capabilities**</u>

•**M18A1 Claymore anti-personnel munitions;**

•**C-4 explosives, demolition munitions, and demolition equipment for obstacle clearing; • Obstacle emplacement equipment;**

•**Counter air defense capability;**

•**More than 100,000 sets of body armor and helmets;**

•**Tactical secure communications systems and support equipment;**

•**Four satellite communications (SATCOM) antennas;**

•**SATCOM terminals and services;**

•**Electronic warfare (EW) and counter-EW equipment;**

•**Commercial satellite imagery services;**

•**Night vision devices, surveillance and thermal imagery systems, optics, and rangefinders;**

•**Explosive ordnance disposal equipment and protective gear;**

•**Chemical, Biological, Radiological, Nuclear protective equipment;**

•**Medical supplies, including first aid kits, bandages, monitors, and other equipment;**

•**Field equipment, cold weather gear, generators, and spare parts;**

•**Support for training, maintenance, and sustainment activities.**

The United States also continues to work with its Allies and partners to provide Ukraine with additional capabilities to defend itself.[259]

THURSDAY, DECEMBER 28, 2023

Lieutenant-General Martin Wijnen, the departing Commander of the Royal Netherlands Army, warned

---

[259] https://media.defense.gov/2023/Dec/27/2003366049/-1/-1/1/UKRAINE-FACT-SHEET-27-DEC.PDF.

citizens of rising tensions with Russia in a public speech. He stated: "*The Netherlands needs to learn that the entire society needs to be prepared for when things go wrong. . . For civilians, that means having food and drinking water in stock for times of emergency. . . The Netherlands should not think our safety is guaranteed because we are 1,500 kilometers away.*"[260]

MONDAY, JANUARY 1, 2024

Sweden warns its citizens to be prepared for the reality of war. Swedish Defense Minister Pål Jonson stated: "An armed attack against Sweden cannot be ruled out. . . War can also come to us. These serious times require clarity of vision, capacity to act and persistence—clarity of vision to understand that Russia's goal remains the eradication of a free Ukraine and creation of a Europe in which 'might is right,' with buffer states and spheres of interest. . . We have already experienced this in the past. We must not go back there and allow our children to grow up in that kind of Europe."[261]

THURSDAY, JANUARY 11, 2024

Tensions in the Middle East escalated as US and British forces launched airstrikes on Yemen after months of Iranian-backed Houthi attacks on commercial and military vessels in the Red Sea.

SATURDAY, JANUARY 13, 2024

Taiwan held elections for its presidency and its 113-seat legislature. Democratic Progressive Party (DPP) candidate William Lai, current vice president, secured the election with 40 percent of the votes. The transition inauguration is scheduled for May 20. Although not a clear majority, Taiwan's election confirms its commitment to remain independent. Beijing's reaction is still to be determined, and their continued insistence to see an eventual unification with Taiwan.[262]

MONDAY, JANUARY 15, 2024

German newspaper *BILD* published German classified documents warning that Russia could expand its war in Ukraine and, in turn, could lead to a wider war with NATO. *Reuters* reported: "The article said that the report it had seen outlined a scenario where emboldened by *faltering Western support for Kyiv, Moscow would launch a spring offensive that could push back the Ukrainian army.*"[263]

TUESDAY, JANUARY 16, 2024

United Kingdom Defense Secretary Grant Shapps warns of the likelihood and need to prepare for wars

---

[260] Martin Wijnen,, *Dutch Army commander warns of a potential war with Russia*, website NL#Times, December 28, 2023, https://nltimes.nl/2023/12/28/dutch-army-commander-warns-potential-war-russia.

[261] Nick Mordowanec, *Sweden Issues Ominous Warning to Citizens*, Newsweek, January 8, 2024, https://www.newsweek.com/sweden-issues-ominous-warning-citizens-1858841.

[262] *Critical Questions* by Brian HartScott KennedyJude Blanchetteand Bonny Lin, website Center For Internal and Strategic Studies, Published January 19, 2024, https://www.csis.org/analysis/taiwans-2024-elections-results-and-implications.

[263] Brendan Cole, *Russia Responds to Report About War With NATO*, Newsweek, January 15, 2024, https://www.newsweek.com/russia- germany-bundeswehr-bild-nato-war-1860675.

with China, Russia, and North Korea.[264]

THURSDAY, JANUARY 18, 2024

NATO's top military authority, the Military Committee, held a meeting in Brussels confirming that it will mobilize for the largest European exercise since the end of the Cold War. The exercise will begin in the amazingly short span of only a few days and be named Operation Steadfast Defender. NATO countries, together with NATO candidate Sweden, will mobilize 90,000 soldiers in total by February. At least 20,000 troops will come from Great Britain, representing a substantial part of their army. The soldiers engaged in the maneuver will train to "deter Russian aggression." Gen Christopher Cavoli, NATO's most senior commander, said, "*. . . the exercises would demonstrate the alliance's ability to quickly 'reinforce' its territory in the event of an attack.*"[265] To the outside observer, it looks clear that NATO has information on Russian intentions, which it is not fully sharing because suddenly putting so many troops into Europe seems just too convenient for a needed preparatory laying of logistical groundwork for a wider war. With the delayed US aid to Ukraine, military observers are concerned that Zelenskyy's defenses could collapse and cause an escalation of NATO/Russian tensions into a hot war.

FRIDAY, JANUARY 19, 2024

The United Nation's nuclear monitoring group, the International Atomic Energy Agency or IAEA, issued a warning that Iran has increased uranium enrichment to 60 percent purity. The IAEA states that Iran now has the material to assemble multiple nuclear devices.[266] On that same date, Estonia's Defense Ministry stated that the Baltic states of Estonia, Latvia, and Lithuania had agreed to build a series of bunkers on their borders with Russia and Belarus to defend against Russian expansion.[267]

SATURDAY, JANUARY 20, 2024

NATO's reports circulate on social media from NATO's Military Committee Chief, Dutch Admiral Rob Bauer, urging that the public be prepared with basic necessities such as flashlights and battery-powered radios. He stated: "*We have to realize it's not a given that we are in peace. That's why we must prepare for a conflict with Russia. In order to be fully effective, also in the future, we need a warfighting transformation of NATO.*"[268] The German Defense Minister Boris Pistorius gave additional warnings that "the war in Ukraine could expand to neighboring countries.

---

[264] Ahtra Elnashar, *British defense chief warns 'pre-war' era has begun*, TND, January 17, 2024, https://thenationaldesk.com/news/americas-news-now/british-defense-chief-warns-pre-war-era-has-begun-world-war-iii-secretary-grant-shapps-united-kingdom-russia-china-north-korea-iran-relationships-nuclear-capabilities-red-sea-houthi-rebels.

[265] Daniel Naupold, *NATO set to mobilize 90,000 soldiers for biggest drill since Cold War*, DPA International, January 18, 2024, https://news.yahoo.com/nato-set-mobilize-90-000-152642019.html.

[266] *Iran has enough highly enriched uranium for several nuclear warheads, reports IAEA*, The New Voice of Ukraine, January 19, 2024, https://news.yahoo.com/iran-enough-highly-enriched-uranium-104800933.html.

[267] *Baltic Nations to Build Defense Network Along Borders With Russia, Belarus*, The Moscow Times, January 19, 2024, ttps://www.themoscowtimes.com/2024/01/19/baltic-nations-to-build-defense-network-along-borders-with-russia-belarus-a83786.

[268] *NATO needs 'warfighting transformation', top military official says*, Reuters, January 17, 2024, https://www.reuters.com/world/europe/nato-needs-warfighting-transformation-top-military-official-says-2024-01-17/.

. . We hear threats from the Kremlin almost every day—most recently again against our friends in the Baltic states."[269]

MONDAY, JANUARY 22, 2024

NATO's Operation Steadfast Defender begins and is planned to run through May. The alliance commander, General Chris Cavoli, stated: ". . . *our unity, our strength, and our determination to protect each other.*"[270] By this point in time, it becomes evident that Western and/or NATO military intelligence knows far more than is being reported. Even the causal observer can see that Europe is increasingly concerned about Russia.

TUESDAY, JANUARY 23, 2024

The Turkish parliament voted for approval of Sweden's NATO membership. Turkey had been the only delaying factor in Sweden's admission.[280][271] NATO Secretary General Jens Stoltenberg and Stacy Cummings, General Manager of the NATO Support and Procurement Agency, announced contracts for the purchase of approximately 220,000 155-millimetre artillery shells worth 1.2 billion US dollars.[272]

WEDNESDAY, JANUARY 24, 2024

Eirik Kristoffersen, the head of the Norwegian Armed Forces, warned of a coming war with Russia. He stated Russia has moved to a war economy and is forming alliances in Iran and North Korea. Kristoffersen believes Russia could invade other countries as it continues to fight Ukraine. "*The general urged Norwegians to stock food and gear up for potential warfare.*"[273]

THURSDAY, JANUARY 25, 2024

United Kingdom General Sir Patrick Sanders warned of the threat of Russian expansionism. He stated: "Our predecessors failed to perceive the implications of the so-called July Crisis in 1914 and stumbled into the most ghastly of wars. We cannot afford to make the same mistake today. Ukraine really matters. . . This war is not merely about the black soil of the Donbas, nor the re-establishment of a Russian empire, it's about defeating our system and way of life politically, psychologically, and symbolically. How we respond as the pre-war generation will reverberate through history. Ukrainian bravery is buying time, for now."[274] On that same

---

[269] *West Must Be Prepared For War With Russia, NATO Official Warns Ahead Of Major Military Drills*, Ukrainian Services, January 20, 2024, https://www.rferl.org/a/ukraine-west-war-russia-nato-admiral-bauer-drills/32783552.html.

[270] *Aljazeera*, https://www.aljazeera.com/news/2024/1/18/biggest-nato-drills-since-cold-war-to-start-next-week.

[271] Isil Sariyuce, Jessie Gretener, and Kristin Wilson, *Turkish parliament approves Sweden's NATO membership bid*, CNN, January 23, 2024, https://www.cnn.com/2024/01/23/europe/turkey-vote-sweden-nato-intl'.

[272] https://www.nato.int/cps/en/natohq/news_221972.htm#:~:text=Since%20Allied%20leaders%20agreed%20NATO%E2%80%99s%20Defence%20Production%20Action,155-mm%20artillery%2C%20anti-tank%20guided%20missiles%20and%20tank%20ammunition

[273] *Norwegian military leader warns of possible Russian invasion, urges Europe to prepare*, Essanews.com, https://www.msn.com/en-us/news/world/norwegian-military-leader-warns-of-possible-russian-invasion-urges-europe-to-prepare/ar-BB1haFu9.

[274] Andrew Naughtie, *UK army chief warns citizens to prepare for massive war with Russia*, Euronews, January 25, 2024,

date, Vladimir Putin flew to the Russian exclave of Kaliningrad to bolster the local populous and armed forces stationed there. This constituted a very provocative move due to the recent arms buildup in that former Prussian port on the Baltic, once called Königsberg. It had been handed down and inherited by modern Russia ever since the days of the Soviet victory over Germany in 1945. Clearly, by this date, the West is aware, as they were in 1938, that a general European war is once again a real possibility.

FRIDAY, JANUARY 26, 2024

"The Director General Rafael Mariano Grossi of the International Atomic Energy Agency (IAEA) told the United Nations Security Council that the nuclear safety and security situation at Ukraine's Zaporizhzhya Nuclear Power Plant (ZNPP) remains extremely fragile with 'very real' potential dangers of a major accident."[275]

SATURDAY, JANUARY 27, 2024

News reports circulate that United States nuclear weapons will be stationed at RAF Lakenheath in Suffolk. Not since the height of the Cold War has a similar move occurred when, in the 1980s, British Prime Minister Margaret Thatcher approved the basing of US nuclear weapons in the UK, which also occurred in Germany. The move spurred highly charged antinuclear protests throughout Europe in the 1980s and created some very memorable cultural influences in music, film, and art, which are still with us today.[276]

WEDNESDAY, JANUARY 31, 2024

Ukraine launches a drone strike on St. Petersburg for a second time in two weeks, targeting the Nevsky Oil Refinery.[277] The world starts to recognize by this date that Russia's industrial infrastructure is becoming threatened just as Ukraine's infrastructure has been threatened by Russian attacks for over a year. The key difference, of course, is that Ukraine is not a nuclear power, and Russia has made it clear that it will not allow its national interest to be threatened. This is a significant red line for Russia. In this chapter, the escalating rhetoric has been often quoted in blue italicized print. The words bear attention because, in contrast, Western countries seem to be, at least publicly, ignoring it.

FRIDAY, FEBRUARY 2, 2024

Two United States Air Force B-1 bombers fly from Dyess Air Force Base to conduct airstrikes with Middle East theater-based aircraft, all striking 85 targets at seven sites in Iraq and Syria. This came in direct response after three US service personnel were killed and 38 wounded in a lethal drone strike. That January 28 attack on "Tower 22" on a US military base in northeastern Jordan was attributed to Iran-supported

---

https://www.euronews.com/2024/01/24/uk-army-chief-warns-citizens-to-prepare-for-massive-war-with-russia.

[275] International Atomic Energy Agency Update 208, https://www.iaea.org/.

[276] Tony Diver, *US to station nuclear weapons in UK to counter threat from Russia*, The Telegraph, January 2024,

https://www.telegraph.co.uk/world-news/2024/01/26/us-nuclear-bombs-lackenheath-raf-russia-threat-hiroshima/.

[277] Euro News report, *Ukraine war: Kyiv strikes St Petersburg oil refinery as prisoner swap concludes,* Euronews, January 31, 2024, https://news.yahoo.com/ukraine-war-russia-faces-icj-100216307.html.

groups.[278]

SATURDAY, FEBRUARY 3, 2024

Ukraine strikes Russia's largest oil refinery in Volgograd with two drones. *Reuters* reported that: "The strike is the latest in a series of Ukrainian drone attacks targeting Russian oil facilities in recent weeks, infrastructure that Kyiv sees as important for the Kremlin's war effort."[281]

MONDAY, FEBRUARY 5, 2024

General Gheorghe Vlad, Romanian military chief, stated that his country must prepare for war with Russia.[279]

WEDNESDAY, FEBRUARY 7, 2024

News stories report that FBI Director Christopher Wray briefed Congress on threats which included the likelihood that the Chinese government is planning cyber attacks on civilian infrastructure, which could cripple our society. China expert Gordon Chang, speaking on *The Steve Malzberg Show*, said that the chances of that happening, on a scale of 1-10, is a 20! He made it clear that this is something that is not a theoretical threat. On November 25th, 2023, exemplifying the threats, Iranian hackers took control of part of the water system in Aliquippa, Pennsylvania.[280]

THURSDAY, FEBRUARY 8, 2024

Ukraine successfully struck a Russian oil refinery in Moscow on the same night of Tucker Carlson's interview with Russian President Vladimir Putin. A significant fire broke out in Moscow, as evidenced by press photos. As Russia's war with Ukraine escalates with the increasingly long-range use of US and NATO munitions provided to Ukraine, analysts fear just how far Putin will allow himself to be repeatedly pushed behind his own redlines.[281]

Fires raged in Moscow, *Reuters*.

TUESDAY, FEBRUARY 13, 2024

Estonia's Foreign Intelligence Service warns that Russia appears ready to increase troop levels near the

---

[278] Tom Bowman, Jane Arraf, *US hits Iranian proxies in Iraq, Syria in retaliation for deadly strikes*, PBS, All Things Considered, February 2, 2024, https://www.npr.org/2024/02/02/1228132782/us-biden-iran-drone-response-strike.

[279] Mark Channer, *NATO military chief is preparing for war with Russia and calls for voluntary conscription*, London Loves Business, February 5, 2024, https://londonlovesbusiness.com/nato-military-chief-is-preparing-for-war-with-russia-and-calls-for-voluntary- conscription/.

[280] Ian Schwartz, *Gordon Chang Warns About Threat Of Crippling Chinese Cyber Attacks: On A Scale Of 1-10, It Is Likely A 20*, Real Clear Politics, January 7, 2024. Suggested reading: Gordon G. Chang, *China Is Going to War*. (New York: Criterion Books, 2023.)

[281] Julian Cruz Lima, Will Stewart, *Night of Fury, Russian oil refinery explodes in drone strike & huge fires burn in Moscow as Putin blitzed on night of Tucker interview*, The Sun, February 9, 2024, https://www.thesun.co.uk/news/25823518/russia-oil-refinery-fire- ukraine-putin-tucker-chat/.

Baltic states' borders as well as Finland.[292282]

WEDNESDAY, FEBRUARY 14, 2024

In 1962, the United States Atomic Energy Commission conducted Starfish Prime, the largest nuclear test in outer space. In that operation, a 1.4-megaton nuclear weapon was detonated 250 miles above the Pacific Ocean.[283] The effects were devastating. Electrons generated by the blast became trapped in the Earth's magnetic field, forming radiation belts that lingered for several months. The belts extended around the planet, destroying or damaging one-third of all satellites in low orbit, including Britain's first satellite called Ariel One.[284] Partially addressing this issue, the Limited Test Ban Treaty followed in August 1963 via President John F. Kennedy's determination to deescalate a dangerous Cold War.[285] That treaty prohibited the testing of nuclear weapons in outer space, underwater, or in the atmosphere. Then came the Treaty on Principles Governing the Activities of States in the Exploration and Use of Outer Space, including the Moon and Other Celestial Bodies. This second attempt at preventing the nuclearization of space became a multilateral treaty drafted under the auspices of the United Nations and signed by the United States, the United Kingdom, and the Soviet Union on January 27th, 1967. Today, 114 countries are parties to the treaty, with another 22 additional signatories. On February 14, 2024, top members of Congress met in closed-door briefings to discuss classified intelligence that suggest that Russia may be in preparation for launching a nuclear-armed antisatellite weapon into orbit.[286] CNN sources described the Russian threat as a *"nuclear-powered and space-based device designed to generate a huge energy wave that would knock out vast numbers of satellites that the world down below depends on for infrastructure."*[287]

FRIDAY, FEBRUARY 16, 2024

---

[282] Kim Hielmgaard, *Putin plans to double Russia's troops along NATO border, Estonian intelligence warns*, USA Today, February 13, 2024, https://www.usatoday.com/story/news/world/2024/02/13/putin-plans-for-russia-troops-nato-border/72580109007/.

[283] An artificial aura was created and visible from Honolulu as the bomb detonated 250 miles above Johnston Island, which is only 717 nautical miles from the Hawaiian Islands. This was one of the first experiences with the phenomenon of an EMP, which was created by the bomb, but little understood at the time. In those days, the effects of EMP were not as disastrous to civilian infrastructure because the vacuum tube technology of the day was not as highly sensitive and vulnerable as modern electronics.

[284] Los Alamos space scientist Greg Cunningham is researching the Starfish Prime data. Up to six months after the tests, residual radiation caused significant damage to satellites' electronic systems and solar arrays. Among the satellites damaged were Telstar, the first commercial relay satellite, as well as the Transit Research and Altitude Control (TRAAC) satellite, Transit 4B. Included was the Ariel 1, the first British-American satellite. Source: https://thedebrief.org/for-half-a-century-our-calculations-on-nuclear-explosions- in-space-have-been-wrong-los-alamos-scientist-reveals/.

[285] See Appendix I.

[286] Fabian Hoffmann, *Why a Russian nuclear weapon in space could be so devastating*, The Telegraph, February 13, 2024, https://www.usatoday.com/story/news/world/2024/02/13/putin-plans-for-russia-troops-nato-border/72580109007/.

[287] Jon Christian, *Russian Nuclear Spacecraft Would Reportedly Attack With Massive Energy Wave*, Futurism, February 17, 2024, https://www.yahoo.com/news/russian-nuclear-spacecraft-reportedly-attack-195524173.html.; and Jacob Phillips, *Russia threatens to unleash 'entire arsenal on London if it loses war in Ukraine'* Evening Standard, February 18, 2024, https://www.msn.com/en-us/news/world/russia-threatens-to-unleash-entire-arsenal-on-london-if-it-loses-war-in-ukraine/ar-BB1itC9r.

It seems too much of a coincidence that on the morning that the wife of Russian opposition leader Alexei Navalny, Yulia Navalnaya, was addressing NATO members and Vice President Kamala Harris at the Munich Security Conference, Russia announced her husband's death. Navalny, a Russian attorney and the most vocal and highly visible opposition leader to Vladimir Putin, had been imprisoned for the past two years in the Russian prison system after returning to his native country. His death occurred in a Siberian penal camp.[288]

SUNDAY, FEBRUARY 18, 2024

Former Russian president Dmitry Medvedev waved the nuclear saber with threatening rhetoric. He stated, ". . . attempting to return Russia to its 1991 borders will only lead to one outcome. *To a global war with Western countries, utilizing our entire strategic arsenal. In Kyiv, Berlin, London, and Washington. For all other beautiful historical sites, which have long been included in our nuclear triad's attack goals. "*[289]

WEDNESDAY, FEBRUARY 21, 2024

CNN reported that:

Chinese companies are doing something rarely seen since the 1970s: setting up their own volunteer armies. At least 16 major Chinese firms, including a privately-owned dairy giant, have established fighting forces over the past year, according to a CNN analysis of state media reports. These units, known as the People's Armed Forces Departments, are composed of civilians who retain their regular jobs. They act as a reserve and auxiliary force for China's military, the world's largest, and are available for missions ranging from responding to natural disasters and helping maintain "social order" to providing support during wartime. The forces, which do not currently operate outside China, have more in common with America's National Guard than its militia movement, which refers to private paramilitary organizations that usually have a right-wing political focus. The establishment of corporate brigades highlights Beijing's growing concerns about potential conflict abroad as well as social unrest at home as the economy stumbles, analysts say.[290]

THURSDAY, FEBRUARY 22, 2024

Reuters reports: "Putin sends a signal to the West by taking a short flight on a modernized Tu-160M nuclear-capable strategic bomber plane."[291]

---

[288] Paul Kirby, *Putin critic Alexei Navalny dies in Arctic Circle jail, says Russia*, BBC, February 16, 2024, https://www.bbc.com/news/world-europe-68315943.

[289] Jacob Phillips, *Russia threatens to unleash 'entire arsenal on London if it loses war in Ukraine'*, Evening Standard, February 18, 2024, https://www.msn.com/en-us/news/world/russia-threatens-to-unleash-entire-arsenal-on-london-if-it-loses-war-in-ukraine/ar- BB1itC9r.

[290] Laura He, *Preparing for war, social unrest or a new pandemic? Chinese companies are raising militias like it's the 1970s*, CNN, February 21, 2024, https://www.yahoo.com/finance/news/preparing-war-social-unrest-pandemic-012931701.html.

[291] Mark Trevelyan, *Putin's nuclear warnings since Russia invaded Ukraine*, Reuters, March 13, 2024, https://www.reuters.com/world/europe/putins-nuclear-warnings-since-russia-invaded-ukraine-2024-03-13/. *NATO allies doing too little as Ukraine runs out of ammunition, Stoltenberg says*, Reuters, March 14, 2024, https://www.yahoo.com/news/nato-allies-doing-too-little-133220865.html.

FRIDAY, FEBRUARY 23, 2024

Newsweek reported that for the second time, Ukrainian forces downed one of Russia's scarce and valuable A-50 military spy planes, which form a vital part of their radar defense network.[292]

SATURDAY, FEBRUARY 24, 2024

On the two-year anniversary of the Russian invasion, Ukrainian drones hit a major steel factory in the city of Lipetsk, 250 miles north of the Ukrainian border.[293] After more than a month of debate over continued funding for Ukraine, lagging US munition shipments cause severe shortages for Zelenskyy's forces. Russian forces simultaneously begin to make significant gains in the land war in Ukraine. As NATO forces begin to worry about a possible collapse of the Ukraine frontline, speculation arises over the possibility of NATO members like France, Poland, and Germany sending either advisors or troops into western Ukraine for added support.

TUESDAY, FEBRUARY 27, 2024

Reuters reports: "The Kremlin warned on Tuesday that *conflict between Russia and the US-led NATO military alliance would be inevitable.*" This comes after dramatic Russian advances on the Ukrainian frontlines stimulated discussions among NATO members, who are considering the options of moving troops from select NATO countries into western Ukraine.[294] This fear of a collapsing front line is very reminiscent of the 1917 collapse of the Russian army, which panicked the allied countries of Britain and France, who were then in a desperate war with Germany, which a two-front war had previously checked.

THURSDAY, FEBRUARY 29, 2024

Reuters reports: "President Vladimir Putin told Western countries on Thursday they risked provoking a nuclear war if they send troops to fight in Ukraine, warning that Moscow had the weapons to strike targets in the West."[295] The Washington Post put it more bluntly: "Russian President Vladimir Putin used his annual State of the Nation address on Thursday to take aim at the West, threatening to use nuclear weapons against NATO countries if they send forces to help defend Ukraine from a Russian victory."[296] Putin proceeded to make this warning both publicly and through back channels communicating to Washington. Since the beginning of the war in Ukraine, Russia has boasted so-called "nuclear blackmail." This involves basically

---

[292] Kaitlin Lewi, *Ukraine Shoots Down Second Highly Advanced Russian A-50 Spy Plane*, Newsweek, February 24, 2024, https://www.msn.com/en-us/news/world/ukraine-shoots-down-second-highly-advanced-russian-a-50-spy-plane/ar-BB1iMWPI.

[293] *Ukraine says it struck big Russian steel plant on invasion anniversary*, Reuters, February 24, 2024, https://www.msn.com/en- us/news/world/ukraine-says-it-struck-big-russian-steel-plant-on-invasion-anniversary/ar-BB1iOT18.

[294] Guy Faulconbridge, *Kremlin warns of conflict with NATO if alliance troops fight in Ukraine*, Reuters, February 27, 2024, https://www.yahoo.com/news/kremlin-warns-conflict-nato-alliance-102317133.html/.

[295] Vladimir Soldatkin and Andrew Osborn, *Putin warns West of risk of nuclear war, says Moscow can strike Western targets, Russian President Putin addresses the Federal Assembly in Moscow*, Reuters, February 29, 2024, https://www.yahoo.com/news/putin-warns- west-risk-nuclear-101243590.html.

[296] [306] Francesca Ebel, *Putin threatens nuclear response to NATO troops if they go to Ukraine*, The Washington Post, February 29, 2024, https://www.washingtonpost.com/world/2024/02/29/putin-russia-state-union-speech-military/.

warning the West, in nonspecific language, that if they support Ukraine to the point of a Russian disadvantage, then they would consider using nuclear weapons. On the other hand, the US has expressed itself much more straightforwardly, indicating both publicly and through private channels that if Russia uses a tactical (theater-based) nuclear weapon, then Russian forces in the region would come under overwhelming conventional retaliation. Of course, the US still maintains the decades-old principle of deterrence with the option of a full retaliatory response if Russia ever threatened to use its strategic (intercontinental) nuclear forces. This recent nuclear saber-rattling by Russia has been the most concerning and specifically aggressive warning yet.

FRIDAY, MARCH 1, 2024

Russia provocatively test-fired an RS-24 Yars ICBM from the Plesetsk state test site in northern Russia to the Kamchatka Peninsula. The Russian defense ministry stated that "The Yars missile, which is capable of carrying multiple nuclear warheads and is normally equipped with a multiple independently targetable reentry vehicle (MIRV), successfully hit its target."[297]

SUNDAY, MARCH 3, 2024

The German healthcare system has begun to prepare to handle a large casualty event. German Health Care Minister Karl Lauterbach said: "*In the event of a crisis, every doctor, every hospital, every health authority must know what to do.*"[298] The evidence becomes overwhelming that the NATO powers are not sharing all they know with the public.

MONDAY, MARCH 4, 2024

More red lines are crossed, and Moscow's nuclear saber-rattling increases as Ukraine continues drone strikes into Russia. On this date, Ukraine destroyed a railway bridge over the Chapaevka River in Russia's Samara region and hit St. Petersburg. Ukrainian naval drones attacked and reportedly sank the *Sergei Kotov* Russian patrol ship near the Kerch Strait.[299] Meanwhile, Russia leaked an intercepted German intelligence communication by high-ranking German military officers who indicated that "Berlin was preparing for war against Russia."[300] In actuality, German military officials in Singapore were carelessly sending communications that were intercepted, stating that German and UK military advisors were aiding the Ukrainians in long-range strikes on targets, possibly including the Crimean Bridge. Adding to the tensions,

---

[297] [307]*Russia tests RS-24 Yars intercontinental missile for nuclear capability*, Defense News, March 4, 2024, https://www.armyrecognition.com/defense_news_march_2024_global_security_army_industry/russia_tests_rs-24_yars_intercontinental_missile_for_nuclear_capability.html.

[298] [308] *German Healthcare System Should Prepare for War: Minister*, Tasnim News Agency, March 3, 2024, https://www.tasnimnews.com/en/news/2024/03/03/3049018/german-healthcare-system-should-prepare-for-war-minister.

[299] *Ukrainian attacks increasingly sap the power of Russia's Black Sea fleet*, Associated Press, March 5, 2024, https://www.yahoo.com/news/ukrainian-attacks-increasingly-sap-power-184937827.html.

[300] James Jackson of the UK *Telegraph* wrote that Germany accidentally leaked British military secrets to Russia by using off-the-shelf video phone technology to discuss missiles in Ukraine. The head of the Luftwaffe told air force officers and a general who dialed in from his hotel room how British and French officials were delivering Storm Shadows to Ukrainian soldiers. He also said British troops were "on the ground," a highly sensitive detail that has already caused division and infighting among NATO allies.

NATO's *Exercise* Steadfast Defender 24, which began in February, is being seen as a convenient way to mass large numbers of troops on the Russian, Belarus, and Ukrainian border should Ukrainian forces collapse. Those maneuvers are augmented with armed forces from other countries under a special NATO designation, Dragon-24.[301] In addition, on this date, NATO supplements their maneuvers Steadfast Defender 24 with another exercise called Nordic Response 24, which places 20,000 NATO troops in Finland, Norway, and Sweden, along with 100 fighter jets, transport aircraft, helicopters, and maritime vessels.[302] Tensions and escalation reach levels exceeding even the darkest Cold War era as it becomes clear that if Ukraine collapses, the West will be forced to fill the vacuum.

TUESDAY, MARCH 5, 2024

China's National People's Congress increased its military budget by 7.2 percent over an already excessive defense budget of $231 billion, the biggest increase in five years. In its mention of Taiwan, the government dropped all reference to the use of the word "peaceful" in its intent for reunification.[303]

WEDNESDAY, MARCH 6, 2024

News outlets reported: "The German Federal Chancellery conducted a surprise emergency drill, relocating to an undisclosed location to test government readiness and inter-agency coordination. This proactive measure underscores Germany's commitment to national security."[304] On this same day, Russia had another national test of its emergency alert system. *Newsweek* reported: "Sirens blared across Russia as part of a national emergency drill on Wednesday, but in Russian President Vladimir Putin's home city, St. Petersburg, speakers played a polonaise from Pyotr Tchaikovsky's opera *Eugene Onegin*."[305]

THURSDAY, MARCH 7, 2024

Sweden is officially declared the 32nd member of NATO. Ironically, Vladimir Putin justified his country's initial 2022 invasion of Ukraine as a response to NATO expansion into Poland, the Baltic States, and southeastern Europe. Since Russia invaded Ukraine, NATO has only expanded further with the addition of Finland and Sweden. Traditionally, these countries have a long history of neutrality, and Sweden has not

---

[301] [311] Kateryna Serohina, *NATO begins large-scale exercises near borders of Russia*, RBC Ukraine, March 4, 2024, https://www.msn.com/en-us/news/world/nato-begins-large-scale-exercises-near-borders-of-russia/ar-BB1jhvTT.

[302] Kevin McSpadden, *NATO Launches Biggest-Ever War Game Exercise on Vladimir Putin's Doorstep With 20,000 Troops*, Knewz, March 4, 2024, https://www.msn.com/en-ca/news/world/nato-launches-biggest-ever-war-game-exercise-on-vladimir-putins-doorstep- with-20000-troops/ar-BB1jniTA.

[303] Mikhaila Friel, *A subtle shift in how China talks about Taiwan suggests it is gearing up, for war*, Business Insider, March 5, 2024, https://www.msn.com/en-us/news/world/a-subtle-shift-in-how-china-talks-about-taiwan-suggests-it-is-gearing-up-for-war/ar- BB1jmdUW; and *China Drops 'Peaceful Reunification' Reference to Taiwan, Raises Defence Spending by 7.2%*, Reuters, March 4, 2024, https://www.usnews.com/news/world/articles/2024-03-04/china-drops-peaceful-reunification-reference-to-taiwan-raises- defence-spending-by-7-2.

[304] Wojciech Zylm, *German Chancellery Conducts Surprise Drill, Relocates to Secret HQ Without Stated Reason*, BNN, March 6, 2024, https://bnnbreaking.com/conflict-defence/security/german-chancellery-conducts-surprise-drill-relocates-to-secret-hq-without-stated- reason.

[305] Isabel van Brugen, *Russia's Nationwide Emergency Alert Drill Brings Surprise Twist*, Newsweek, March 6, 2024, https://www.msn.com/en- us/news/world/russias-nationwide-emergency-alert-drill-brings-surprise-twist/ar-BB1jqGLH.

been involved in a war since 1814.[306]

FRIDAY, MARCH 8, 2024

By this date, it becomes obvious that for weeks, Russia has been using and experimenting with the tactical use of GPS (global positioning system) jamming on NATO countries in northern Europe. *Newsweek* has reported that this "has become a daily occurrence near strategic sites, particularly since Russia invaded Ukraine."[307] The jamming is designed to disable or confuse military tracking and navigation equipment; however, fears are rising that this could present a safety hazard to civilian air traffic. *Newsweek* stated: "GPS jamming reduces positioning accuracy and could cause the receivers in the cockpit of civilian planes to lose positioning, although experts have said the recent spike in disturbances over the Baltic Sea does not threaten aircraft safety."[308]

SATURDAY, MARCH 9, 2024

CNN reported a story detailing two senior White House officials who confirmed that *"In late 2022, the US began "preparing rigorously" for Russia potentially striking Ukraine with a nuclear weapon. . . Multiple senior administration officials took part in an urgent outreach. Secretary of State Antony Blinken communicated US concerns 'very directly' with Russian foreign minister Sergey Lavrov, according to senior administration officials. Joint Chiefs Chairman General Mark Milley called his Russian counterpart, General Valery Gerasimov, chief of the general staff of the Russian Armed Forces. According to a senior US official, President Joe Biden sent CIA Director Bill Burns to speak to Sergey Naryshkin, the head of Russia's foreign intelligence service, in Turkey to communicate US concerns about a nuclear strike taking place and gauge Russian intentions."[309]*

TUESDAY, MARCH 12-13, 2024

Drone attacks on Russia have increased. The Associated Press reported that "one Ukrainian drone struck an oil refinery in the Nizhny Novgorod region, according to regional governor Gleb Nikitin. That region is 775 kilometers (480 miles) from the Ukraine border."[310] Another drone was reported to have been shot

---

[306] Emily Rauhala, *Sweden finally joins NATO in expansion spurred by Putin's Ukraine war*, The Washington Post, March 7. 2024, https://www.msn.com/en-us/news/world/sweden-finally-joins-nato-in-expansion-spurred-by-putin-s-ukraine-war/ar-BB1jv7C2; and Christian Edwards, Radina Gigova, Jennifer Hansler and Mariya Knight, *Sweden officially joins NATO, becoming alliance's 32nd member*, CNN, https://www.msn.com/en-us/news/world/sweden-officially-joins-nato-becoming-alliance-s-32nd-member/ar- BB1juWU6.

[307] *NATO Nations Hit by GPS Attack Blamed on Russia*, Newsweek, March 2, 2024, https://www.newsweek.com/nato-gps-russia-attack- blamed-1875276.

[308] Ibid.

[309] [319] Jim Sciutto, Exclusive: *US prepared 'rigorously' for potential Russian nuclear strike in Ukraine in late 2022, officials say,* CNN March 9, 2024, https://www.cnn.com/2024/03/09/politics/us-prepared-rigorously-potential-russian-nuclear-strike-ukraine.

[310] Jim Heintz and Hanna Arhirova, *Russia reports Ukrainian drone strikes on targets deep inside its territory and a border incursion*, Associate Press, March 12, 2024, https://www.yahoo.com/news/russia-reports-ukrainian-drone-strikes-094644024.html.

down near Moscow's Zhukovsky Airport, one of the city's four international airports. Then, a drone hit a Russian oil depot in Oryol, 116 kilometers (95 miles) from Ukraine. *The Associate Press* added, "The Russian Defense Ministry said Ukrainian drones were also intercepted Tuesday over the Belgorod, Bryansk, Kursk, Leningrad, and Tula regions of Russia."[311] *CNN* reports Ukrainian drones have targeted "three Russian oil refineries were targeted in the cities of Ryazan, about 130 miles southeast of Moscow; Kstovo, in the Nizhny Novgorod region, nearly 300 miles east of the capital; and Kirishi in Russia's northwest. The source said that the trio of facilities are among Russia's largest refineries. A fourth facility–the Novoshakhtinsky oil refinery in Rostov-on-Don–was also hit, a representative of the Defense Intelligence of Ukraine said Wednesday."[312] Russian oil production falls by ten percent, according to analysts.

WEDNESDAY, MARCH 13, 2024

Reuters reports on continued nuclear saber-rattling from Vladimir Putin: "Putin said on Wednesday that "*From a military-technical point of view, we are, of course, ready.*" - the latest in a series of warnings he has delivered to the West in the two years since he launched his invasion of Ukraine. Western security analysts say Putin's statements are designed to deter and intimidate."[313]

THURSDAY, MARCH 14, 2024

Reuters reports that NATO Secretary General Jens Stoltenberg publicly warned that "Ukraine is running out of ammunition in its war against Russia's invasion and NATO members are not doing enough to help Kyiv."[314]

FRIDAY, MARCH 15, 2024

NATO confirms Russia has moved tactical nuclear weapons into Belarus. It is believed this has been underway since June of 2023, clearly because Vladimir Putin has often said as much. However, Arvydas Anusauskas, Lithuania's defense minister, has become the first top official within the NATO alliance to confirm the news of the deployments. He also confirmed that nuclear weapons were being moved into the Russian exclave of Kaliningrad.[315] If tactical nuclear weapons are used on Ukraine and launched from

---

[311] Ibid.

[312] Rob Picheta, Victoria Butenko, Martin Goillandeau, Josh Pennington, Olga Voitovych and Anna Chernova, *Ukraine hits oil refineries deep inside Russian territory, as Kyiv steps up drone attacks before Putin's likely re-election*, CNN, March 13, 2024, https://www.cnn.com/2024/03/13/europe/ukraine-russia-drone-strikes-putin-intl.

[313] Mark Trevelyan, *Putin's nuclear warnings since Russia invaded Ukraine*, Reuters, March 13, 2024, https://www.reuters.com/world/europe/putins-nuclear-warnings-since-russia-invaded-ukraine-2024-03-13/.

[314] *NATO allies doing too little as Ukraine runs out of ammunition, Stoltenberg says*, Reuters, March 14, 2024, https://www.yahoo.com/news/nato-allies-doing-too-little-133220865.html.

[315] Jack Detsch and Robbie Gramer, *Russia's Nuclear Weapons Are Now in Belarus*, Foreign Policy, March 14, 2024, https://foreignpolicy.com/2024/03/14/russia-nuclear-weapons-belarus-putin/.; *Foreign Policy* reported: "Western intelligence officials and open-source sleuths have spent months tracking the status of the Russian deployment to Belarus, which Putin himself framed as a warning to the West. The movement of the weapons to Belarus marks one of the westernmost deployment points of the Kremlin's nuclear arsenal. The movement of its nuclear weapons has clear political signaling, but some experts downplayed the military significance of the move—arguing that the weapons don't pose a higher or lower threat to the alliance simply by being

Belarus, it would conceivably give Vladamir Putin some latitude as Belarus is not technically part of the Russian Federation, although highly dependent on. Putin, therefore, could feasibly claim Belarusian President Alexander Lukashenko launched the nuclear weapons to defend his own country. It would be a weak argument, but it is considered a possible political ploy by Putin to avert a direct retaliation on Russia. It is also apparent from various media sources that during the past week, both Western and Russian nuclear forces have been on high alert, and a large part of the Russian submarine fleet has been put to sea.

SUNDAY, MARCH 17, 2024

Reuters reported: "President Vladimir Putin won a record post-Soviet landslide in Russia's election on Sunday, cementing his grip on power, though thousands of opponents staged a noon protest at polling stations and Western countries said the vote was neither free nor fair."[316326] The fear by Western experts is that Putin will now feel emboldened by this election outcome and more aggressively challenge Ukraine and NATO.

TUESDAY, MARCH 19, 2024

The Greek City Times reported that: "The American cargo ship *Leroy A. Mendonica* delivered dozens of Bradley and Abrams infantry fighting vehicles to the port of Alexandroupolis in northern Greece as part of NATO's military reinforcement in Europe. . . in support of the deployment of the [US] 3rd Armored Brigade Combat Team, 4th Infantry Division"[327317] *The DC Weekly* provided additional details stating: "It is reported that over 3,000 units of weapons and heavy transport vehicles have already been unloaded at the port of Alexandroupolis, which were transported by road and rail to Eastern Europe bypassing the Dardanelles Strait."[328318]

---

moved several hundred miles closer to NATO territory."

[316] *Putin's election win tells us about Russia today*, Reuters, March 18, 2024, https://www.yahoo.com/finance/news/takeaways-putin- election-win-tells-160807168.html.

[317] *The Greek City Times* website, March 19, 2024, https://greekcitytimes.com/2024/03/19/us-sends-huge-shipment-of-military- equipment-to-greece/.

[318] *The DC Weekly* website, March 18, 2024, https://dcweekly.org/2024/03/18/us-tanks-and-infantry-fighting-vehicles-arrive-in-greece- strengthening-natos-presence-in-europe/.

U.S. Army Sgt. 1st Class Terysa King reported: "The port of Alexandroupolis in Greece made sustainment history as it facilitates movement of the first "heavy" infantry brigade on March 11, 2024. The 624th Movement Control Team, 598th Transportation Brigade, 3rd Brigade Combat Team, 4th Infantry Division, the 21st Theater Sustainment Command, and Hellenic Forces downloaded approximately 3,000 pieces of equipment to include Bradley Fighting Vehicles, Joint Light Tactical Vehicles, Mine Resistant Ambush Protected All-Terrain Vehicles, and M1 Abrams Tanks from the vessel Leroy A. Mendonca. The equipment will be transported to multiple locations to generate rapid combat power to deter potential aggression and reassure NATO allies." https://www.dvidshub.net/video/915895/21st-theater-sustainment-command-and-598th-transportation- brigade-makes-history-port-alexandrouplis.

WEDNESDAY, MARCH 20, 2024

The Institute for The Study of War reports: "Several Russian financial, economic, and military indicators suggest that Russia is preparing for a large-scale conventional conflict with NATO, not imminently but likely on a shorter timeline than what some Western analysts have initially posited. The Russian military continues to undertake structural reforms to simultaneously support the war in Ukraine while expanding Russia's conventional capabilities in the long term in preparation for a potential future large-scale conflict with NATO."[319]

THURSDAY-TUESDAY, MARCH 21-26, 2024

A tragic series of events unfolded beginning Thursday night when, after a long series of hard-hitting Ukrainian drone strikes on Russian oil refineries, Russia launched 60 drones and 90 missiles in their largest attack to date on Ukrainian energy infrastructure, which included the DniproHES dam, in the southern city of Zaporizhzhia. The *Associated Press* characterized this as a significant escalation.[320] Then, on Friday night, Islamic State-Khorasanor IS-K gunmen committed a humanitarian massacre, killing over 140 civilians at a concert at the Crocus City Hall, on the outskirts of Moscow just 12 miles from the Kremlin. The US had intelligence that warned Moscow about just such an attack weeks prior. Despite that, Vladimir Putin falsely linked Ukraine, the UK, and even American intelligence to the tragedy.[321] *Reuters* reported that: "The director of Russia's FSB security agency said on Tuesday that he believed Ukraine, along with the United States and Britain, were involved in the Moscow attack."[322] Ukraine continues to deny any involvement in the attack as British Foreign Secretary David Cameron insists that the Russian allegations are "utter nonsense."[323] Ukrainian and Russian attacks on each other's infrastructure continue as both sides experience problems related to losses in their energy sectors. As Russia moves trainloads of armaments to

---

[319] *The Institute For The Study Of War* website, March 20, 2024, https://www.understandingwar.org/.

[320] Hanna Arhirova and Jim Heintz, *Russia launches sweeping attack on Ukraine's power sector, a sign of possible escalation*, Associated Press, March 22, 2024,vhttps://www.yahoo.com/news/russia-attacks-ukrainian-electrical-power-081145116.html.

[321] Reshad Hudson,*Moscow attack is country's deadliest in years*, WDVNCBSNews, March25, https://www.yahoo.com/news/moscow-attack-country-deadliest-years-132908155.html.

[322] *Russia says it's hard to believe Islamic State could have launched Moscow attack*, Reuters, March 27, 2024, https://www.yahoo.com/news/russia-says-hard-believe-islamic-135347339.html.

[323] Ibid.

its western borders, Poland and France warn of regional instability. French President Emmanuel Macron sent 2,000 of his country's troops to Romania in preparation for an intervention in Ukraine be needed. In addition, French Defense Minister Sébastien Lecornu said France will soon be able to deliver 78 Caesar howitzers to Ukraine and will boost its supply of shells to meet Kyiv's urgent needs for ammunition to fight Russia's full-scale invasion." As American aid to Ukraine continues to be delayed, real worries arise as to how long the Russians can be held back.

MONDAY, MARCH 25, 2024

The Joint Base Charleston public affairs website announced: "Army vehicles and equipment are staged, ready to be loaded onto the USNS Charlton at Joint Base Charleston, South Carolina, March 25, 2024. The proposition, maintenance, and availability of this equipment enable the rapid deployment of a brigade combat team, ensuring quick response in the event of any possible contingency."324 The USNS Charlton (T-AKR-314) is one 19 "roll on, roll off" vehicle cargo ships in the Military Sealift Command. Indications are it is going to sea with a battalion size force in early April.

U.S. Air Force photo by Tech. Sgt. Alex Fox Echols III.

THURSDAY, APRIL 4, 2024

On the 75th anniversary of the founding of NATO, Kremlin spokesman Dmitry Peskov told reporters: "*In fact, relations have now slipped to the level of direct confrontation.*"325

---

324 *DoD, contracted personnel load hundreds of U.S. Army equipment aboard USNS Charlton*, Defense Visual Information Distribution Service, March 24, 2024, https://www.dvidshub.net/image/8313542/dod-contracted-personnel-load-hundreds-us-army-equipment- aboard-usns-charlton.

325 *Kremlin says Russia and NATO are now in "direct confrontation,"* Reuters, April 4, 2024, https://www.yahoo.com/news/russia- nato-relations-level-direct-093001322.html.

MONDAY, APRIL 22, 2024

As the US moves closer to renewing military funding to Ukraine, Russian Foreign Minister Sergei Lavrov states that Western support for Ukraine has put the *"United States and its allies on the verge of a direct military clash with Russia. . . The Westerners are teetering dangerously on the brink of a direct military clash between nuclear powers, which is fraught with catastrophic consequences."*[326] Meanwhile, the United Nations International Atomic Energy Commission or IAEA convened a meeting of Russian and Ukrainian officials pleading for "sanity" after drones struck the Zaporizhzhia nuclear power plant the previous week.[327] During a time of escalating tensions with Israel and Iran, following the first "direct" exchange of missiles and drones between the two states, world tensions skyrocket. During the week of April 14-20, tensions were at an extreme level. On the night of April 17th into the early morning of the 18th, short-wave radio operators overhead an incredible number of Emergency Action Messages or EMAs.[328] These were overlapped with even higher priority messages known as "Skyking Messages."[329]

---

[326]. Guy Faulconbridge, Russia warns of direct clash with West over Ukraine, Reuters, April 22, 2024, https://www.yahoo.com/news/russia-says-west-teetering-brink-074546436.html.

[327] *337 The Threat of Nuclear Terror Looms Over Ukraine, Daily Beast, April 21, 2024, https://www.yahoo.com/news/threat-nuclear-terror- looms-over-010706110.html..*

[328] *According to Numbers Station and Information Center (https://www.numbers-stations.com/usa/hfgcs/), EAMs are frequently read on the HF-GCS frequencies and usually won't take you long to hear one. The United States Air Force uses the HF-GCS to send instructions for their operations through messages and, most commonly, send Emergency Action Messages (EAMs). The HF-GCS is not exclusive to the USAF and is used by other countries, too, but not as often. They also send higher priority messages known as "Skyking Messages," which will even be read over-top and interrupt an EAM to be read. These messages are time-sensitive and read live in NATO Phonetic letters. They begin with a 6-letter header. This preamble could have a few different uses, but NPS states that for Minuteman Missile launches, "The preamble told the crew which edition and page number of a non-sealed authentication to use. Once at the right page, the crew would know what message checklist to use." The receiving crew can access an emergency action checklist binder where the message and instructions are copied. Then, the message continues afterward and is repeated. A typical EAM message is 30 characters long but can be different. There have been EAMs over 200 characters long before. The message usually ends with "Mainsail Out," but it can change based on where it is being sent from, for example, (Offutt out). Mainsail is the collective calling for all ground stations in the network. From 2013 to April 2015, "Mainsail Out" was a much more commonly used ending than the originating base. Since then, you will now mostly hear a callsign being used for the sender/recipient. Another unique callsign is SkyMaster – the collective callsign for all USSTRATCOM airborne command units. The primary HF-GCS frequency is 11175 kHz, and it transmits 24/7 on that frequency as well as 8992 kHz. The HF-GCS also transmits on 4724, 6712 (Croughton), 6739, 13200, and 15016 during their scheduled times, which is still most of the day. The High Frequency (HF) Global Communications System (HFGCS) supports war plans and operational requirements for the following organizations: White House Communications Agency (WHCA), Joint Chiefs of Staff (JCS), Air Mobility Command (AMC), Air Combat Command (ACC), AF Air Intelligence Agency (AIA), Air Force Materiel Command (AFMC), Air Force Space Command (AFSPC), United States Air Forces in Europe (USAFE), Pacific Air Forces (PACAF), and Air Weather Service (AWS).*

[329] *According to Numbers Station and Information Center (https://www.numbers-stations.com/usa/hfgcs/), "Foxtrot Broadcasts" are Skyking messages. These are higher-priority messages and are sent in a format different from EAMs. Skyking messages sometimes interrupt an ongoing EAM since it's the highest priority. "Skyking" is the collective callsign for sending messages to all Single Integrated Operation Plan (SIOP) aircraft and missile Ops which are also responsible for deploying strategic bombers, reconnaissance aircraft, and various support aircraft. A Skyking message begins with the reader speaking, "Skyking Skyking do not answer," followed by a codeword, then two numbers for the time of the hour, and ends with a two letter authentication string. The message is then repeated. Skyking messages have the same ending as a regular EAM and can also*

The level of these coded signals proved unprecedented, indicating either a high-level exercise was going on or a significant military development had transpired. Stories have circulated that on that evening, Israel had planned a massive attack on Iran, but due to one of its F-35s being shot down, it canceled the operation and followed up the next night of the 18th with a very low-key attack on a Russian-made S-300 air defense missile battery near the Iranian central city of Isfahan.340 That area is also home to sites associated with Iran's nuclear program. Overall, military escalation continues around the world. During this week, Bill Nelson, NASA Administrator, told Congress that he has serious concerns that China is using its civilian launches as cover for military space-based weapons systems.341 Meanwhile, reports circulate that France has placed troops in Odesa, and Iran indicates it may be moving toward a nuclear test.

THURSDAY, APRIL 24, 2024

President Biden signs into law a $95 billion war aid package for assistance to Ukraine, Israel, and Taiwan. Simultaneously the Russian Federation drafts legislation of an unknown nature for "retaliatory measures" if the West pursues proposed plans to use $300 billion of seized Russian assets to fund aid to Ukraine, which is assumed to be linked to the recent aid bill legislation.[330]

TUESDAY, APRIL 30, 2024

News reports highlight a pattern in China's stockpiling of assets such as gold and copper and selling US investments. *The* (UK) *Telegraph* stated that "China has built up a $170bn (£135bn) stockpile of gold after a record buying spree, in a move that has raised fears Beijing is preparing its economy for a possible conflict over Taiwan. The People's Bank of China (PBOC) bought 27 tons of gold in the first three months of the year, taking its reserves to a record high of 2,262 tons, according to data from the World Gold Council.[331]

WEDNESDAY, MAY 1, 2024

On this date the US discussed the threat of Russian anti-satellite capability for the first time in an open congressional hearing. NBC reported: "A senior Defense Department official told lawmakers Wednesday that Russia is developing an 'indiscriminate' anti-satellite nuclear device that would pose a threat to all satellites operated by countries and companies around the world."[332] Indiscriminate means a nuclear weapon in space, as John Plumb, the assistant secretary of defense for space policy told the House Armed Services subcommittee.[333] The Institute For The Study of War confirmed that Ukrainian drone attacks on Russian oil facilities have been renewed. They report: "Ukrainian forces struck an oil refinery in Ryazan Oblast for the

---

change depending on where they are sent from. Until 2016, Skyking began with a 3 letter trigraph instead of a codeword.

[330] *Russia warns Europe: if you take our assets, we have a response that will hurt*, Reuters, April 24, 2024, https://www.yahoo.com/news/russia-warns-europe-assets-response-061530314.html.

[331] Melissa Lawford, *China's $170bn gold rush triggers Taiwan invasion fears*, The (UK) Telegraph, April 30, 2024, https://www.yahoo.com/finance/news/china-launches-record-gold-buying-060000862.html. (The article added: Beijing's stockpile is dwarfed by the holdings of the US, which has the largest reserves in the world. US holdings are worth $602bn, while the UK owns $23bn of the precious metal.)

[332] Dan De Luce, *Pentagon official warns Russian anti-satellite nuclear weapon could be devastating*, NBC News, May 1, 2024, https://www.yahoo.com/news/pentagon-official-warns-russian-anti-015113620.html.

[333] Ibid.

second time in less than a month on the night of April 30 to May 1."[334] On this same date China rejected arms control talks after the Biden administration proposed "common-sense steps that addressed fundamental risks for conflict and uncontrolled escalation in the nuclear and space domains."[335]

THURSDAY, MAY 2, 2024

The (UK) *Telegraph* reports opensource intelligence indicating that Russia's war effort in Ukraine is receiving significant assistance from China, India, and North Korea. The accounts state: "UN investigators have concluded that debris from a missile that landed in the Ukrainian city of Kharkiv on January 2 was from a North Korean Hwasong-11 series ballistic missile[336]. . . The missile, moreover, will have been part of the massive consignment of weapons North Korea has sent to Russia in recent months, with South Korean security officials estimating that nearly 7,000 shipping containers filled with missiles, artillery shells and other ammunition have been dispatched since December. . . Iranian drones, too, have become a familiar feature of the Ukrainian battlespace, with Russian forces using them regularly to conduct swarm attacks against key Ukrainian infrastructure. China's support for Putin's war effort is, by contrast, focused more on helping Russia to rebuild its military industrial base to a level not seen since the Soviet era. . . Secretary of State Antony Blinken commented following his inconclusive visit, China is 'providing components that are powering Russia's brutal war of aggression against Ukraine.'"[337] On this same day the US State Department accused Russia of using chloropicrin, a chemical weapon, in Ukraine.

THURSDAY, MAY 2, 2024

The US government's Department of Energy (DOE) suspends plans to restock the Strategic Petroleum Reserve after heavily drawing on it in past years. Oilprice.com reports: "Following Russia's invasion of Ukraine, the Biden Administration made the largest withdrawal in SPR history to curb the oil price spikes that happened in the wake of the invasion. Despite indicating they would refill the Strategic Petroleum Reserve (SPR) by the end of this year, the Department of Energy has now cancelled solicitations. The DOE has consistently promised to refill the SPR as market conditions allow. The SPR is the world's largest supply of emergency crude oil. It was established primarily to reduce the impact of disruptions in supplies of petroleum products and to carry out obligations of the United States under the international energy program. The SPR is maintained by the U.S. DOE and its oil stocks are stored in huge underground salt caverns at four sites along the coastline of the Gulf of Mexico."[338]

---

[334] ISW website: https://www.understandingwar.org/backgrounder/ukraine-conflict-updates; and Charles Kennedy, *Ukrainian Drones Hit Major Rosneft Refinery in Russia*, Oil Price.Com, May 1, 2024, https://oilprice.com/Latest-Energy-News/World-News/Ukrainian-Drones-Hit-Major-Rosneft-Refinery-in-Russia.html.

[335] Mathias Hammer, *China declines to meet with US on nuclear arms control, US official says*, Semafor, May 1, 2024, https://www.yahoo.com/news/china-declines-meet-us-nuclear-211255064.html.

[336] This is North Korea's version of the Russian Iskander short range ballistic missile.

[337] Con Coughlin, *Putin's crushing new offensive could be the end of Ukraine*, The (UK) Telegraph, May 2, 2024, https://www.yahoo.com/news/putin-crushing-offensive-could-end-052400437.html.

[338] Robert Rapier, *High Oil Prices Force Biden Admin to Halt SPR Refill Plans*, Oil Price.Com, May 3, 2024, https://oilprice.com/Energy/Crude-Oil/High-Oil-Prices-Force-Biden-Admin-to-Halt-SPR-Refill-Plans.html. The size of the SPR (authorized storage capacity of 714 million barrels) makes it a significant deterrent to oil import cutoffs.

SATURDAY, MAY 4, 2024

In a major shift in Russian policy, Russia named Ukrainian President Volodymyr Zelenskyy as a wanted criminal. Also on that list are Estonia and Lithuania cabinet ministers. This is extremely significant as Russia goes from defense to offense on the Ukraine battlefield, putting the Ukrainian leader on a wanted list now inhibits any effort for a negotiated truce or peace. It also indicates Russia's increasing and concerning preparations for a possible military action in the Baltic states.[339]

MONDAY, MAY 6, 2024

After British Foreign Secretary David Cameron insisted that Ukraine had the right to use British weapons to strike Russia, *Reuters* reports: "Russia warned Britain on Monday that if British weapons were used by Ukraine to strike Russian territory then Moscow could hit back at British military installations and equipment both inside Ukraine and elsewhere."[340] *Reuters* also reported that as a response Russia would hold drills simulating the use of battlefield nuclear weapons. This may not be an empty threat because after the UK provided Ukraine with armor-piercing shells containing depleted uranium in March of 2023, Putin deployed tactical nuclear weapons to Belarus in response. Escalating tensions further, French President Emmanuel Macron renewed his pledge to send troops to Ukraine if needed. This only further infuriated Vladimir Putin who considers NATO intervention in the form of troops to be his most sacred redline. Putin also continues to be concerned about the soon expected deployment of NATO donated, American built, F-16s to Ukraine whose pilots have been training to fly for over a year. Putin's main concern is that the aircraft was designed, in part, as a platform to carry American B61 nuclear bombs.[341]

TUESDAY, MAY 7, 2024

The *BBC* reported: "Russia has started preparations for missile drills near Ukraine simulating the use of tactical nuclear weapons in response to 'threats' by Western officials. Kremlin spokesman Dmitry Peskov said recent statements by French President Emmanuel Macron and the British Foreign Secretary David Cameron constituted a '*completely new round of escalation of tension.*" NATO spokesperson Farah Dakhlallah said it was "dangerous and irresponsible" talk.[342] Belarus, where a number of Russia's tactical nuclear missiles were deployed last year, also joined the drills. Pictured below is a nuclear tipped Iskander missile, apparently, being loaded into a mobile launching vehicle for those drills. Since the beginning of war in Ukraine two years ago the US and NATO has maintained surveillance of nuclear sites, fearing a tactical nuclear weapon ever being removed from its storage facility.

---

[339] *Ukrainian President Zelenskyy is on Russia's wanted list, state media*, Associated Press, May 4, 2024, https://www.pbs.org/newshour/world/ukrainian-president-zelenskyy-is-on-russias-wanted-list-state-media.

[340] Guy Faulconbridge, *Russia warns Britain it could strike back after Cameron remark on Ukraine*, Reuters, May 6, 2024, https://www.yahoo.com/news/russia-warns-strike-british-military-135226225.html.

[341] Laura Gozzi, Russia to hold nuclear drills following 'threats' from West, BBC News, May 6, 2024, https://www.yahoo.com/news/world/.

[342] Laura Gozzi, *Russia to hold nuclear drills following "threats" from West*, BBC, May 7, 2024, https://www.bbc.com/news/articles/cq5npwdv3wzo.

Russian Ambassador-at-Large Grigory Mashkov stated to press that "Moscow would have to increase its entire missile arsenal as a deterrent following what was framed as the worst breakdown in relations since the 1962 Cuban Missile Crisis. We are now at the stage of open confrontation, which, I hope, will not result in a direct armed conflict."[343]

WEDNESDAY, MAY 8, 2024

Russia dealt a heavy attack on Ukrainian energy sites with missile and drone strikes. This appeared the worst blow to Ukrainian infrastructure yet.[344]

THURSDAY, MAY 9, 2024

Despite warnings of escalating an already dangerous situation with Russia, Ukraine hit an oil depot with drones 746 miles beyond their own lines into the Russian republic of Bashkortostan.[345] The Institute for The Study of War reports on this date: "Russian President Vladimir Putin used his May 9 Victory Day speech to relitigate his belief that the West is attempting to erase the Soviet Union's contributions to defeating Nazi Germany during the Great Patriotic War (Second World War), a grievance that is at the core of Russia's adversarial perceptions of the West. Putin claimed during the Victory Day parade, which is held to commemorate the Soviet Union's victory and sacrifices during the Second World War, that 'they,' referring to the West, are attempting to "distort" the truth about the Second World War and 'demolish' the memory of Soviet heroism and sacrifice."[346] Putin warned during his speech that Russia's *"nuclear forces were always on alert and added that Moscow would not tolerate any Western threats."*[347]

---

[343] *Belarus stages nuclear drills after Putin ordered nuke strike practice*, Daily Mail, 7, 2024, https://www.msn.com/en-za/news/other/belarus-stages-nuclear-drills-after-putin-ordered-nuke-striyoutube.come-practice/ss-BB1lYFK4#image=2.

[344] The Institute For The Study Of War website, May 8, 2024, https://www.understandingwar.org/.

[345] Verity Bowman, *Ukraine strikes refinery 746 miles inside Russia*, The (UK) Telegraph, May 9, 2024, https://www.yahoo.com/news/ukraine-drone-hits-refinery-nearly-140115545.html.

[346] The Institute For The Study Of War website, May 9, 2024, https://www.understandingwar.org/.

[347] *Putin says nuclear forces 'always' on alert in Victory Day speech*, AFP, May 9, 2024, https://www.yahoo.com/news/putin-

FRIDAY, MAY 10-19, 2024

Russia launches the most significant offensive since the beginning of the war. Ukraine follows with the largest drone attack on Russia to date as NATO considers sending troops as "trainers."[348] Later, the next week, Vladimir Putin travels to Beijing to visit Xi Jinping. They "declare a deepening of their 'strategic partnership' while scolding the United States for a series of moves that they said threatened their countries."[349] Russian Foreign Ministry Spokeswoman Maria Zakharova said Russian President Vladimir Putin's recent visit to China is "*set to determine the future of the world.*"[350]

As this book moves to publication, escalation continues. Not since the early days of the Second World War has there been an analogous situation of Western nations facing two peer level powers with associated axis allies threatening military action in three separate geographic theaters. The difference of course is in that example, nuclear weapons did not yet exist nor did mankind's ability to destroy itself. This time, history must not repeat itself into another world war.

---

mark-victory-day-emboldened-015221818.html?fr=yhssrp_catchall.

[348] *As Russia Advances, NATO Considers Sending Trainers Into Ukraine*, New York Times, May 18, 2024, https://www.nytimes.com/2024/05/16/us/politics/nato-ukraine.html.

[349] *What is Putin and Xi's 'new era' strategic partnership?*, Reuters, May 16, 2024, https://www.reuters.com/world/what-is-putin-xis-new-era-strategic-partnership-2024-05-16/.

[350] *Russian MFA Describes Putin's China Trip as Defining for World's Future*, Tasnim News Agency, May 19, 2024, https://www.tasnimnews.com/en/news/2024/05/19/3088290/russian-mfa-describes-putin-s-china-trip-as-defining-for-world-s-future.

CONCLUSION

AMERICA'S STRATEGIC POSTURE

The Final Report of the Congressional Commission on the Strategic Posture of the United States

Madelyn R. Creedon, Chair	Jon L. Kyl, Vice Chair
Marshall S. Billingslea	Gloria C. Duffy
Rose E. Gottemoeller	Lisa E. Gordon-Hagerty
Rebeccah L. Heinrichs	John E. Hyten
Robert M. Scher	Matthew H. Kroenig
Franklin C. Miller	Leonor A. Tomero

Reuters recently reported that a congressionally appointed bipartisan panel concluded that: "The United States must prepare for possible simultaneous wars with Russia and China by expanding its conventional forces, strengthening alliances, and enhancing its nuclear weapons modernization program."[351] The report comes from the most recent 2023 Strategic Posture Commission called *Final Report of the Congressional Commission on the Strategic Posture of the United States*.[352] The report elaborates:

China's rapid military build-up, **including the unprecedented growth of its nuclear forces, Russia's diversification and expansion of its theater-based nuclear systems, the invasion of Ukraine in 2014 and subsequent full-scale invasion in February 2022, have all fundamentally altered the geopolitical landscape. As a result of China's and Russia's growing competition with the United States and its Allies and partners, and the increasing risk of military conflict with one or both, as well as concerns about whether the United States would be prepared to deter two nuclear peers, Congress determined it was time for a new look at US strategic policy, strategy, and force structure. . . The United States faces a strategic challenge requiring urgent action. Given current threat trajectories, our nation will soon encounter a fundamentally different global setting than it has ever experienced: we will face a world where two nations possess nuclear arsenals on par with our own. In addition, the risk of conflict with these two nuclear peers is increasing. It is an existential challenge for which the United States is ill-prepared, unless its leaders make decisions now to adjust the US strategic posture. . . The US theater nuclear force posture should be urgently modified to: Provide the President a range of militarily effective nuclear response options to deter or counter Russian or Chinese limited nuclear use in theater.[353]**

After almost eight decades, the sword of Damocles still hangs over us all. During much of that time, there has been a basic justifying assumption (or excuse) that nuclear weapons form effective deterrence. There is now, however, for the first time, a new phrase increasingly being heard which is called "deterrence failure." That term can be found in the 2023 *Final Report of the Congressional Commission on the Strategic*

---

[351] Jonathan Landay, *US must be ready for simultaneous wars with China, Russia, report says*, Reuters, October 12, 2023, https://www.reuters.com/world/us-must-be-ready-simultaneous-wars-with-china-russia-report-says-2023-10-12/.

[352] 344 *The Final Report of the Congressional Commission on the Strategic Posture of the United States*, https://www.ida.org/-/media/feature/publications/A/Am/Americas%20Strategic%20Posture/Strategic-Posture-Commission-Report.pdf.

[353] Ibid, pp. V-VIII

Posture of the United States.[354]

Deterrence failure, we believe as authors, is the growing realization that Russia, China, or both countries combined are ready to use nuclear weapons to further their strategic goals. If and when that happens, once even one nuclear weapon is used, the whole principle of deterrence is gone.

We both believe that the concept of nuclear deterrence was doomed to fail anyway because it can only work so long as all leaders concerned remain rational. Yes, so far this has remained true, and a third world war has most likely been prevented by nuclear deterrence, but this is because reason has held. Even Nikita Khrushchev, banging his shoe at the delegate desk of the UN in 1960, was still as rational as he was later during the Cuban Missile Crisis when he and Kennedy kept their cool. Khrushchev was seasoned because he participated in the Soviet defense of Stalingrad and saw firsthand how terrible modern war is.[355]

My son James and I had the great pleasure to talk at length with Nikita Khrushchev's son Sergei about his father in 2017 while hosting his visit to a lecture in Las Vegas, Nevada with Dr. Linda Miller.

Few people who have personal experience of the First or Second World Wars are alive now. Almost no world leader currently has a firsthand concept of what a global war is really like. No one has ever seen a true nuclear war. The last person alive who we knew to have witnessed an actual above-ground nuclear explosion in an atomic test died in 2022 at the age of 92. His name was Ernest Williams, and he had been interviewed in numerous documentaries. He was also a chief tour guide for many years at the Smithsonian-affiliated National Atomic Testing Museum (now called Atomic Museum) when I served there as Executive Director. He often said to us that the day will come when there will be no one still living who has personally witnessed the horrors of an actual nuclear blast. His point was that without that knowledge, it is impossible to appreciate or respect how bad a nuclear war would be.

The previously mentioned *Final Report of the Congressional Commission on the Strategic Posture of the United States* provides a very good summary of how we came to this precarious situation as traced from the days of the Cold War.

After the Cold War, the United States sought to build cooperative relationships with China and Russia and to reduce the role of nuclear weapons. Despite the US best efforts to create the conditions for a just and prosperous world, both Russia and China chose a different path, one of military build-ups, aggression, and extortion. Russia's criminal and unjust war against Ukraine, backed by nuclear threats, is the most potent demonstration of its belligerency. Russia brought trenches and tanks back to Europe, making it an epicenter of conflict once more. At the same time, China is violating its neighbors' territorial sovereignty and by its aggressive behavior is making Asia a military

---

[354] Ibid. p. 76

[355] Personal conversation with Sergei Khrushchev, November 3, 2017.

flashpoint.[356]

An October 2023 Foreign Affairs article by Keir A. Lieber and Daryl G. Press has put it even more succinctly:

Nuclear weapons once again loom large in international politics, and a dangerous pattern is emerging. In the regions most likely to draw the United States into conflict—the Korean Peninsula, the Taiwan Strait, eastern Europe, and the Persian Gulf—US adversaries appear to be acquiring, enhancing, or threatening to use nuclear weapons. North Korea is developing intercontinental ballistic missiles that can reach the United States; China is doubling the size of its arsenal; Russia is threatening to use nuclear weapons in its war in Ukraine; and according to US officials, Iran has amassed enough fissile material for a bomb. Many people hoped that once the Cold War ended, nuclear weapons would recede into irrelevance. Instead, many countries are relying on them to make up for the weakness of their conventional military force.[357]

A recent book further explaining this dynamic is called *From Cold War to Hot Peace* by former US Ambassador to the Russian Federation Michael McFaul. It provides insight into how the old Cold War fits into the broader picture of history in what is now clearly an extended age, or new hybrid Cold War, still caught in the shadow of nuclear weapons.[358] And as before, it is those nuclear weapons which make the current situation so dangerous. UN Secretary General Antonio Guterres put it very clearly. He stated in August of 2022 that humanity is just one misunderstanding away from nuclear annihilation."[359]

As we conclude, we realize that our book has not sought to give a definitive picture of the history of nuclear issues but rather to speak to some of the rarely detailed stories. It is an interesting subject that my son and I became immersed in for eight years during what was part of my long 40-year museum career. I had the opportunity to meet a great deal of people involved in this topic, from retired National Lab directors to those who still work in nuclear security to those who actually detonated the many atomic and hydrogen bombs during the heyday of nuclear testing.

I very much identify with their stories, having grown up during the Cold War. My son, who assisted me with this book, is now living through very similar times. We continue to repeat and stress that all of us still live under the "sword of Damocles."

Our conclusion is not just about the past but the future. Looking back does give us that insight into the

---

[356] *The Final Report of the Congressional Commission on the Strategic Posture of the United States*, https://www.ida.org/-   /media/feature/publications/A/Am/Americas%20Strategic%20Posture/Strategic-Posture-Commission-Report.pdf, p. 87.

[357] Keir A. Lieber and Daryl G. Press, *The Return of Nuclear Escalation: How America's Adversaries Have Hijacked Its Old Deterrence Strategy*, Foreign Affairs, November/December 2023, 32-41, https://www.foreignaffairs.com/united-states/return-nuclear-escalation.

[358] McFaul, Michael McFaul, *From Cold War to Hot Peace: An American Ambassador in Putin's Russia*, (New York: Houghton Mifflin Harcourt, 2018).

[359] The UN chief was speaking at the opening of the Tenth Review Conference of the Parties to the Treaty on the Non-Proliferation of Nuclear Weaponshttps://news.un.org/en/story/2022/08/1123752.

future. Robert Oppenheimer's formative years were immersed as a student in the foundational stage of quantum physics in Europe in the late 1920s and early 1930s. He then fathered that progressive discipline in this country. The team of students and international colleagues he proceeded to mentor introduced us, for better or worse, to the atomic age. Oppenheimer and Niels Bohr then sincerely hoped against hope that fission weapons would make war too costly to ever wage again. That proved naive. Sadly, an inevitable arms race and the mastery of the fusion bomb then made war apocalyptic. By the end of his life, Oppenheimer, like his friend Albert Einstein, may have thought future peace and prosperity hopeless.

Yet not all technological advancements from our early 20th century proved forlorn. Quantum physics, thanks to many talented scientists and engineers, has created our present and will continue to make our future. Most of what drives our modern economies is thanks to quantum discoveries, and yes, notable practical business applications came out of the Manhattan Project following the war via its brilliant physicists. Many of those hesitant weaponeers left Los Alamos and went on to build modern marvels like advanced computing and telecommunications.

Even though we are still left with their nuclear legacy, the world continues to advance. So, what will our future be? Only H.G. Wells could make a good guess at that, although our own guess would go like this: Quantum principles will exponentially accelerate quantum computing, which will exponentially accelerate artificial intelligence. Once AI fully integrates with robotics, the world's economy and politics will change so dramatically that we cannot currently imagine it. Human beings may own their own robots as we once did slaves, and our wealth and productivity will be based on our accumulated collection of these bots. Yet it is hoped these will only resemble human servants and function merely as non-conscious machines. The earlier industrial revolution largely enslaved us as laborers, but perhaps the future will free us all as our more innovative technology allows us equal participation in a new economy.

Technology may also accelerate our learning from a young age and focus our energies not on repetitive tasks but on improving our social conditions. A better and cleaner world could come out of these frightening and life-changing technological advancements. This assumes we as a civilization do not destroy ourselves or at the very least, revert to a new dark age. Ironically, new technologies may make nuclear weapons as obsolete as traditional tanks are now being proved to be on the Ukrainian battlefields or as obsolete as our giant aircraft carriers will soon become to even non-peer-level countries like Iran and North Korea. Modern warfare is changing so fast that we cannot even appreciate this phenomenon, as cheap drones and computer systems are making traditional weapon systems unwieldy. We are at the point now that our tech is so advanced, yet so efficient, that only computers are able to fix computers, and we as humans ignore just how dependent we are becoming on a system that is beginning to manage itself. We are not losing control of our own technology; we have already lost it. Maybe though that is a natural part of the evolution of an advancing civilization.

The future, in fact, may hold even more disturbing concepts and could foretell a weapon much more advanced than thermonuclear bombs. New advances in AI that can facilitate genetic engineering will one day free us from most diseases, cancers, viruses, and harmful bacteria. However, that same technology could also easily create new designer viruses that, unlike nuclear weapons, do not destroy property or the environment but simply remove an adversary's enemy cheaply and easily. The fear is that this technology will be so easy to wield with AI assistance that even third-world countries or terrorist organizations could access it and use it before countermeasures could meet the threat. That could be the future, and such a

MICHAEL AND JAMES HALL

scenario would be potentially so surreal that even war itself, along with its human enabler, would not survive.

History shows us that it is not the weapon that makes war. It only makes war worse. Unfortunately, the most significant weapons that our advancing technology has created are almost all eventually used at some point. So, we, as authors, tell the world that it is war itself that must be eliminated. Communication would be the best start to reach that goal.

The newest emerging theory in quantum mechanics is that human consciousness may be the central constant that animates our physical world. Bohr and Oppenheimer had similar assumptions in their early days as quantum theorists. Maybe our blue world as a collective spirit can make the future better and more peaceful. It could only take our conscious will to accomplish that dream if we all collectively focus on peace in our actions and our prayers.

Michael and James Hall

APPENDIX I
JOHN F. KENNEDY'S "WORLD PEACE"

On June 10, 1963, President John F. Kennedy delivered a speech at American University in Washington, DC. In this address, he laid the groundwork for what became The Limited Test Ban Treaty, which was eventually ratified by the Senate on September 24, 1963, less than a year after the Cuban Missile Crisis. The speech is considered one of his finest moments, and it helped deescalate US Soviet tensions and secure the peace of what became a long Cold War.

President Anderson, members of the faculty, board of trustees, distinguished guests . . .

I have, therefore, chosen this time and this place to discuss a topic on which ignorance too often abounds and the truth is too rarely perceived—yet it is the most important topic on earth: world peace.

What kind of peace do I mean? What kind of peace do we seek? Not a Pax Americana enforced on the world by American weapons of war. Not the peace of the grave or the security of the slave. I am talking about genuine peace, the kind of peace that makes life on earth worth living, the kind that enables men and nations to grow and to hope and to build a better life for their children— not merely peace for Americans but peace for all men and women—not merely peace in our time but peace for all time.

I speak of peace because of the new face of war. Total war makes no sense in an age when great powers can maintain large and relatively invulnerable nuclear forces and refuse to surrender without resort to those forces. It makes no sense in an age when a single nuclear weapon contains almost ten times the explosive force delivered by all the allied air forces in the Second World War. It makes no sense in an age when the deadly poisons produced by a nuclear exchange would be carried by wind and water and soil and seed to the far corners of the globe and to generations yet unborn.

Today the expenditure of billions of dollars every year on weapons acquired for the purpose of making sure we never need to use them is essential to keeping the peace. But surely the acquisition of such idle stockpiles—which can only destroy and never create—is not the only, much less the most efficient, means of assuring peace.

I speak of peace, therefore, as the necessary rational end of rational men. I realize that the pursuit of peace is not as dramatic as the pursuit of war—and frequently the words of the pursuer fall on deaf ears. But we have no more urgent task.

Some say that it is useless to speak of world peace or world law or world disarmament—and that it will be useless until the leaders of the Soviet Union adopt a more enlightened attitude. I hope they do. I believe we can help them do it. But I also believe that we must reexamine our own attitude--as individuals and as a Nation—for our attitude is as essential as theirs. And every graduate of this school, every thoughtful citizen who despairs of war and wishes to bring peace, should begin by looking inward—by examining his own attitude toward the possibilities of peace, toward the Soviet Union, toward the course of the cold war and toward freedom and peace here at home.

First: Let us examine our attitude toward peace itself. Too many of us think it is impossible. Too

many think it unreal. But that is a dangerous, defeatist belief. It leads to the conclusion that war is inevitable—that mankind is doomed—that we are gripped by forces we cannot control.

We need not accept that view. Our problems are manmade—therefore, they can be solved by man. And man can be as big as he wants. No problem of human destiny is beyond human beings. Man's reason and spirit have often solved the seemingly unsolvable—and we believe they can do it again.

I am not referring to the absolute, infinite concept of peace and good will of which some fantasies and fanatics dream. I do not deny the value of hopes and dreams but we merely invite discouragement and incredulity by making that our only and immediate goal.

Let us focus instead on a more practical, more attainable peace—based not on a sudden revolution in human nature but on a gradual evolution in human institutions—on a series of concrete actions and effective agreements which are in the interest of all concerned. There is no single, simple key to this peace—no grand or magic formula to be adopted by one or two powers. Genuine peace must be the product of many nations, the sum of many acts. It must be dynamic, not static, changing to meet the challenge of each new generation. For peace is a process—a way of solving problems.

With such a peace, there will still be quarrels and conflicting interests, as there are within families and nations. World peace, like community peace, does not require that each man love his neighbor-- it requires only that they live together in mutual tolerance, submitting their disputes to a just and peaceful settlement. And history teaches us that enmities between nations, as between individuals, do not last forever. However fixed our likes and dislikes may seem, the tide of time and events will often bring surprising changes in the relations between nations and neighbors.

So let us persevere. Peace need not be impracticable, and war need not be inevitable. By defining our goal more clearly, by making it seem more manageable and less remote, we can help all peoples to see it, to draw hope from it, and to move irresistibly toward it.

Second: Let us reexamine our attitude toward the Soviet Union. It is discouraging to think that their leaders may actually believe what their propagandists write. It is discouraging to read a recent authoritative Soviet text on Military Strategy and find, on page after page, wholly baseless and incredible claims—such as the allegation that "American imperialist circles are preparing to unleash different types of wars . . . that there is a very real threat of a preventive war being unleashed by American imperialists against the Soviet Union . . . [and that] the political aims of the American imperialists are to enslave economically and politically the European and other capitalist countries . . . [and] to achieve world domination . . . by means of aggressive wars."

Truly, as it was written long ago: "The wicked flee when no man pursueth." Yet it is sad to read these Soviet statements—to realize the extent of the gulf between us. But it is also a warning—a warning to the American people not to fall into the same trap as the Soviets, not to see only a distorted and desperate view of the other side, not to see conflict as inevitable, accommodation as impossible, and communication as nothing more than an exchange of threats.

No government or social system is so evil that its people must be considered as lacking in virtue. As Americans, we find communism profoundly repugnant as a negation of personal freedom and dignity. But we can still hail the Russian people for their many achievements—in science and space, in economic and industrial growth, in culture and in acts of courage.

Among the many traits the peoples of our two countries have in common, none is stronger than our mutual abhorrence of war. Almost unique among the major world powers, we have never been at war with each other. And no nation in the history of battle ever suffered more than the Soviet Union suffered in the course of the Second World War. At least 20 million lost their lives. Countless millions of homes and farms were burned or sacked. A third of the nation's territory, including nearly two thirds of its industrial base, was turned into a wasteland—a loss equivalent to the devastation of this country east of Chicago.

Today, should total war ever break out again—no matter how—our two countries would become the primary targets. It is an ironic but accurate fact that the two strongest powers are the two in the most danger of devastation. All we have built, all we have worked for, would be destroyed in the first 24 hours. And even in the cold war, which brings burdens and dangers to so many nations, including this Nation's closest allies--our two countries bear the heaviest burdens. For we are both devoting massive sums of money to weapons that could be better devoted to combating ignorance, poverty, and disease. We are both caught up in a vicious and dangerous cycle in which suspicion on one side breeds suspicion on the other, and new weapons beget counterweapons.

In short, both the United States and its allies, and the Soviet Union and its allies, have a mutually deep interest in a just and genuine peace and in halting the arms race. Agreements to this end are in the interests of the Soviet Union as well as ours—and even the most hostile nations can be relied upon to accept and keep those treaty obligations, and only those treaty obligations, which are in their own interest.

So, let us not be blind to our differences—but let us also direct attention to our common interests and to the means by which those differences can be resolved. And if we cannot end now our differences, at least we can help make the world safe for diversity. For, in the final analysis, our most basic common link is that we all inhabit this small planet. We all breathe the same air. We all cherish our children's future. And we are all mortal.

Third: Let us reexamine our attitude toward the cold war, remembering that we are not engaged in a debate, seeking to pile up debating points. We are not here distributing blame or pointing the finger of judgment. We must deal with the world as it is, and not as it might have been had the history of the last 18 years been different.

We must, therefore, persevere in the search for peace in the hope that constructive changes within the Communist bloc might bring within reach solutions which now seem beyond us. We must conduct our affairs in such a way that it becomes in the Communists' interest to agree on a genuine peace. Above all, while defending our own vital interests, nuclear powers must avert those confrontations which bring an adversary to a choice of either a humiliating retreat or a nuclear war. To adopt that kind of course in the nuclear age would be evidence only of the bankruptcy of our policy—or of a collective death—wish for the world.

To secure these ends, America's weapons are nonprovocative, carefully controlled, designed to deter, and capable of selective use. Our military forces are committed to peace and disciplined in self-restraint. Our diplomats are instructed to avoid unnecessary irritants and purely rhetorical hostility.

For we can seek a relaxation of tension without relaxing our guard. And, for our part, we do not need to use threats to prove that we are resolute. We do not need to jam foreign broadcasts out of fear our faith will be eroded. We are unwilling to impose our system on any unwilling people—but we are willing and able to engage in peaceful competition with any people on earth.

Meanwhile, we seek to strengthen the United Nations, to help solve its financial problems, to make it a more effective instrument for peace, to develop it into a genuine world security system—a system capable of resolving disputes on the basis of law, of insuring the security of the large and the small, and of creating conditions under which arms can finally be abolished.

At the same time we seek to keep peace inside the non-Communist world, where many nations, all of them our friends, are divided over issues which weaken Western unity, which invite Communist intervention or which threaten to erupt into war. Our efforts in West New Guinea, in the Congo, in the Middle East, and in the Indian subcontinent, have been persistent and patient despite criticism from both sides. We have also tried to set an example for other—by seeking to adjust small but significant differences with our own closest neighbors in Mexico and in Canada.

Speaking of other nations, I wish to make one point clear. We are bound to many nations by alliances. Those alliances exist because our concern and theirs substantially overlap. Our commitment to defend Western Europe and West Berlin, for example, stands undiminished because of the identity of our vital interests. The United States will make no deal with the Soviet Union at the expense of other nations and other peoples, not merely because they are our partners, but also because their interests and ours converge.

Our interests converge, however, not only in defending the frontiers of freedom, but in pursuing the paths of peace. It is our hope—and the purpose of allied policies—to convince the Soviet Union that she, too, should let each nation choose its own future, so long as that choice does not interfere with the choices of others. The Communist drive to impose their political and economic system on others is the primary cause of world tension today. For there can be no doubt that, if all nations could refrain from interfering in the self- determination of others, the peace would be much more assured.

This will require a new effort to achieve world law—a new context for world discussions. It will require increased understanding between the Soviets and ourselves. And increased understanding will require increased contact and communication. One step in this direction is the proposed arrangement for a direct line between Moscow and Washington, to avoid on each side the dangerous delays, misunderstandings, and misreadings of the other's actions which might occur at a time of crisis.

We have also been talking in Geneva about the other first-step measures of arms control designed to limit the intensity of the arms race and to reduce the risks of accidental war. Our primary long range interest in Geneva, however, is general and complete disarmament—designed to take place by stages, permitting parallel political developments to build the new institutions of peace which would take the place of arms. The pursuit of disarmament has been an effort of this Government since the 1920s. It has been urgently sought by the past three administrations. And however dim the prospects may be today, we intend to continue this effort—to continue it in order that all countries, including our own, can better grasp what the problems and possibilities of disarmament are.

The one major area of these negotiations where the end is in sight, yet where a fresh start is badly needed, is in a treaty to outlaw nuclear tests. The conclusion of such a treaty, so near and yet so far, would check the spiraling arms race in one of its most dangerous areas. It would place the nuclear powers in a position to deal more effectively with one of the greatest hazards which man faces in 1963, the further spread of nuclear arms. It would increase our security—it would decrease the prospects of war. Surely this goal is sufficiently important to require our steady pursuit, yielding neither to the temptation to give up the whole effort nor the temptation to give up our insistence on vital and responsible safeguards.

I am taking this opportunity, therefore, to announce two important decisions in this regard.

First: Chairman Khrushchev, Prime Minister Macmillan, and I have agreed that high-level discussions will shortly begin in Moscow looking toward early agreement on a comprehensive test ban treaty. Our hopes must be tempered with the caution of history—but with our hopes go the hopes of all mankind.

Second: To make clear our good faith and solemn convictions on the matter, I now declare that the United States does not propose to conduct nuclear tests in the atmosphere so long as other states do not do so. We will not be the first to resume. Such a declaration is no substitute for a formal binding treaty, but I hope it will help us achieve one. Nor would such a treaty be a substitute for disarmament, but I hope it will help us achieve it.

Finally, my fellow Americans, let us examine our attitude toward peace and freedom here at home. The quality and spirit of our own society must justify and support our efforts abroad. We must show it in the dedication of our own lives—as many of you who are graduating today will have a unique opportunity to do, by serving without pay in the Peace Corps abroad or in the proposed National Service Corps here at home.

But wherever we are, we must all, in our daily lives, live up to the age-old faith that peace and freedom walk together. In too many of our cities today, the peace is not secure because the freedom is incomplete.

It is the responsibility of the executive branch at all levels of government--local, State, and National—to provide and protect that freedom for all of our citizens by all means within their authority. It is the responsibility of the legislative branch at all levels, wherever that authority is not now adequate, to make it adequate. And it is the responsibility of all citizens in all sections of this country to respect the rights of all others and to respect the law of the land.

All this is not unrelated to world peace. "When a man's ways please the Lord," the Scriptures tell us, "he maketh even his enemies to be at peace with him." And is not peace, in the last analysis, basically a matter of human rights—the right to live out our lives without fear of devastation—the right to breathe air as nature provided it--the right of future generations to a healthy existence?

While we proceed to safeguard our national interests, let us also safeguard human interests. And the elimination of war and arms is clearly in the interest of both. No treaty, however much it may be to the advantage of all, however tightly it may be worded, can provide absolute security against the risks of deception and evasion. But it can—if it is sufficiently effective in its enforcement and if it is sufficiently in the interests of its signers—offer far more security and far fewer risks than an unabated, uncontrolled, unpredictable arms race.

The United States, as the world knows, will never start a war. We do not want a war. We do not now expect a war. This generation of Americans has already had enough--more than enough—of war and hate and oppression. We shall be prepared if others wish it. We shall be alert to try to stop it. But we shall also do our part to build a world of peace where the weak are safe and the strong are just. We are not helpless before that task or hopeless of its success. Confident and unafraid, we labor on—not toward a strategy of annihilation but toward a strategy of peace.[360]

---

[360] John F. Kennedy Presidential Library and Museum, https://www.jfklibrary.org/archives/other-resources/john-f-kennedy- speeches/american-university-19630610.

APPENDIX II B61-13

B61-13 illustrative fact sheet, whatsnew2day, https://whatsnew2day.com/maps-show-devastating- impact-of-americas-new-super-nuke-on-russian-capital-b61-13-gravity-bomb-has-24-times-more-power-than-the-one-dropped-on-hiroshima/.

The B61 gravity bomb was conceived after the Cuban Missile Crisis as a device capable of being carried by a fast, low-flying fighter or bomber. It typically has an adjustable yield capability ranging from one to 100 kilotons, which was designed to give a US president more flexible options in the event of a nuclear conflict. The following Department of Defense press release illuminates one of the most significant proposed future upgrades to that weapon, which will dramatically increase its yield by up to 360 kilotons:

Today, (October 27, 2023) the Department of Defense (DoD) announced that the United States will pursue a modern variant of the B61 nuclear gravity bomb, designated the B61-13, pending Congressional authorization and appropriation.

The Department of Energy's National Nuclear Security Administration (NNSA) would produce the B61-13. The decision to pursue this capability, which was undertaken in close collaboration with the NNSA, responds to the demands of a rapidly evolving security environment as described in the 2022 Nuclear Posture Review.

"Today's announcement is reflective of a changing security environment and growing threats from potential adversaries," said Assistant Secretary of Defense for Space Policy John Plumb. "The United States has a responsibility to continue to assess and field the capabilities we need to credibly deter and, if necessary, respond to strategic attacks, and assure our allies."

The B61-13 would be deliverable by modern aircraft, strengthening deterrence of adversaries and assurance of allies and partners by providing the President with additional options against certain harder and large-area military targets. It would replace some of the B61-7s in the current nuclear stockpile and have a yield similar to the B61-7, which is higher than that of the B61-12.

"The B61-13 represents a reasonable step to manage the challenges of a highly dynamic security environment," said Plumb. "While it provides us with additional flexibility, production of the B61-13 will not increase the overall number of weapons in our nuclear stockpile."

The B61-13 would take advantage of the current, established production capabilities supporting the B61-12, and would include the modern safety, security, and accuracy features of the B61-12.

This initiative follows several months of review and consideration. The fielding of the B61-13 is not in response to any specific current event; it reflects an ongoing assessment of a changing security environment.[361]

---

[361] Department of Defense web site: https://www.defense.gov/News/Releases/Release/Article/3571660/department-of-defense- announces-pursuit-of-b61-gravity-bomb-variant/.

APPENDIX III IN MEMORIAM

"Chuck" Costa was a great mentor and leader to me. He was the first person I met when I came to interview in Las Vegas for the position of executive director of the Smithsonian-affiliated National Atomic Testing Museum. He took such care and interest in giving me an initial tour of the Museum. I could tell from that first meeting how much he cared about this subject and his passion for educating the public on the science and history of nuclear testing. As years followed, he and Nelson Cochran (another key mentor) made sure to have me accompany them on many tours of the historic Nevada Test Site, now called Nevada National Security Site. Chuck and Nelson took me places many may never see or even hear about. They knew so many intricate stories about the days of nuclear testing. I recall on one of those tours, I had off-handedly mentioned something about that famous and now iconic film clip of the pine trees being bent to and fro by the blast of a nuclear test. The 145 Ponderosa Pines had been cut down in the

mountains and placed in holes drilled in the desert floor for the Annie test in Operation Upshot-Knothole. Millions and millions have seen the film over the years in a multitude of disaster movies, but when I mentioned it on that day, Chuck spoke up and said: "I bet no one knows those tree trunks are still around." He and Nelson then drove around for a very long time, trying to remember where exactly they were stored, but they found them and, with great interest, showed the amazing artifacts to me. Not very many people have been as lucky as I have been to experience such things, thanks to the passion of such atomic veterans and educators.

"Charles (Chuck) Francis Costa, beloved father and grandfather and faithful government servant, died in the arms of his daughters on Sunday, August 29, 2021, in Dallas, Texas. Born on March 20, 1939, in Medford, Massachusetts, Chuck moved in 1955 to Tewksbury, Massachusetts, graduating from Tewksbury High School in 1957. The first of his family to attend college, he earned a Civil Engineering degree from the University of Massachusetts in Amherst in 1961 and a graduate degree in Radiation Health Physics from the University of Michigan in 1968. After graduating from the University of Massachusetts, Chuck migrated to Las Vegas, Nevada, and never looked back. He was fascinated by the nuclear weapons testing program and wanted to be part of it. He ultimately realized his dream and worked from 1962 to 1992 for the United States Public Health Service, assigned to the Environmental Protection Agency from 1970 to 1992. In this capacity, he helped establish a radiation monitoring program for northern Nevada and Utah. In the process of doing so, he developed lifelong friendships with coworkers and some of the early Nevada ranching families, especially those near Ely, Nevada. Upon cessation of the above-ground testing program, Chuck continued his work monitoring the radiation from below-ground weapons testing, ultimately working his way up from Field Monitor to becoming a member of the Advisory Panel for the weapons testing program. Following his retirement from the EPA in 1992, Chuck joined the Los Alamos National Laboratory from 1992-2008, serving as the Los Alamos Liaison at the Nevada Test Site on the Advisory Board and later as a Test Director. During his years with the PHS and the EPA, he had many cherished experiences, including radiation monitoring in Amchitka in the Aleutian Islands as well as in Eniwetok in the Marshall Islands. He also participated in the cleanup of the Three-Mile Island nuclear accident in Pennsylvania and

the Exxon Valdez oil spill in Alaska. Chuck loved every minute of his work, including the lifelong friendships he made, the places he saw, and the mission of nuclear weapons testing. After retiring from Los Alamos, Chuck worked tirelessly to support and promote the National Atomic Testing Museum in Las Vegas.

APPENDIX IV MAUD REPORT

1.General Statement

Work to investigate the possibilities of utilizing the atomic energy of uranium for military purposes has been in progress since 1939, and a stage has now been reached when it seems desirable to report progress.

At the beginning of this report, we should emphasize that we entered the project with more skepticism than belief, though we felt it was a matter that had to be investigated. As we proceeded, we became more and more convinced that releasing atomic energy on a large scale is possible and that conditions can be chosen, making it a very powerful weapon of war. We have now concluded that it will be possible to make an effective uranium bomb which, containing some 25 lb of active material, would be equivalent as regards destructive effect to 1,800 tons of T.N.T. and would also release large quantities of radioactive substance, which would make places near to where the bomb exploded dangerous to human life for a long period. The bomb would be composed of an active constituent (referred to in what follows as ^{235}U) present to the extent of about a part in 140 in ordinary Uranium. Owing to the very small difference in properties (other than explosive) between this substance and the rest of the Uranium, its extraction is a matter of great difficulty, and a plant to produce 2–4 lb (1 kg) per day (or three bombs per month) is estimated to cost approximately £5,000,000, of which sum a considerable proportion would be spent on engineering, requiring the labor of the same highly skilled character as is needed for making turbines.

In spite of this very large expenditure, we consider that the destructive effect, both material and moral, is so great that every effort should be made to produce bombs of this kind. As regards the time required, Imperial Chemical Industries, after consultation with Dr. Guy of Metropolitan-Vickers, estimated that the material for the first bomb could be ready by the end of 1943. This of course, assumes that no major difficulty of an entirely unforeseen character arises. Dr. Ferguson of Woolwich estimates that the time required to work out the method of producing high velocities required for fusing (see paragraph 3) is 1–2 months. As this could be done concurrently with the production of the material, no further delay is to be anticipated in this score. Even if the war should end before the bombs are ready, the effort would not be wasted, except in the unlikely event of complete disarmament, since no nation would care to risk being caught without a weapon of such decisive possibilities.

We know that Germany has had a great deal of trouble securing supplies of the substance known as heavy water. In the earlier stages, we thought this substance might be very important to our work. It appears, in fact, that its usefulness in the release of atomic energy is limited to processes that are not likely to be of immediate war value, but the Germans may by now have realized this, and it may be mentioned that the lines on which we are now working are such as would be likely to suggest themselves to any capable physicist.

By far, the largest supplies of Uranium are in Canada and the Belgian Congo. Since it has been actively looked for because of the radium that accompanies it, it is unlikely that any considerable quantities exist which are unknown except possibly in unexplored regions.

2.Principle Involved

This type of bomb is possible because of the enormous store of energy resident in atoms and because of the special properties of the active constituent of uranium. The explosion is very different in its mechanism from the ordinary chemical explosion, for it can occur only if the quantity of 235U is greater than a certain critical amount. Quantities of the material that are less than the critical amount are quite stable. Such quantities are, therefore, perfectly safe, and this is a point that we wish to emphasize. On the other hand, if the amount of material exceeds the critical value, it is unstable, and a reaction will develop and multiply itself with enormous rapidity, resulting in an explosion of unprecedented violence. Thus, all that is necessary to detonate the bomb is to bring together two pieces of the active material, each less than the critical size but which, when in contact, form a mass exceeding it.

3.Method of Fusing

In order to achieve the greatest efficiency in an explosion of this type, it is necessary to bring the two halves together at high velocity, and it is proposed to do this by firing them together with charges of ordinary explosives in the form of a double gun.

The weight of this gun will, of course, greatly exceed the weight of the bomb itself but should not be more than 1 ton, and it would certainly be within the carrying capacity of a modern bomber. It is suggested that the bomb (contained in the gun) should be dropped by parachute, and the gun should be fired by means of a percussion device when it hits the ground. The drop time can be long enough to allow the airplane to escape from the danger zone, and as this is very large, great accuracy of aim is not required.

4.Probable Effect

The best estimate of the kind of damage likely to be produced by the explosion of 1,800 tons of T.N.T. is afforded by the great explosion at Halifax N.S. in 1917. The following account is from the *History of Explosives*. "The ship contained 450,000 lb. of T.N.T., 122,960 lb. of guncotton, and 4,661,794 lb. of picric acid wet and dry, making a total of 5,234,754 lb. The zone of the explosion extended for about $3/4$ mile in every direction; in this zone, the destruction was almost complete. Severe structural damage extended generally for a radius of $1^{1}/_{8}$ to $1^{1}/4$ miles and in one direction up to 1-3/4 miles from the origin. Missiles were projected to 3–4 miles, window glass broken up to 10 miles generally, and in one instance up to 61 miles." In considering this description, it is worth remembering that part of the explosives cargo was situated below water level and part above.

5.Preparation of Material and Cost

We have thoroughly considered the possible methods of extracting the ^{235}U from ordinary uranium and have conducted several experiments. The scheme which we recommend is described in Part 11 of this report and in greater detail in Appendix IV. It involves essentially the gaseous diffusion of a compound of uranium through gauzes of very fine mesh.

In the estimates of size and cost accompanying this report, we have only assumed the types of gauze that are presently in existence. A comparatively small amount of development would probably enable the making of gauzes of smaller mesh, and this would allow the construction of a somewhat smaller and, consequently, cheaper separation plant for the same output.

Although the cost per lb. of this explosive is so great, it compares very favorably with ordinary explosives in terms of energy released and damage done. It is considerably cheaper, but the points which we regard as of overwhelming importance are the concentrated destruction which it would produce the large moral effect, and the saving in air effort the use of this substance would allow, as compared with bombing with ordinary explosives.

6.Discussion

One outstanding difficulty of the scheme is that the main principle cannot be tested on a small scale. Even to produce a bomb of the minimum critical size would involve a great expenditure of time and money. We are, however, convinced that the principle is correct, and whilst there is still some uncertainty as to the critical size, it is most unlikely that the best estimate we can make is so far in error that it invalidates the general conclusions. We feel that the present evidence is sufficient to justify the strongly-pressed scheme.

As regards the manufacture of the 235U, we have gone nearly as far as we can on a laboratory scale. The principle of the method is certain, and the application does not appear unduly difficult as a piece of chemical engineering. The need to work on a larger scale is now very apparent, and we are beginning to have difficulty in finding the necessary scientific personnel. Further, if the weapon is to be available in, say, two years from now, it is necessary to start plans for the erection of a factory, though no really large expenditure will be needed till the 20-stage model has been tested. It is also important to begin training men who can ultimately act as supervisors of the manufacturer. There are a number of auxiliary pieces of apparatus to be developed, such as those for measuring the concentration of the 235U. In addition, work on a fairly large scale is needed to develop the chemical side for the production of the bulk of uranium hexafluoride, the gaseous compound we propose to use.

It will be seen from the foregoing that a stage in the work has now been reached at which it is important that a decision should be made as to whether the work is to be continued on the increasing scale which would be necessary if we are to hope for it as an effective weapon for this war. Any considerable delay now would retard by an equivalent amount the date by which the weapon could come into effect.

7.Action in U.S.

We are informed that while the Americans are working on the uranium problem, the bulk of their effort has been directed to the production of energy, as discussed in our report on uranium as a source of power rather than to the production of a bomb. We are cooperating with the United States to the extent of exchanging information, and they have undertaken one or two pieces of laboratory work for us. We feel that it is important and desirable that development work should proceed on both sides of the Atlantic irrespective of where it may be finally decided to locate the plant for separating the ^{235}U, and for this purpose, it seems desirable that certain members of the committee should visit the United States. We are informed that such a visit would be welcomed by the members of the United States committees dealing with this matter.

8.Conclusions and Recommendations

(i).The committee considers that the scheme for a uranium bomb is practicable and likely to lead to decisive results in the war.

(ii).It is recommended that this work be continued as the highest priority and on the increasing scale necessary to obtain the weapon in the shortest possible time.

(iii).The present collaboration with America should be continued and extended, especially in the experimental work area.[362]

---

[362] Atomic Heritage Foundation, https://ahf.nuclearmuseum.org/ahf/key-documents/maud-committee-report/.

SUGGESTED READING

Bernstein, Jeremy. *Plutonium, A History of the World's Most Dangerous Element*. Ithaca and London: Cornell University Press, 2009.

Bird, Kai, and Martin J. Sherwin. *American Prometheus: The Triumph and Tragedy of J. Robert Oppenheimer*. New York: A.A. Knopf, 2005.

Blanchard, B. Wayne. *American Civil Defense 1945-1984: The Evolution of Programs and Policies*. Washington, District of Columbia: United States of America, U. S. Government Printing Office, 1986.

Blanchard, B. Wayne. *The Official History of Civil Defense*. Emmitsburg, MD: National Emergency Training Center, 1985.

Bouverie, Tim. *Appeasement*. London: Vintage, 2019.

Brinkley, William. *The Last Ship*. New York: Penguin Books, 1988.

Carothers, Jim. Gaging the Dragon, *The Containment of Underground Nuclear Explosives.* Washington DC: United States Department of Energy, DOE/NV-388, DNA TR 95-74, 1995.

Chang, Gordon G. *China Is Going to War*. New York: Criterion Books, 2023.

Church, Bruce W. *The Fallout Story*. National Atomic Testing Museum, Las Vegas, Nevada: Blurb Publishing, 2022.

Churchill, Winston S. *The Aftermath*. Vol. 4 of *The World Crisis*. London: Thornton Butterworth, 1929.

Conroy, Robert. *1901*. New York: Ballantine Books, 2010.

Coulthart, Ross. *In Plain Sight: A Fascinating Investigation into UFOs and Alien Encounters*. New York: HarperCollins, 2021.

Defense Threat Reduction Agency. *Defense's Nuclear Agency, 1947-1997*. Washington DC: US Department of Defense, 2002.

Forstche, William R. *One Second After*. New York: Saint Martin's Press, 2009.

Frisch, Otto. *What Little I Remember*. Cambridge: Cambridge University Press, 1979.

Glasstone, Samuel. *The Effects of Nuclear Weapons*. US Government Printing Office: United States Department of Defense, 1957.

Hacker, Barton C. *Elements of Controversy: the Atomic Energy Commission and Radiation Safety in Nuclear Weapons Testing, 1947-1974*. Berkeley, CA: University of California Press, 1994.

Hager, Thomas. *The Alchemy of Air: A Jewish Genius, a Doomed Tycoon, and the Scientific Discovery that Fed the World but Fueled the Rise of Hitler*. 1st ed. New York: Harmony Books, 2008.

Hall, Michael. *A Century of Sightings*. Hendersonville, North Carolina: Galde Press, 1999.

Hall, Michael, Connors, Wendy. *Captain Edward J. Ruppelt, Summer of the Saucers-1952.*

Albuquerque, New Mexico: Rose Press, 2000.

Hawkins, David and Truslow, Edith C. *Project Y: The Los Alamos Story*. Los Angeles: Tomash Publisgers,1983.

Hecker, Siegfried S. *Doomed to Cooperate*. Two Volumes. Los Alamos, New Mexico: Bathtub Row Press, 2016. Holloway, David. *Stalin And The Bomb*. New Haven and London: Yale University Press, 1994.

Hopkins, John C. and Killian, Barbara Germain. *Nuclear Weapons Testing At The Nevada Test* Site, *The First Decade*. 2013.

Jacobsen, Annie. *Nuclear War: A Scenario*. Boston, Massachusetts: Dutton, 2024.

Keller, Jonathan F. *Why The Nazi Atomic Bomb Never Happened*. https://www.mdpi.com/2673-4362/2/1/2/htm.

Khrushchev, Sergei N. *Nikita Khrushchev, and the Creation of Superpower*. University Park, PA: The Pennsylvania State University Press, 2000.

King, J. E. Cicero. *Tusculan Disputations*. Williamsburg, Virginia: Loeb Classical Library. Harvard University Press, 1927.

Lanouette, William. *Genius in the Shadow: A Biography of Leo Szilard, the Man Behind the Bomb, with Bela A. Szilard and Jonas Salk*. New York: Scribner's Sons, 1992.

Lieber, Keir A. and Daryl G. Press. *The Return of Nuclear Escalation: How America's Adversaries Have Hijacked Its Old Deterrence Strategy*. Foreign Affairs. November/December 2023, 32-41, https://www.foreignaffairs.com/united-states/return-nuclear-escalation.

Manchester, William. *The Last Lion: Winston Spencer Churchill*, Volume 1-3. Boston: Little, Brown, 1983.

Massie, Robert K. Dreadnought. New York: Random House, 1991.

Massie, Robert K. Castles of Steel. New York: Random House, 2003.

McFaul, Michael. *From Cold War to Hot Peace: An American Ambassador in Putin's Russia*. New York: Houghton Mifflin Harcourt, 2018.

Monk, Ray. *Robert Oppenheimer: A Life Inside the Center*. New York: Knopf Doubleday Publishing Group, 2013.

Office of The Historian Joint Task Force One. *Operation Crossroads, The Official Pictorial History*. New York: Wm. H. Wise. Co., Inc, 1946.

O'Keefe, Bernard J. *Nuclear Hostages*. Boston: Houghton Mifflin Co. 1983.

O'Keefe, Bernard J. *Shooting Ourselves In The Foot*. Boston: Houghton Mifflin Co. 1985.

Pat, Frank. *Alas Babylon*. New York; London; Toronto; Sydney: Harper Perennial Modern Classics, 1959. Pringle, Peter, Spigelman, James. *The Nuclear Barons*. New York: Holt, Rinehart and Winston, 1981.

Reed, Thomas, and Stillman, Danny. *The Nuclear Express, A Political History of The Bomb and*

Its Proliferation. Minneapolis, MN: Zenith Press, 2009.

Rhodes, Richard. *Dark Sun, The Making of the Hydrogen Bomb*. New York: Simon & Schuster, 1996.

Rhodes, Richard. *The Making of the Atomic Bomb*. New York: Simon & Schuster, 1986.

Rose, Kenneth D. *One Nation Underground: The Fallout Shelter in American Culture*. New York: New York University Press, 2001.

Ruane, Kevin. *Churchill and the Bomb in War and the Cold War*. London: Bloomsbury Publishing, 2016.

Schneider, Dr. Mark. *The Emerging EMP Threat to The United States*. Fairfax, VA: National Institute Press, National Institute for Public Policy, foreword by Congressman Roscoe Bartlett. US House of Representatives, 2007.

Sime, Ruth Lewin. *Lise Meitner*. Berkeley, CA: University of California Press, 1996. Stavridis, James, Ackerman, Elliot. *2034*. New York: Penguin Press, 2021.

Stewart, Graham. *Buring Caesar, The Churchill-Chamberlain Rivalry*. Woodstock, New York: The Overlook Press, 2001.

Szilárd, Leo. *The Voice of the Dolphins and Other Stories*. New York: Simon and Schuster, 1961.

Vale, Lawrence J. *The Limits of Civil Defense in the USA*, Switzerland; Britain; Soviet Union; New York: St. Martin's Press, 1987.

Wells, H.G. *The World Set Free*. London: W. Collins Sons, 1924. Online version at https://novelonline.ir/read/1781/The-World-Set-Free.

Wilson, David. *Rutherford, Simple Genius*. London: Hodder & Stoughton, 1983.

Zavattaro, Peter. *EG&G*. Las Vegas, Nevada: Nevada Test Site Historical Foundation, 2007.

ABOUT THE AUTHORS

Michael Hall has worked for over four decades in the museum field as a managing director in five different accredited culturally and history-focused organizations nationwide. His son James grew up beside him in many of those assignments and has become his father's chief assistant.

While serving as Executive Director of the Smithsonian-affiliated National Atomic Testing Museum in Las Vegas, Nevada, Michael and James drafted a manuscript based on their many experiences working with veterans of the Cold War and nuclear testing. Their story is very relevant to today's concerns over nuclear conflict. In fact, the threat of nuclear war has never been higher since the days of the Cuban Missile Crisis in October of 1962. That is why the lessons of the past are important today. Michael and James provide those insights through stories of our nuclear age.

Today Michael has retired from the sciences and has returned to a personal love of art and is managing with his son, James, a regional art museum. Michael studied both disciplines and is a graduate of Spoon River College, Illinois College, Western Illinois University, and alumni of Purdue University. James studied at The Las Vegas Art Institute for three years. Today Michael and James are constantly reminded of their years being so close to the nuclear weapons field and understand how dangerous the current and rapidly escalating international tensions are to us all and our fragile planet. They feel the current world situation is clearly unprecedented.

The authors can be reached at mike19001901@gmail.com for your questions.

www.ingramcontent.com/pod-product-compliance
Lightning Source LLC
Chambersburg PA
CBHW080953120626
46546CB00010B/2877